前言
PREFACE

Python 作为人工智能和大数据的主要开发语言,具有灵活性强、扩展性好、应用面广、可移植、可扩展、可嵌入等特点,近几年发展迅速,热度上涨,人才需求量逐年攀升,相关课程已经成为高等院校的专业课程。

为适应当前教育教学改革的要求,更好地践行人工智能模型与算法应用,作者以实践教学与创新能力培养为目标,采取了创新方式,从不同难度、不同类型、不同算法,融合同类图书的优点,对实际智能应用案例进行总结。

本书主要内容和素材来自开源网站的人工智能经典模型算法、信息工程专业创新课程内容、作者近几年承担的科研项目成果、作者指导学生所完成的创新项目,学生不仅学到了知识,提高了能力,而且为本书提供了第一手素材和相关资料。

本书内容由总到分、先思考后实践、注重整体架构、系统流程与代码实现相结合。对于从事人工智能开发、机器学习和算法实现的专业技术人员,本书可以作为技术参考书、提高工程创新实践手册;也可以作为信息通信工程及相关领域的本科生参考书,为机器学习模型分析、算法设计、应用实现提供帮助。

本书的编写得到了教育部高等学校电子信息类专业教学指导委员会、信息工程专业国家第一类特色专业建设项目、信息工程专业国家第二类特色专业建设项目、教育部 CDIO 工程教育模式研究与实践项目、教育部本科教学工程项目、信息工程专业北京市特色专业建设、北京市教育教学改革项目和北京邮电大学教育教学改革项目(编号为 2019TD01)的大力支持,在此表示感谢!

由于作者经验与水平有限,书中难免存在疏漏及不当之处,衷心地希望各位读者多提宝贵意见及具体整改措施,以便作者进一步修改和完善。

作 者

2020 年 12 月

目 录
CONTENTS

项目 1　文章辅助生成系统 ··· 1
  1.1　总体设计 ··· 1
    1.1.1　系统整体结构图 ··· 2
    1.1.2　系统流程图 ·· 2
  1.2　运行环境 ··· 3
    1.2.1　Python 环境 ·· 3
    1.2.2　TextRank 环境 ·· 3
    1.2.3　TensorFlow 环境 ··· 3
    1.2.4　PyQt5 及 Qt Designer 运行环境 ·· 4
  1.3　模块实现 ··· 4
    1.3.1　数据预处理 ·· 6
    1.3.2　抽取摘要 ··· 8
    1.3.3　模型搭建与编译 ··· 9
    1.3.4　模型训练与保存 ··· 12
    1.3.5　图形化界面的开发 ··· 21
    1.3.6　应用封装 ··· 21
  1.4　系统测试 ··· 22
    1.4.1　训练困惑度 ·· 22
    1.4.2　测试效果 ··· 24
    1.4.3　模型应用 ··· 29

项目 2　Trump 推特的情感分析 ··· 29
  2.1　总体设计 ··· 29
    2.1.1　系统整体结构图 ··· 30
    2.1.2　系统流程图 ·· 30
  2.2　运行环境 ··· 30
    2.2.1　Python 环境 ·· 30
    2.2.2　TensorFlow 环境 ··· 31
    2.2.3　工具包 ·· 31
  2.3　模块实现 ··· 31
    2.3.1　准备数据 ···

　　　　2.3.2　数据预处理 ·· 32
　　　　2.3.3　模型构建 ·· 32
　　　　2.3.4　模型测试 ·· 33
　　2.4　系统测试 ··· 35
　　　　2.4.1　模型效果 ·· 42
　　　　2.4.2　模型应用 ·· 42

项目3　基于LSTM的影评情感分析 ···························· 42
　　3.1　总体设计 ··· 46
　　　　3.1.1　系统整体结构图 ·· 46
　　　　3.1.2　系统前后端流程图 ·· 46
　　3.2　运行环境 ··· 47
　　　　3.2.1　Python 环境 ·· 47
　　　　3.2.2　TensorFlow 环境 ·· 47
　　　　3.2.3　Android 环境 ·· 48
　　3.3　模块实现 ··· 48
　　　　3.3.1　数据预处理 ·· 48
　　　　3.3.2　模型构建及训练 ·· 48
　　　　3.3.3　模型保存 ·· 51
　　　　3.3.4　词典保存 ·· 53
　　　　3.3.5　模型测试 ·· 54
　　3.4　系统测试 ··· 55
　　　　3.4.1　数据处理 ·· 59
　　　　3.4.2　模型训练 ·· 59
　　　　3.4.3　词典保存 ·· 60
　　　　3.4.4　模型效果 ·· 60

项目4　Image2Poem——根据图像生成古体诗句 ·········· 61
　　4.1　总体设计 ··· 63
　　　　4.1.1　系统整体结构图 ·· 63
　　　　4.1.2　系统流程图 ·· 63
　　4.2　运行环境 ··· 64
　　　　4.2.1　Python 环境 ·· 64
　　　　4.2.2　TensorFlow 安装 ·· 65
　　　　4.2.3　其他 Python 模块的安装 ·· 65
　　　　4.2.4　百度通用翻译 API 开通及使用 ···································· 66
　　4.3　模块实现 ··· 66
　　　　4.3.1　数据准备 ·· 66
　　　　4.3.2　Web 后端准备 ·· 66
　　　　4.3.3　百度通用翻译 ·· 72
　　　　4.3.4　全局变量声明 ·· 73
　　　73

            4.3.5 创建模型 ··· 75
            4.3.6 模型训练及保存 ·· 83
            4.3.7 模型调用 ··· 86
      4.4 系统测试 ·· 88
            4.4.1 训练准确率 ·· 88
            4.4.2 模型效果 ··· 90
            4.4.3 整合应用 ··· 91

项目 5 歌曲人声分离 ··· 93
      5.1 总体设计 ·· 93
            5.1.1 系统整体结构图 ·· 94
            5.1.2 系统流程图 ·· 94
      5.2 运行环境 ·· 94
            5.2.1 Python 环境 ··· 95
            5.2.2 TensorFlow 环境 ··· 95
            5.2.3 Jupyter Notebook 环境 ·· 96
      5.3 模块实现 ·· 97
            5.3.1 数据准备 ··· 97
            5.3.2 数据预处理 ··· 101
            5.3.3 模型构建 ··· 105
            5.3.4 模型训练及保存 ·· 107
            5.3.5 模型测试 ·· 111
      5.4 系统测试 ··· 111
            5.4.1 训练准确率 ·· 112
            5.4.2 测试效果 ·· 112
            5.4.3 模型应用 ·· 115

项目 6 基于 Image Caption 的英语学习 ·· 115
      6.1 总体设计 ··· 115
            6.1.1 系统整体结构图 ·· 116
            6.1.2 系统流程图 ·· 116
      6.2 运行环境 ··· 116
            6.2.1 Python 环境 ·· 117
            6.2.2 TensorFlow 环境 ·· 117
            6.2.3 微信开发者工具 ·· 117
      6.3 模块实现 ··· 117
            6.3.1 准备数据 ·· 119
            6.3.2 模型构建 ·· 121
            6.3.3 模型训练及保存 ·· 123
            6.3.4 模型调用 ·· 127
            6.3.5 模型测试

6.4 系统测试
6.4.1 训练准确率 ... 136
6.4.2 测试效果 ... 136
6.4.3 模型应用 ... 137

项目7 智能聊天机器人 ... 137
7.1 总体设计 ... 140
7.1.1 系统整体结构图 ... 140
7.1.2 系统流程图 ... 140
7.2 运行环境 ... 141
7.2.1 Python 环境 ... 141
7.2.2 TensorFlow 环境 ... 142
7.3 模块实现 ... 142
7.3.1 数据预处理 ... 143
7.3.2 模型构建 ... 143
7.3.3 模型测试 ... 146
7.4 系统测试 ... 153
7.4.1 训练损失 ... 153
7.4.2 测试效果 ... 153
7.4.3 模型应用 ... 154

项目8 说唱歌词创作应用 ... 154
8.1 总体设计 ... 157
8.1.1 系统整体结构图 ... 157
8.1.2 系统流程图和前端流程图 ... 157
8.2 运行环境 ... 158
8.2.1 Python 环境 ... 158
8.2.2 TensorFlow 环境 ... 158
8.2.3 其他环境 ... 158
8.3 模块实现 ... 159
8.3.1 数据预处理与加载 ... 159
8.3.2 模型构建 ... 159
8.3.3 模型训练及保存 ... 164
8.3.4 模型测试 ... 169
8.4 系统测试 ... 172
8.4.1 模型困惑度 ... 178
8.4.2 模型应用 ... 178

项目9 基于 LSTM 的语音/文本/情感识别系统 ... 178
9.1 总体设计 ... 180
9.1.1 系统整体结构图 ... 180
9.1.2 系统流程图 ... 180
... 181

人工智能科学与技术丛书

AI SOURCE CODE ANALYZING
RECURRENT NEURAL NETWORK (RNN)
DEEP LEARNING CASES WITH PYTHON

AI源码解读
循环神经网络（RNN）
深度学习案例
（Python版）

李永华　曲宗峰　李红伟◎编著
Li Yonghua　Qu Zongfeng　Li Hongwei

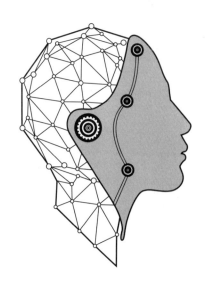

清华大学出版社
北京

内 容 简 介

本书以人工智能发展为时代背景,通过20个案例,应用机器学习模型和算法,为工程技术人员提供较为详细的实战方案,以便深度学习。

在编排方式上,全书侧重对创新项目的过程进行介绍。分别从整体设计、系统流程、实现模块等角度论述数据处理、模型训练及模型应用,并剖析模块的功能、使用和程序代码。为便于读者高效学习,快速掌握人工智能开发方法,本书配套提供项目设计工程文档、程序代码、实现过程中出现的问题,并给予解决方法,可供读者举一反三,二次开发。

本书将系统设计、代码实现以及运行结果展示相结合,语言简洁,深入浅出,通俗易懂,不仅适合Python 编程爱好者,而且可作为高等院校教材,还可作为从事智能应用创新开发专业人员的参考书。

本书封面贴有清华大学出版社防伪标签,无标签者不得销售。
版权所有,侵权必究。举报: 010-62782989,beiqinquan@tup.tsinghua.edu.cn。

图书在版编目(CIP)数据

AI 源码解读: 循环神经网络(RNN)深度学习案例: Python 版/李永华,曲宗峰,李红伟编著. —北京: 清华大学出版社,2021.7
(人工智能科学与技术丛书)
ISBN 978-7-302-57909-0

Ⅰ. ①A… Ⅱ. ①李… ②曲… ③李… Ⅲ. ①机器学习 ②软件工具-程序设计 Ⅳ. ①TP181 ②TP311.561

中国版本图书馆 CIP 数据核字(2021)第 060943 号

责任编辑: 曾　珊
封面设计: 李召霞
责任校对: 胡伟民
责任印制: 刘海龙

出版发行: 清华大学出版社
网　　址: http://www.tup.com.cn, http://www.wqbook.com
地　　址: 北京清华大学学研大厦 A 座
邮　　编: 100084
社 总 机: 010-62770175
邮　　购: 010-83470235
投稿与读者服务: 010-62776969, c-service@tup.tsinghua.edu.cn
质量反馈: 010-62772015, zhiliang@tup.tsinghua.edu.cn
课件下载: http://www.tup.com.cn, 010-83470236

印 装 者: 北京国马印刷厂
经　　销: 全国新华书店
开　　本: 186mm×240mm
印　　张: 24.5
字　　数: 610 千字
版　　次: 2021 年 7 月第 1 版
印　　次: 2021 年 7 月第 1 次印刷
印　　数: 1~1500
定　　价: 89.00 元

产品编号: 090184-01

目 录
CONTENTS

项目 1　文章辅助生成系统 ··· 1
 1.1　总体设计 ··· 1
  1.1.1　系统整体结构图 ··· 2
  1.1.2　系统流程图 ··· 2
 1.2　运行环境 ··· 3
  1.2.1　Python 环境 ·· 3
  1.2.2　TextRank 环境 ·· 3
  1.2.3　TensorFlow 环境 ··· 3
  1.2.4　PyQt5 及 Qt Designer 运行环境 ··· 4
 1.3　模块实现 ··· 4
  1.3.1　数据预处理 ··· 6
  1.3.2　抽取摘要 ·· 8
  1.3.3　模型搭建与编译 ··· 9
  1.3.4　模型训练与保存 ··· 12
  1.3.5　图形化界面的开发 ·· 21
  1.3.6　应用封装 ·· 21
 1.4　系统测试 ··· 22
  1.4.1　训练困惑度 ··· 22
  1.4.2　测试效果 ·· 24
  1.4.3　模型应用 ·· 29

项目 2　Trump 推特的情感分析 ·· 29
 2.1　总体设计 ··· 29
  2.1.1　系统整体结构图 ··· 30
  2.1.2　系统流程图 ··· 30
 2.2　运行环境 ··· 30
  2.2.1　Python 环境 ·· 30
  2.2.2　TensorFlow 环境 ··· 31
  2.2.3　工具包 ··· 31
 2.3　模块实现 ··· 31
  2.3.1　准备数据 ·· 31

2.3.2　数据预处理 ··· 32
　　　2.3.3　模型构建 ··· 32
　　　2.3.4　模型测试 ··· 33
　2.4　系统测试 ·· 35
　　　2.4.1　模型效果 ··· 42
　　　2.4.2　模型应用 ··· 42

项目3　基于LSTM的影评情感分析 ··· 42
　3.1　总体设计 ·· 46
　　　3.1.1　系统整体结构图 ·· 46
　　　3.1.2　系统前后端流程图 ··· 46
　3.2　运行环境 ·· 47
　　　3.2.1　Python 环境 ·· 47
　　　3.2.2　TensorFlow 环境 ··· 47
　　　3.2.3　Android 环境 ··· 48
　3.3　模块实现 ·· 48
　　　3.3.1　数据预处理 ··· 48
　　　3.3.2　模型构建及训练 ·· 48
　　　3.3.3　模型保存 ·· 51
　　　3.3.4　词典保存 ·· 53
　　　3.3.5　模型测试 ·· 54
　3.4　系统测试 ·· 55
　　　3.4.1　数据处理 ·· 59
　　　3.4.2　模型训练 ·· 59
　　　3.4.3　词典保存 ·· 60
　　　3.4.4　模型效果 ·· 60

项目4　Image2Poem——根据图像生成古体诗句 ································· 61
　4.1　总体设计 ·· 63
　　　4.1.1　系统整体结构图 ·· 63
　　　4.1.2　系统流程图 ·· 63
　4.2　运行环境 ·· 64
　　　4.2.1　Python 环境 ·· 64
　　　4.2.2　TensorFlow 安装 ·· 65
　　　4.2.3　其他 Python 模块的安装 ··· 65
　　　4.2.4　百度通用翻译 API 开通及使用 ······································ 66
　4.3　模块实现 ·· 66
　　　4.3.1　数据准备 ·· 66
　　　4.3.2　Web 后端准备 ··· 66
　　　4.3.3　百度通用翻译 ··· 72
　　　4.3.4　全局变量声明 ··· 73
　　　　　　　　　　　　　　　　　　　　　　　　　　　　　　　　　73

		4.3.5 创建模型	83
		4.3.6 模型训练及保存	86
		4.3.7 模型调用	88
	4.4	系统测试	88
		4.4.1 训练准确率	90
		4.4.2 模型效果	91
		4.4.3 整合应用	93

项目 5　歌曲人声分离

- 5.1 总体设计　93
  - 5.1.1 系统整体结构图　94
  - 5.1.2 系统流程图　94
- 5.2 运行环境　94
  - 5.2.1 Python 环境　95
  - 5.2.2 TensorFlow 环境　95
  - 5.2.3 Jupyter Notebook 环境　96
- 5.3 模块实现　97
  - 5.3.1 数据准备　97
  - 5.3.2 数据预处理　101
  - 5.3.3 模型构建　105
  - 5.3.4 模型训练及保存　107
  - 5.3.5 模型测试　111
- 5.4 系统测试　111
  - 5.4.1 训练准确率　112
  - 5.4.2 测试效果　112
  - 5.4.3 模型应用　115

项目 6　基于 Image Caption 的英语学习

- 6.1 总体设计　115
  - 6.1.1 系统整体结构图　116
  - 6.1.2 系统流程图　116
- 6.2 运行环境　116
  - 6.2.1 Python 环境　117
  - 6.2.2 TensorFlow 环境　117
  - 6.2.3 微信开发者工具　117
- 6.3 模块实现　117
  - 6.3.1 准备数据　119
  - 6.3.2 模型构建　121
  - 6.3.3 模型训练及保存　123
  - 6.3.4 模型调用　127
  - 6.3.5 模型测试

6.4 系统测试 ... 136
6.4.1 训练准确率 ... 136
6.4.2 测试效果 ... 136
6.4.3 模型应用 ... 137

项目7 智能聊天机器人 ... 137
7.1 总体设计 ... 140
7.1.1 系统整体结构图 ... 140
7.1.2 系统流程图 ... 140
7.2 运行环境 ... 141
7.2.1 Python 环境 ... 141
7.2.2 TensorFlow 环境 ... 142
7.3 模块实现 ... 142
7.3.1 数据预处理 ... 143
7.3.2 模型构建 ... 143
7.3.3 模型测试 ... 146
7.4 系统测试 ... 153
7.4.1 训练损失 ... 153
7.4.2 测试效果 ... 153
7.4.3 模型应用 ... 154

项目8 说唱歌词创作应用 ... 154
8.1 总体设计 ... 157
8.1.1 系统整体结构图 ... 157
8.1.2 系统流程图和前端流程图 ... 157
8.2 运行环境 ... 158
8.2.1 Python 环境 ... 158
8.2.2 TensorFlow 环境 ... 158
8.2.3 其他环境 ... 158
8.3 模块实现 ... 159
8.3.1 数据预处理与加载 ... 159
8.3.2 模型构建 ... 159
8.3.3 模型训练及保存 ... 164
8.3.4 模型测试 ... 169
8.4 系统测试 ... 172
8.4.1 模型困惑度 ... 178
8.4.2 模型应用 ... 178

项目9 基于LSTM的语音/文本/情感识别系统 ... 178
9.1 总体设计 ... 180
9.1.1 系统整体结构图 ... 180
9.1.2 系统流程图 ... 180
... 181

前言
PREFACE

  Python作为人工智能和大数据的主要开发语言，具有灵活性强、扩展性好、应用面广、可移植、可扩展、可嵌入等特点，近几年发展迅速，热度上涨，人才需求量逐年攀升，相关课程已经成为高等院校的专业课程。

  为适应当前教育教学改革的要求，更好地践行人工智能模型与算法应用，作者以实践教学与创新能力培养为目标，采取了创新方式，从不同难度、不同类型、不同算法，融合同类图书的优点，对实际智能应用案例进行总结。

  本书主要内容和素材来自开源网站的人工智能经典模型算法、信息工程专业创新课程内容、作者近几年承担的科研项目成果、作者指导学生所完成的创新项目，学生不仅学到了知识，提高了能力，而且为本书提供了第一手素材和相关资料。

  本书内容由总到分、先思考后实践、注重整体架构、系统流程与代码实现相结合。对于从事人工智能开发、机器学习和算法实现的专业技术人员，本书可以作为技术参考书、提高工程创新实践手册；也可以作为信息通信工程及相关领域的本科生参考书，为机器学习模型分析、算法设计、应用实现提供帮助。

  本书的编写得到了教育部高等学校电子信息类专业教学指导委员会、信息工程专业国家第一类特色专业建设项目、信息工程专业国家第二类特色专业建设项目、教育部CDIO工程教育模式研究与实践项目、教育部本科教学工程项目、信息工程专业北京市特色专业建设、北京市教育教学改革项目和北京邮电大学教育教学改革项目(编号为2019TD01)的大力支持，在此表示感谢！

  由于作者经验与水平有限，书中难免存在疏漏及不当之处，衷心地希望各位读者多提宝贵意见及具体整改措施，以便作者进一步修改和完善。

作 者
2020年12月

9.1.3 网页端配置流程图 ... 181

9.2 运行环境 ... 182
9.2.1 Python 环境 ... 182
9.2.2 TensorFlow 环境 ... 182
9.2.3 网页端环境框架——Django ... 183

9.3 模块实现（服务器端） ... 183
9.3.1 数据处理 ... 184
9.3.2 调用 API ... 184
9.3.3 模型构建 ... 187
9.3.4 模型训练及保存 ... 188

9.4 网页实现（前端） ... 188
9.4.1 Django 的管理脚本 ... 188
9.4.2 Django 的核心脚本 ... 190
9.4.3 网页端模板的组成 ... 198
9.4.4 Django 的接口验证脚本 ... 198
9.4.5 Django 中 URL 模板的连接器 ... 199
9.4.6 Django 中 URL 配置 ... 199

9.5 系统测试 ... 199
9.5.1 训练准确率 ... 200
9.5.2 效果展示 ... 202

项目 10 基于人脸检测的表情包自动生成器 ... 202

10.1 总体设计 ... 202
10.1.1 系统整体结构图 ... 203
10.1.2 系统流程图 ... 203
10.1.3 文件结构 ... 204

10.2 运行环境 ... 204
10.2.1 Python 环境 ... 204
10.2.2 TensorFlow 环境 ... 204
10.2.3 OpenCV 环境 ... 204
10.2.4 Pillow 环境 ... 205

10.3 模块实现 ... 205
10.3.1 图形用户界面 ... 206
10.3.2 人脸检测与标注 ... 207
10.3.3 人脸朝向识别 ... 211
10.3.4 人脸处理与表情包合成 ... 215

10.4 系统测试 ... 215
10.4.1 确定运行环境符合要求 ... 215
10.4.2 应用使用说明

项目 11　AI 作曲 ... 218
11.1　总体设计 ... 218
11.1.1　系统整体结构图 ... 218
11.1.2　系统流程图 ... 218
11.2　运行环境 ... 219
11.2.1　Python 环境 ... 219
11.2.2　虚拟机环境 ... 219
11.2.3　TensorFlow 环境 ... 219
11.2.4　Python 类库及项目软件 ... 222
11.3　模块实现 ... 222
11.3.1　数据预处理 ... 222
11.3.2　信息提取 ... 222
11.3.3　模型构建 ... 222
11.3.4　模型训练及保存 ... 225
11.3.5　音乐生成 ... 227
11.4　系统测试 ... 229
11.4.1　模型训练 ... 231
11.4.2　测试效果 ... 231

项目 12　智能作文打分系统 ... 231
12.1　总体设计 ... 233
12.1.1　系统整体结构图 ... 233
12.1.2　系统流程图 ... 233
12.1.3　前端流程图 ... 233
12.2　运行环境 ... 233
12.2.1　Python 环境 ... 235
12.2.2　Keras 环境 ... 235
12.2.3　Django 环境 ... 235
12.3　模块实现 ... 236
12.3.1　数据预处理 ... 236
12.3.2　模型构建 ... 236
12.3.3　模型训练及保存 ... 240
12.3.4　模型测试 ... 241
12.4　系统测试 ... 242
12.4.1　训练准确率 ... 249
12.4.2　模型应用 ... 249
12.4.3　测试效果 ... 250

项目 13　新冠疫情舆情监督 ... 251
13.1　总体设计 ... 252
13.1.1　系统整体结构图 ... 252
... 252

　　　　13.1.2　系统流程图 ··· 253
13.2　运行环境 ··· 253
　　　　13.2.1　Python 环境 ··· 254
　　　　13.2.2　PaddlePaddle 环境 ·· 254
13.3　模块实现 ··· 254
　　　　13.3.1　准备预处理 ··· 260
　　　　13.3.2　模型构建 ·· 263
　　　　13.3.3　模型训练 ·· 265
　　　　13.3.4　模型评估 ·· 265
　　　　13.3.5　模型预测 ·· 267
13.4　系统测试 ··· 267
　　　　13.4.1　训练准确率 ··· 267
　　　　13.4.2　测试效果 ·· 269
　　　　13.4.3　模型应用 ·· 270

项目14　语音识别——视频添加字幕 ·· 270

14.1　总体设计 ··· 270
　　　　14.1.1　系统整体结构图 ··· 271
　　　　14.1.2　系统流程图 ··· 271
14.2　运行环境 ··· 271
14.3　模块实现 ··· 272
　　　　14.3.1　分离音频 ·· 272
　　　　14.3.2　分割音频 ·· 273
　　　　14.3.3　提取音频 ·· 275
　　　　14.3.4　模型构建 ·· 279
　　　　14.3.5　识别音频 ·· 279
　　　　14.3.6　添加字幕 ·· 280
　　　　14.3.7　GUI 界面 ·· 282
14.4　系统测试 ··· 285

项目15　人脸识别与机器翻译小程序 ·· 285

15.1　总体设计 ··· 285
　　　　15.1.1　系统整体结构图 ··· 286
　　　　15.1.2　系统流程图 ··· 286
15.2　运行环境 ··· 287
　　　　15.2.1　Python 环境 ··· 287
　　　　15.2.2　TensorFlow-GPU/CPU 环境 ·· 287
　　　　15.2.3　OpenCV2 库 ··· 287
　　　　15.2.4　Dlib 库 ·· 287
　　　　15.2.5　Flask 环境 ··· 287
　　　　15.2.6　TensorFlow-SSD 目标（人脸）检测框架

15.2.7　TensorFlow-FaceNet 人脸匹配框架 ······ 287
　　　15.2.8　微信小程序开发环境 ······ 288
　　　15.2.9　JupyterLab ······ 288
　15.3　模块实现 ······ 288
　　　15.3.1　数据预处理 ······ 288
　　　15.3.2　创建模型 ······ 288
　15.4　系统测试 ······ 291

项目 16　基于循环神经网络的机器翻译 ······ 297
　16.1　总体设计 ······ 299
　　　16.1.1　系统整体结构图 ······ 299
　　　16.1.2　系统流程图 ······ 299
　16.2　运行环境 ······ 300
　　　16.2.1　Python 环境 ······ 300
　　　16.2.2　PyTorch 环境 ······ 300
　　　16.2.3　Flask 环境 ······ 300
　16.3　模块实现 ······ 301
　　　16.3.1　数据预处理 ······ 301
　　　16.3.2　模型构建 ······ 301
　　　16.3.3　模型训练及保存 ······ 303
　　　16.3.4　模型测试 ······ 306
　16.4　系统测试 ······ 307
　　　16.4.1　训练准确率 ······ 311
　　　16.4.2　模型应用 ······ 311

项目 17　基于 LSTM 的股票预测 ······ 311
　17.1　总体设计 ······ 314
　　　17.1.1　系统整体结构图 ······ 314
　　　17.1.2　系统流程图 ······ 314
　17.2　运行环境 ······ 315
　　　17.2.1　Python 环境 ······ 315
　　　17.2.2　TensorFlow 环境 ······ 315
　　　17.2.3　Numpy 环境 ······ 315
　　　17.2.4　Pandas 环境 ······ 316
　　　17.2.5　Keras 环境 ······ 316
　　　17.2.6　Matplotlib 环境 ······ 316
　17.3　模块实现 ······ 316
　　　17.3.1　数据预处理 ······ 317
　　　17.3.2　模型构建 ······ 317
　　　17.3.3　模型保存及输出预测 ······ 318
　　　17.3.4　模型测试 ······ 320
　　　　　　　　　　　　　　　　　　　　　　　　　　　　322

17.4 系统测试 330
    17.4.1 训练准确率 330
    17.4.2 模型效果 331

项目 18 基于 LSTM 的豆瓣影评分类情感分析 333
18.1 总体设计 333
    18.1.1 系统整体结构图 334
    18.1.2 系统流程图 334
18.2 运行环境 334
    18.2.1 Python 环境 334
    18.2.2 TensorFlow 环境 335
    18.2.3 Keras 环境 336
18.3 模块实现 336
    18.3.1 数据收集 337
    18.3.2 数据处理 339
    18.3.3 Word2Vec 模型 339
    18.3.4 LSTM 模型 342
    18.3.5 完整流程 343
    18.3.6 模型测试 346
18.4 系统测试 346
    18.4.1 训练准确率 346
    18.4.2 应用效果 349

项目 19 AI 写诗机器人 349
19.1 总体设计 349
    19.1.1 系统整体结构图 350
    19.1.2 系统流程图 350
19.2 运行环境 350
    19.2.1 Python 环境 351
    19.2.2 TensorFlow 环境 351
    19.2.3 Qt Creator 下载与安装 351
19.3 模块实现 352
    19.3.1 语料获取和整理 352
    19.3.2 特征提取与预训练 353
    19.3.3 构建模型 356
    19.3.4 模型训练 359
    19.3.5 结果预测 360
    19.3.6 设置诗句评分标准 361
    19.3.7 界面设计 362
19.4 系统测试 ....

项目 20 基于 COCO 数据集的自动图像描述 ································· 365
20.1 总体设计 ·· 365
20.1.1 系统整体结构图 ·· 365
20.1.2 系统流程图 ··· 365
20.2 运行环境 ·· 366
20.3 模块实现 ·· 366
20.3.1 数据准备 ·· 366
20.3.2 模型创建及保存 ·· 366
20.3.3 模型训练及保存 ·· 369
20.3.4 界面设置及演示 ·· 370
20.4 系统测试 ·· 373

项目 1 文章辅助生成系统
PROJECT 1

本项目基于学术论文、维基百科等数据集,通过 TextRank 和 Seq2Seq 算法对模型进行优化和改进,构建一体化的文章摘要、标题和关键词辅助生成系统,设计、对接可视化界面,将程序封装为可执行文件并在 PC 端直接运行。

1.1 总体设计

本部分包括系统整体结构图和系统流程图。

1.1.1 系统整体结构图

系统整体结构如图 1-1 所示。

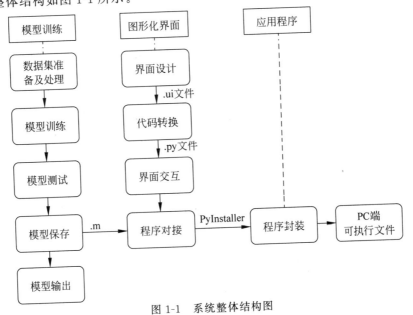

图 1-1 系统整体结构图

1.1.2 系统流程图

系统流程如图 1-2 所示。

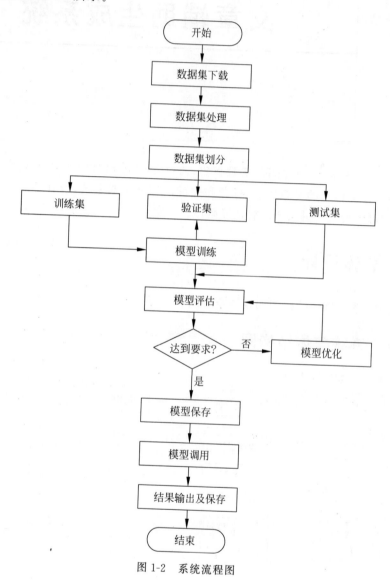

图 1-2 系统流程图

1.2 运行环境

本部分包括 Python 环境、TextRank 环境、TensorFlow 环境、PyQt5 及 Qt Designer 运行环境。

1.2.1　Python 环境

版本：Python 3.5。

1.2.2　TextRank 环境

从清华仓库镜像中下载 numpy-1.9.3.tar.gz、networkx-2.4.tar.gz、math-0.5.tar.gz 文件，在本地解压后，使用 cmd 命令行进入控制台，切换到对应目录中，执行 python.exe setup.py install 命令，完成安装。

1.2.3　TensorFlow 环境

下载 tensorflow-1.0.1-cp35-cp35m-win_amd64.whl，使用 cmd pip 命令进行安装，使用 pip 命令安装 tarfile、matplotlib、jieba 依赖包，实现 TensorFlow 平台相关模型的准备。

1.2.4　PyQt5 及 Qt Designer 运行环境

使用 pip 命令安装与 Python 语言对应版本的 PyQt5 工具包，同时在环境中配置 PyUIC5 和 PyQt5-tools，用于图形化界面的快速开发及转换。将上述工具添加至 PyCharm 编辑器的 ExternalTools 中。

从 PyCharm 编辑器的 Tools-External Tools 中打开 Qt Designer，如图 1-3 所示，表明 Qt Designer 安装成功。

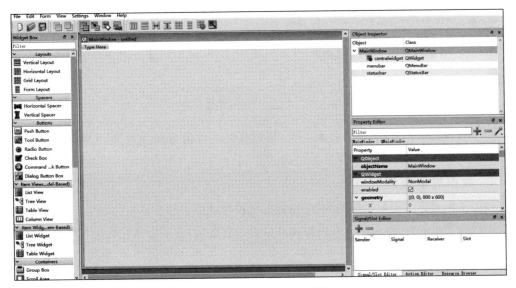

图 1-3　Qt Designer 工具图

1.3 模块实现

本项目包括6个模块：数据预处理、抽取摘要、模型搭建与编译、模型训练与保存、图形化界面的开发和应用封装。下面分别给出各模块的功能介绍及相关代码。

1.3.1 数据预处理

数据预处理下载地址为 http://www.sogou.com/labs/resource/cs.php，未经处理的原始数据图如图1-4所示。

```
<doc>
<url>http://gongyi.sohu.com/20120706/n347457739.shtml</url>
<docno>98590b972ad2f0ea-34913306c0bb3300</docno>
<contenttitle>深圳地铁将设立ＶＩＰ头等车厢　买双倍票可享坐票</contenttitle>
<content>南都讯　记者刘凡　周昌和　任笑一　继推出日票后，深圳今后将设地铁ＶＩＰ头等车厢，设坐票制。昨日，《南都ＭＥＴＲＯ》创刊仪式暨２０１２年深港地铁圈高峰论坛上透露，在未来的１１号线上将增
</doc>
<doc>
<url>http://gongyi.sohu.com/20120724/n348878190.shtml</url>
<docno>5fa7926d2cd2f0ea-34913306c0bb3300</docno>
<contenttitle>爸爸为女儿百万建幼儿园　消防设施３年仍不过关</contenttitle>
<content></content>
</doc>
<doc>
<url>http://gongyi.sohu.com/s2008/sourceoflife/</url>
<docno>f2467af22cd2f0ea-34913306c0bb3300</docno>
<contenttitle>中国西部是地球上主要干旱带之一，妇女是当地劳动力．．．</contenttitle>
<content>同心县地处宁夏中部干旱带的核心区，　冬寒长，春暖迟，夏热短，秋凉早，干旱少雨，蒸发强烈，风大沙多。主要自然灾害有沙尘暴、干热风、霜冻、冰雹等，其中以干旱危害最为严重。　由于生态环境的
</doc>
<doc>
<url>http://gongyi.sohu.com/20120612/n345424232.shtml</url>
<docno>0dadd5002ed2f0ea-34913306c0bb3300</docno>
<contenttitle>思源焦点公益基金救助孩子：永康</contenttitle>
<content>不满一岁的永康是个饱经病痛折磨的孩子，２０１１年７月５日出生的他，患有先天性心脏病、疝气，一出生便被遗弃。２０１２年１月８日，才５个月大的永康被发现呼吸困难，随后送往医院进行抢救治疗
</doc>
<doc>
<url>http://gongyi.sohu.com/20120629/n346847569.shtml</url>
<docno>be6bcb252fd2f0ea-34913306c0bb3300</docno>
<contenttitle>康师傅回应转卖废弃茶叶：下家承诺用废料做枕头</contenttitle>
<content>就废弃茶叶被转手事件发声明　本报讯（记者刘俊）　“我们也是受害者！”昨日，有媒体报道称康师傅的废弃茶叶被转手卖给不良商家、冒充名茶流入市场，康师傅的一位联系人这样说。康师傅昨日晚间发
</doc>
<doc>
<url>http://gongyi.sohu.com/s2009/gongyidream/</url>
<docno>66c756872fd2f0ea-34913306c0bb3300</docno>
<contenttitle>活动时间：</contenttitle>
<content>·奖励办法：率先提交的前１００个创意项目，经评估，可优先资助实施。　·咨询电话：０１０－６７７８４７１０，０１０－６７７８４７２０。　·报名方式：先下载报名表填写完整，网上直接上传项目概
</doc>
<doc>
<url>http://gongyi.sohu.com/20120730/n349358066.shtml</url>
<docno>fdaa73d52fd2f0ea-34913306c0bb3300</docno>
<contenttitle>失独父母中年遇独子夭折　称不怕死亡怕养老生病</contenttitle>
<content></content>
</doc>
<doc>
<url>http://gongyi.sohu.com/s2009/xianxue/</url>
```

图1-4 未经处理的原始数据图

对于其编码形式，由于文件过大，无法通过打开文件的方式获取编码，采用GBK18030可对其进行编码。处理过程如下。

1. 数据提取及划分

使用正则表达式提取数据的内容，按照比例进行训练集和验证集的划分，去除文本内容长度不符合要求的数据，对划分后的标题和内容进行存储，生成的文件使用软件改变编码格式为utf-8。相关代码如下：

1）正则表达式匹配

```
for_title = '<contenttitle>(.*)</contenttitle>'    #筛选标题，去除标签
for_content = '<content>(.*)</content>'            #筛选内容，去除标签
```

```python
p1 = re.compile(for_title)
p2 = re.compile(for_content)
```

2)将数据筛选进行写入过程

```python
for i in range(4,len(data.values) + 1,6):       #针对位置选择相应的数据
    n = p2.findall(str(data.values[i]))
    text = n[0]
    word = text
    result = ''
    for w in word:
        result = result + str(w.word) + ' '
#对意外的情况进行替换
    result = result.replace(u'\u3000','').replace(u'\ue40c','')
#检查数据长度是否符合需求,太长或者太短,都要舍弃
    if len(result)>= 1024 or len(result) == 0:
        id.append(i)
        continue
    if i < for_train:
        f_content_train.write(result + '\n')
    else:
        f_content_test.write(result + '\n')
    print((i/6)/len(range(3,len(data.values) + 1,6)))
```

2. 替换和分词

对获得的文本进行标签替换,完成分词操作。相关代码如下:

```python
def token(self, sentence):
    words = self.segmentor.segment(sentence)         #分词
    words = list(words)
    postags = self.postagger.postag(words)           #词性标注
    postags = list(postags)
    netags = self.recognizer.recognize(words, postags) #命名实体识别
    netags = list(netags)
    result = []
    for i, j in zip(words, netags):
        if j in ['S-Nh', 'S-Ni', 'S-Ns']:
            result.append(j)
            continue
        result.append(i)
    return result
```

使用上述代码后,得到4个文件——2个训练集和2个测试集,对应文件的同一行分别为标题和内容。对所有的文字都进行标签替换,完成分词。

3. 数据读取

根据得到的文件进行数据读取,相关代码如下:

```python
data_set = [[] for _ in buckets]
with tf.gfile.GFile(source_path, mode = "r") as source_file:
    with tf.gfile.GFile(target_path, mode = "r") as target_file:
        source, target = source_file.readline(), target_file.readline()
        counter = 0                                              #源文件和目标文件
        while source and target and (not max_size or counter < max_size):
            counter += 1
            if counter % 10000 == 0:
                print("reading data line % d" % counter)         #输出信息
                sys.stdout.flush()
            source_ids = [int(x) for x in source.split()]
            target_ids = [int(x) for x in target.split()]
            target_ids.append(data_utils.EOS_ID)                 #添加标识
            for bucket_id, (source_size, target_size) in enumerate(buckets):
                if len(source_ids)< source_size and len(target_ids)< target_size:
                    data_set[bucket_id].append([source_ids, target_ids])
                    break
            source, target = source_file.readline(), target_file.readline()
return data_set
```

该代码会自动读取信息,并将读取的数据存入 bucket 内,深度学习模型通过学习内容与标题生成,数据预处理在准备数据时进行,使用 Python 的 codecs 模块读取即可。

1.3.2 抽取摘要

大多数论文的篇幅都是数万字,直接使用模型对数据进行训练与测试会耗费计算资源。因此,通过文本排序对数据进行重要性提取,算法如下。

1. 排序迭代算法

首先,获得二位列表,句子为子列表,元素是单词;其次,通过判断两个单词是否同时出现在同一个时间窗口内确定链接。将所有词添加到图的链接后,使用 PageRank 算法进行迭代,获得平稳的单词 PR 值;最后,获取重要的单词列表。

```
def sort_words(vertex_source, edge_source, window = 2, pagerank_config = {'alpha': 0.85, }):
    #对单词的关键程度进行排序
    """
    vertex_source: 二维列表,子列表代表句子,其元素是单词,用来构造 PageRank 算法中的节点
    edge_source: 二维列表,子列表代表句子,其元素为单词,根据单词位置关系构造 PageRank 算法中的
    边窗口,一个句子中相邻的单词,两两之间认为有边
    pagerank_config: PageRank 算法的设置
    """
    sorted_words = []
    word_index = {}
    index_word = {}
    _vertex_source = vertex_source
```

```
_edge_source = edge_source
words_number = 0
for word_list in _vertex_source: #对每个句子进行处理,提取包含单词的列表
    for word in word_list:
        if not word in word_index:
#更新 word_index,假如字典中没有单词,将这个单词与索引添加到字典中
            word_index[word] = words_number
            index_word[words_number] = word  #对 word 进行反向映射
            words_number += 1
graph = np.zeros((words_number, words_number))
#构建 word_number * word_number 的矩阵,实现图计算
for word_list in _edge_source:
    for w1, w2 in combine(word_list, window):
        if w1 in word_index and w2 in word_index:
            index1 = word_index[w1]
            index2 = word_index[w2]
            graph[index1][index2] = 1.0
            graph[index2][index1] = 1.0
#根据窗口判断其连接
nx_graph = nx.from_numpy_matrix(graph)
#构成邻接矩阵
scores = nx.pagerank(nx_graph, ** pagerank_config)
#使用 PageRank 算法进行迭代
sorted_scores = sorted(scores.items(), key = lambda item: item[1], reverse = True)
for index, score in sorted_scores:
    item = AttrDict(word = index_word[index], weight = score)
    sorted_words.append(item)
return sorted_words
```

2. 句子相似度算法

在用 TextRank 算法对句子进行输出时,使用的默认节点是句子,两个节点相互连接的权重使用句子的相似度。相关代码如下:

```
def get_similarity(word_list1, word_list2): #计算两个句子的相似程度
    """默认用于计算两个句子相似度的函数
    word_list1, word_list2 分别代表两个句子,都是由单词组成的列表
    """
    words = list(set(word_list1 + word_list2))
    vector1 = [float(word_list1.count(word)) for word in words]
#统计某个单词在句子中的频率
    vector2 = [float(word_list2.count(word)) for word in words]
    vector3 = [vector1[x] * vector2[x] for x in range(len(vector1))]
    vector4 = [1 for num in vector3 if num > 0.]
    co_occur_num = sum(vector4)          #分子
    if abs(co_occur_num) <= 1e - 12:
        return 0.
```

```
            denominator = math.log(float(len(word_list1))) + math.log(float(len(word_list2)))
                                        #分母
            if abs(denominator) < 1e-12:
                return 0.
        return co_occur_num / denominator            #返回句子的相似度
```

1.3.3 模型搭建与编译

完成数据集制作后,进行模型搭建、定义模型输入、确定损失函数。

1. 模型搭建

以 TensorFlow 提供的模型为基础,参数使用类进行传递:

```
class LargeConfig(object):                    #定义网络结构
            learning_rate = 1.0               #学习率
            init_scale = 0.04
            learning_rate_decay_factor = 0.99 #学习率下降
            max_gradient_norm = 5.0
            num_samples = 4096                #采样 Softmax
            batch_size = 64
            size = 256                        #每层节点数
            num_layers = 4                    #层数
    vocab_size = 50000
    #模型构建
    def seq2seq_f(encoder_inputs, decoder_inputs, do_decode):
            return tf.contrib.legacy_seq2seq.embedding_attention_seq2seq(
                encoder_inputs,                          #输入的句子
                decoder_inputs,                          #输出的句子
                cell,                                    #使用的 cell、LSTM 或者 GRU
                num_encoder_symbols = source_vocab_size, #源字典的大小
                num_decoder_symbols = target_vocab_size, #转换后字典的大小
                embedding_size = size,                   #embedding 的大小
                output_projection = output_projection,   #看字典大小
                feed_previous = do_decode,               #进行训练还是测试
                dtype = tf.float32)
```

2. 定义模型输入

在模型中,bucket 承接输入的字符,所以须为 bucket 的每个元素构建一个占位符。

```
    #输入
    self.encoder_inputs = []
    self.decoder_inputs = []
    self.target_weights = []
    for i in xrange(buckets[-1][0]):
        self.encoder_inputs.append(tf.placeholder(tf.int32, shape = [None],
                                    name = "encoder{0}".format(i)))
```

```python
# 为列表对象中的每个元素构建一个占位符,名称分别为 encoder0、encoder1…
    for i in xrange(buckets[-1][1] + 1):
        self.decoder_inputs.append(tf.placeholder(tf.int32, shape=[None],
                                                  name="decoder{0}".format(i)))
        self.target_weights.append(tf.placeholder(tf.float32, shape=[None],
                                                  name="weight{0}".format(i)))
# target_weights 是一个与 decoder_outputs 大小一样的矩阵
# 该矩阵将目标序列长度以外的其他位置填充为标量值 0
# 目标是将解码器输入移位 1
    targets = [self.decoder_inputs[i + 1]
                for i in xrange(len(self.decoder_inputs) - 1)]
# 将 decoder input 向右平移一个单位
```

3. 确定损失函数

在损失函数上,使用 TensorFlow 中的 sampled_softmax_loss() 函数。

```python
def sampled_loss(labels, inputs):          # 使用候选采样损失函数
    labels = tf.reshape(labels, [-1, 1])
# 需要使用 32 位浮点数计算 sampled_softmax_loss,以避免数值不稳定性
    local_b = tf.cast(b, tf.float32)
    local_inputs = tf.cast(inputs, tf.float32)
    return tf.cast(
        tf.nn.sampled_softmax_loss(          # 损失函数
            weights=local_w_t,
            biases=local_b,
            labels=labels,
            inputs=local_inputs,
            num_sampled=num_samples,
            num_classes=self.target_vocab_size), tf.float32)
```

1.3.4 模型训练与保存

设定模型结构后,定义模型训练函数,以导入及调用模型。

1. 定义模型训练函数

定义模型训练函数及相关操作。

```python
def train():
    # 准备标题数据
    print("Preparing Headline data in %s" % FLAGS.data_dir)
    src_train, , dest_train, src_dev, dest_dev, _, _ = data_utils.prepare_headline_data(FLAGS.
data_dir, FLAGS.vocab_size)
# 将获得的数据进行处理,包括:构建词典、根据词典单词 ID 的转换、返回路径
    config = tf.ConfigProto(device_count={"CPU": 4},
                            inter_op_parallelism_threads=1,
                            intra_op_parallelism_threads=2)
```

```python
with tf.Session(config = config) as sess:
    print("Creating %d layers of %d units." % (FLAGS.num_layers, FLAGS.size))
    model = create_model(sess, False)
    # 创建模型
    print ("Reading development and training data (limit: %d)."
        % FLAGS.max_train_data_size)
    dev_set = read_data(src_dev, dest_dev)
    train_set = read_data(src_train, dest_train, FLAGS.max_train_data_size)
    train_bucket_sizes = [len(train_set[b]) for b in xrange(len(buckets))]
    train_total_size = float(sum(train_bucket_sizes))
    trainbuckets_scale = [sum(train_bucket_sizes[:i + 1]) / train_total_size
                          for i in xrange(len(train_bucket_sizes))]
    # 进行循环训练
    step_time, loss = 0.0, 0.0
    current_step = 0
    previous_losses = []
    while True:
        random_number_01 = np.random.random_sample()
        bucket_id = min([i for i in xrange(len(trainbuckets_scale))
                        if trainbuckets_scale[i] > random_number_01])
        # 随机选择一个bucket进行训练
        start_time = time.time()
        encoder_inputs, decoder_inputs, target_weights = model.get_batch(
            train_set, bucket_id)
        _, step_loss, _ = model.step(sess, encoder_inputs, decoder_inputs,
                                     target_weights, bucket_id, False)
        step_time += (time.time() - start_time)/FLAGS.steps_per_checkpoint
        loss += step_loss / FLAGS.steps_per_checkpoint
        current_step += 1
        if current_step % FLAGS.steps_per_checkpoint == 0:
            perplexity = math.exp(float(loss)) if loss < 300 else float("inf")
            print ("global step %d learning rate %.4f step-time %.2f perplexity "
                "%.2f" % (model.global_step.eval(), model.learning_rate.eval(),
                          step_time, perplexity))   # 输出参数
            if len(previous_losses)> 2 and loss > max(previous_losses[-3:]):
                sess.run(model.learning_rate_decay_op)
            previous_losses.append(loss)
            checkpoint_path = os.path.join(FLAGS.train_dir, "headline_large.ckpt")
            model.saver.save(sess, checkpoint_path, global_step = model.global_step)   # 检查点输出路径
            step_time, loss = 0.0, 0.0
            for bucket_id in xrange(len(buckets)):
                if len(dev_set[bucket_id]) == 0:
```

```python
                    print(" eval: empty bucket % d" % (bucket_id))
                    continue
                    encoder_inputs,decoder_inputs,target_weights = 
model.get_batch(dev_set, bucket_id)                         #编解码及目标加权
                    _,eval_loss,_ = model.step(sess,encoder_inputs, 
decoder_inputs, target_weights, bucket_id, True)
                    eval_ppx = math.exp(float(eval_loss))   #计算损失
if eval_loss < 300
else float("inf")
print("eval:bucket % dperplexity % .2f" % (bucket_id, eval_ppx))     #输出困惑度
                    sys.stdout.flush()
```

2. 模型导入及调用

将生成的模型放在/ckpt文件夹内部,运行的过程中加载该模型。当程序获取句子之后,进行以下处理:

```python
while sentence:
    sen = tf.compat.as_bytes(sentence)
    sen = sen.decode('utf-8')
    token_ids = data_utils.sentence_to_token_ids(sen, vocab, 
normalize_digits = False)
    print (token_ids)          #打印 ID
    #选择合适的 bucket
    bucket_id = min([b for b in xrange(len(buckets)) if buckets[b][0] > len(token_ids)])
    print ("current bucket id" + str(bucket_id))
    encoder_inputs, decoder_inputs, target_weights = model.get_batch(
        {bucket_id: [(token_ids, [])]}, bucket_id)
            #获得模型的输出
    _, _, output_logits_batch = model.step(sess, encoder_inputs, 
decoder_inputs, target_weights,
            bucket_id, True)
    #贪婪解码器
    output_logits = []
    for item in output_logits_batch:
        output_logits.append(item[0])
    print (output_logits)
    print (len(output_logits))
    print (output_logits[0])
    outputs = [int(np.argmax(logit)) for logit in output_logits]
    print(output_logits)
    #剔除程序对文本进行的标记
    if data_utils.EOS_ID in outputs:
        outputs = outputs[:outputs.index(data_utils.EOS_ID)]
    print(" ".join([tf.compat.as_str(rev_vocab[output]) for output in outputs]))
```

1.3.5　图形化界面的开发

为提高可用性,将面向代码进行操作的环境转变为面向界面的操作,通过 Python 提供的 Qt Designer 及 PyQt5 环境完成项目的图形化界面。

1. 界面设计

从 PyCharm 的 External Tools 中打开配置好的 Qt Designer,创建主窗口,并使用 WidgetBox 进行组件的布局,如图 1-5 所示。

图 1-5　使用 Qt Designer 进行界面布局设计图

原生组件的美观性不足,需对各组件进行样式自定义。在监控窗口中选择修改对应组件的样式表,通过添加 CSS(Cascading Style Sheets,层叠样式表)完成各组件和界面的美化。如图 1-6 所示的美化图展示了"进入程序"按钮的 CSS 代码,此处分别设置了基础样式、点按样式及鼠标悬浮样式,使按钮的逻辑更贴近真实的使用场景,提升用户体验。

对主窗口和各组件的样式表修改后进行预览,如图 1-7 所示。

2. 代码转换

将上述界面设计保存为 .ui 文件,使用配置好的 PyUIC5 工具进行处理,得到转换后的 .py 代码。

```
QPushButton#pushButton_openfile{
    border: 1px solid #9a8878;
    background-color:#ffffff;
    border-style: solid;
    border-radius:0px;
    width: 40px;
    height:20px;
    padding:0 0px;
    margin:0 0px;
}
QPushButton#pushButton_openfile:pressed{
    background-color:#FBF7F6;
    border:0.5px solid #DDCFC2;
}
QPushButton#pushButton_openfile:hover{
    border:0.5px solid #DDCFC2;
}
```

图 1-6　美化图

图 1-7　主页设计预览(按钮样式为指针悬浮)图

对代码中无法通过程序运行渲染出的组件进行调整,例如,通过.qrc 文件指定关联引入的 icon 修改为引用相对地址。相关代码如下:

```python
from PyQt5 import QtCore, QtGui, QtWidgets              # 引入所需的库
from PyQt5 import QtCore, QtGui, QtWidgets, Qt
from PyQt5.QtWidgets import *
import PreEdit
class Ui_MainWindow_home(QtWidgets.QMainWindow):        # 定义界面类
    def __init__(self):
        super(Ui_MainWindow_home,self).__init__()
        self.setupUi(self)
        self.retranslateUi(self)
    def setupUi(self, MainWindow_home):                  # 设置界面
        MainWindow_home.setObjectName("MainWindow_home")
        MainWindow_home.resize(900, 650)
        MainWindow_home.setMinimumSize(QtCore.QSize(900, 650))
        MainWindow_home.setMaximumSize(QtCore.QSize(900, 650))
        MainWindow_home.setBaseSize(QtCore.QSize(900, 650))
        font = QtGui.QFont()
        font.setFamily("黑体")
        font.setPointSize(12)
        MainWindow_home.setFont(font)                    # 设置字体
        MainWindow_home.setStyleSheet("QMainWindow#MainWindow_home{\n"
            "background:#FFFEF8\n}")
        self.centralwidget = QtWidgets.QWidget(MainWindow_home)
        self.centralwidget.setStyleSheet("")             # 设置表单风格
        self.centralwidget.setObjectName("centralwidget")
        self.pushButton_openfile = QtWidgets.QPushButton(self.centralwidget)
        self.pushButton_openfile.setGeometry(QtCore.QRect(320,328,258,51))
        font = QtGui.QFont()
        font.setFamily("等线")
        font.setPointSize(11)
        font.setBold(True)
        font.setWeight(75)
        self.pushButton_openfile.setFont(font)           # 单击按钮,自动浏览文件设置
        self.pushButton_openfile.setCursor(QtGui.QCursor(QtCore.Qt.PointingHandCursor))
        self.pushButton_openfile.setStyleSheet("QPushButton#pushButton_openfile{  \n"
            "border: 1px solid #9a8878;  \n"
            "background-color:#ffffff;\n"
            "border-style: solid;  \n"
            "border-radius:0px;  \n"
            "width: 40px; \n"
            "height:20px;  \n"
            "padding:0 0px;  \n"
            "margin:0 0px;  \n"
            "} \n"
            "\n"
            "QPushButton#pushButton_openfile:pressed{\n"
            "background-color:#FBF7F6;\n"
            "border:0.5px solid #DDCFC2;\n"
```

```python
            "}\n"
            "\n"
            "QPushButton#pushButton_openfile:hover{\n"
            "border:0.5px solid #DDCFC2;\n"
            "}")
        icon = QtGui.QIcon()                                    #设置图标
        icon.addPixmap(QtGui.QPixmap(r".\icon\enter2.png"),
QtGui.QIcon.Normal, QtGui.QIcon.Off)
        icon.addPixmap(QtGui.QPixmap(r".\icon\enter2.png"),
QtGui.QIcon.Normal, QtGui.QIcon.On)
        self.pushButton_openfile.setIcon(icon)
        self.pushButton_openfile.setCheckable(False)
        self.pushButton_openfile.setObjectName("pushButton_openfile")
        self.label_maintitle_shadow = QtWidgets.QLabel(self.centralwidget)
        self.label_maintitle_shadow.setGeometry(QtCore.QRect(331, 188, 241, 61))
        font = QtGui.QFont()                                    #设置图形界面的字体
        font.setFamily("微软雅黑")
        font.setPointSize(36)
        font.setBold(True)
        font.setWeight(75)
        self.label_maintitle_shadow.setFont(font)
        self.label_maintitle_shadow.setStyleSheet("QLabel#label_maintitle_shadow{\n"
            " color:#847c74\n"
            "}")                                                #设置表单的风格
        self.label_maintitle_shadow.setAlignment(QtCore.Qt.AlignCenter)
        self.label_maintitle_shadow.setObjectName("label_shadow")
        self.label_format = QtWidgets.QLabel(self.centralwidget)
        self.label_format.setGeometry(QtCore.QRect(325, 395, 251, 20))
        font = QtGui.QFont()
        font.setFamily("黑体")
        font.setPointSize(10)
        self.label_format.setFont(font)                         #设置表单的格式字体
        self.label_format.setStyleSheet("QLabel#label_format{\n"
            "color:#3A332A\n"
            "}")
        self.label_format.setObjectName("label_format")
        self.label_maintitle = QtWidgets.QLabel(self.centralwidget)
        self.label_maintitle.setGeometry(QtCore.QRect(331, 189, 241, 61))
        font = QtGui.QFont()
        font.setFamily("微软雅黑")
        font.setPointSize(35)
        font.setBold(True)
        font.setWeight(75)
        self.label_maintitle.setFont(font)
        self.label_maintitle.setStyleSheet("QLabel#label_maintitle{\n"
            "color:#3A332A\n"
            "}")                                                #设置主题标签的风格
```

```python
        self.label_maintitle.setAlignment(QtCore.Qt.AlignCenter)
        self.label_maintitle.setObjectName("label_maintitle")
        self.label_author = QtWidgets.QLabel(self.centralwidget)
        self.label_author.setGeometry(QtCore.QRect(328, 600, 251, 20))
        font = QtGui.QFont()
        font.setFamily("等线")
        font.setPointSize(8)
        self.label_author.setFont(font)
        self.label_author.setStyleSheet("QLabel#label_author{\n"
            "color:#97846c\n"                    #设置表单风格
            "}")
        self.label_author.setAlignment(QtCore.Qt.AlignCenter)
        self.label_author.setObjectName("label_author")
        MainWindow_home.setCentralWidget(self.centralwidget)
        self.menubar = QtWidgets.QMenuBar(MainWindow_home)
        self.menubar.setGeometry(QtCore.QRect(0, 0, 900, 23))
        self.menubar.setObjectName("menubar")
        MainWindow_home.setMenuBar(self.menubar)       #设置主窗口菜单栏
        self.statusbar = QtWidgets.QStatusBar(MainWindow_home)
        self.statusbar.setObjectName("statusbar")
        MainWindow_home.setStatusBar(self.statusbar)   #设置主窗口状态栏
        self.retranslateUi(MainWindow_home)
        QtCore.QMetaObject.connectSlotsByName(MainWindow_home)
    def retranslateUi(self, MainWindow_home):
        _translate = QtCore.QCoreApplication.translate
        MainWindow_home.setWindowTitle(_translate("MainWindow_home",
"MainWindow"))                                 #设置主窗口标题
        self.pushButton_openfile.setText(_translate("MainWindow_home",
"进入程序"))
        self.label_maintitle_shadow.setText(_translate("MainWindow_home",
"论文助手"))
        self.label_format.setText(_translate("MainWindow_home",
"支持扩展名:.pdf  .doc  .docx  .txt"))
        self.label_maintitle.setText(_translate("MainWindow_home",
"论文助手"))
        self.label_author.setText(_translate("MainWindow_home",
"Designed by Hu Tong & Li Shuolin"))
    def openfile(self):
        openfile_name = QFileDialog.getOpenFileName(self,'选择文件',
'','files(*.doc,*.docx,*.pdf,*.txt)')
```

3. 界面交互

完成设计后,在界面之间建立交互关系。此处尝试两种方式:一是定义跳转函数;二是绑定按钮的槽函数,以完成跳转。

1) 定义跳转函数

定义跳转函数及相关操作。

```
#Jumpmain2pre.py
from PyQt5 import QtCore, QtGui, QtWidgets
from home import Ui_MainWindow_home      #跳转按钮所在界面
from PreEdit import Ui_Form               #跳转到的界面
    class Ui_PreEdit(Ui_Form):            #定义跳转函数的名字
        def __init__(self):
            super(Ui_PreEdit,self).__init__()  #跳转函数类名
            self.setupUi(self)
    #主界面
class Mainshow(Ui_MainWindow_home):
    def __init__(self):
        super(Mainshow,self).__init__()
        self.setupUi(self)
    #定义按钮功能
    def loginEvent(self):
        self.hide()
        self.dia = Ui_PreEdit()           #跳转到的界面类名
        self.dia.show()
    def homeshow():                       #调用函数
        import sys
        app = QtWidgets.QApplication(sys.argv)
        first = Mainshow()
        first.show()
        first.pushButton_openfile.clicked.connect(first.loginEvent)
    #绑定跳转功能的按钮
        sys.exit(app.exec_())
```

2）绑定按钮的槽函数

给需要完成跳转功能的按钮定义Click（单击）事件，并使用槽函数绑定事件。此处以绑定showwaiting()函数为例：

```
self.pushButton_create.clicked.connect(self.showwaiting)
```

将按钮要绑定的事件单独定义为函数：

```
def showwaiting(self):
    import sys
    self.MainWindow = QtWidgets.QMainWindow()
    self.newshow = Ui_MainWindow_sumcreating()    #创建图形界面
    self.newshow.setupUi(self.MainWindow)          #设置界面
    self.hide()
    self.MainWindow.show()
    print('生成中…')
```

3）示例：在图形化界面中读取本地文件

将打开动作、保存动作、保存内容分别定义为三个函数,便于单击按钮时通过槽进行调用,相关代码如下:

```python
def open_event(self):  #打开文件事件
    _translate = QtCore.QCoreApplication.translate
    directory1 = QFileDialog.getOpenFileName(None, "选择文件", "C:/",
"Word文档(*.docx;*.doc);;文本文件(*.txt);;pdf(*.pdf);;")
    print(directory1)      #输出路径
    path = directory1[0]
    self.open_path_text.setText(_translate("Form", path))
    if path is not None:
        with open(file = path, mode = 'r+', encoding = 'utf-8') as file:
            self.text_value.setPlainText(file.read())
def save_event(self):      #保存事件
    global save_path
    _translate = QtCore.QCoreApplication.translate
    fileName2, ok2 = QFileDialog.getSaveFileName(None, "文件保存", "C:/",
"Text Files (*.txt)")
    print(fileName2)       #打印保存文件的全部路径(包括文件名和后缀名)
    save_path = fileName2
    self.save_path_text.setText(_translate("Form", save_path))
def save_text(self):       #保存文本
    global save_path
    if save_path is not None:
        with open(file = save_path, mode = 'a+', encoding = 'utf-8') as file:
            file.write(self.text_value.toPlainText())
        print('已保存!')
```

给打开、保存按钮绑定单击动作,并通过槽调用上面定义的相应函数。同时,使用相同的方式将 open_event() 和 save_event() 函数中获取的路径与已定义的两个路径显示框进行关联,相关代码如下:

```python
def retranslateUi(self, Form):
    _translate = QtCore.QCoreApplication.translate
    Form.setWindowTitle(_translate("Form", "Form"))
    self.label_preview.setText(_translate("MainWindow_preview","预览"))
    self.open_path_text.setPlaceholderText(_translate("Form","打开"))
    self.open_path_but.setText(_translate("Form", "浏览"))
    self.save_path_but.setText(_translate("Form", "浏览"))
    self.save_path_text.setPlaceholderText(_translate("Form","保存"))
    self.save_but.setText(_translate("Form", "保存"))
    self.open_path_but.clicked.connect(self.open_event)
    self.save_path_but.clicked.connect(self.save_event)
    self.save_but.clicked.connect(self.save_text)
    self.pushButton_create.clicked.connect(self.showwaiting)
    self.pushButton_create.setText(_translate("Main_preview","生成"))
```

程序运行效果分别如图 1-8 和图 1-9 所示。

图 1-8　程序运行效果一：在图形化界面中打开文件

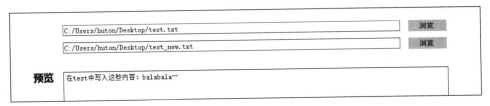

图 1-9　程序运行效果二：显示选定的打开和保存路径

4. 程序对接

为完成模型与图形化界面的联合，需要在代码中留出相应的对接接口。在本项目中，界面与模型主体的重点对接部分共有四处：调用模型、模型处理结束、结果展示和输出保存。

1) 调用模型

为完成单击生成按钮后开始调用模型进行处理的功能，在 PreEdit.py 最后的 showWaiting() 函数中写入调用接口，在摘要生成页弹出的同时调用模型。相关代码如下：

```python
def showWaiting(self):
    import sys
    self.MainWindow = QtWidgets.QMainWindow()
    self.newshow = Ui_MainWindow_sumcreating()      # 创建
    self.newshow.setupUi(self.MainWindow)           # 设置
    self.hide()
    # 待对接程序,读取前面保存的文件(文件的路径在 save_event 函数里)
    # 调用模型进行输出并保存
    self.MainWindow.show()
    print('生成中…')
```

2) 模型处理结束

在模型处理结束后需继续运行结果展示页,故在 PaperMain.py 的主函数里加一个判断,在判断模型处理完毕后,调用 resultShow() 函数,继续运行后续的结果展示。相关代码如下:

```python
def main():
    homeShow()
    # 待对接程序在 PreEdit.py 最后的 showWaiting() 函数里调用模型
    # 待对接程序判断处理完成后继续运行结果展示页
    resultShow()
```

3) 结果展示

模型处理完毕后,需要在结果展示页显示得到摘要、标题、关键词。结果展示页对应 result.py 文件,对接前,界面显示的结果为固定的字符串;对接时仅需将模型运行的结果存为字符串形式,替换之前固定的内容即可。相关代码如下:

```python
# 待对接程序模型运行的结果存为几个字符串后替换下面的文字即可
# 替换摘要
self.plainTextEdit_summary.setPlainText(_translate("MainWindow_result", "生成的摘要"))
# 替换标题1
self.lineEdit_title1.setText(_translate("MainWindow_result","标题1"))
# 替换标题2
self.lineEdit_title2.setText(_translate("MainWindow_result","标题2"))
# 替换标题3
self.lineEdit_title3.setText(_translate("MainWindow_result","标题3"))
# 替换关键词
self.lineEdit_keywords.setText(_translate("MainWindow_result","关键词"))
```

4) 输出保存

单击"保存"按钮时,将模型的输出直接保存到本地。该功能由 result.py 中 save_text() 函数的 f.write() 完成,在对接之前,f.write() 的输出为固定字符串,对接时替换为模型输出的内容即可。相关代码如下:

```python
def save_text(self):
    global save_path
    if save_path is not None:
```

```
with open(file = save_path, mode = 'a + ', encoding = 'utf - 8') as file:
# 对接 file.write,这里直接把程序里的字符串加起来写入保存的结果
    file.write("hello,Tibbarr")
print('已保存!')
```

1.3.6 应用封装

为提高使用的便捷性,降低用户的使用门槛,本项目需要进行一体化封装。考虑到论文的撰写大多是在 PC 端完成的,故使用 PyInstaller 将项目封装为.exe 应用程序。

1. 安装 PyInstaller

从清华仓库镜像中下载 PyInstaller-3.6.tar.gz,在本地解压后,使用 cmd 进入控制台,切换到解压后的对应目录中,执行命令 python.exe setup.py install,即可完成安装。

2. 将程序打包为.exe 文件

打开命令窗口,将目录切换到 papermain.py 路径下,输入命令 pyinstaller -F -w papermain.py,如图 1-10 所示。

图 1-10 使用 PyInstaller 命令进行程序打包

使用 PyInstaller 命令打包成功,如图 1-11 所示。

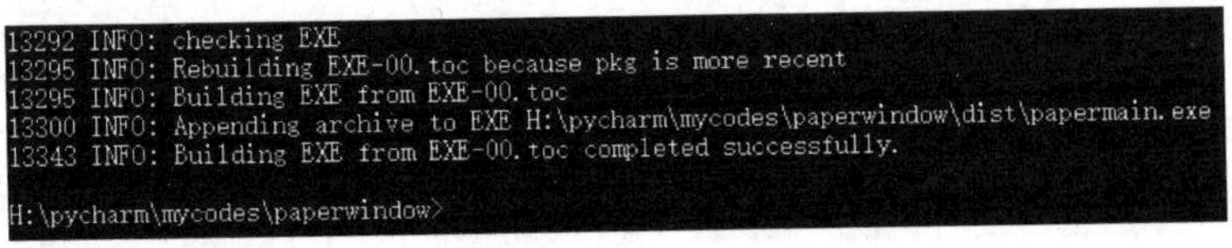

图 1-11 使用 PyInstaller 命令打包成功

3. 查看.exe 文件

成功打包程序后,在 papermain.py 文件目录下生成 dist 文件夹,里面有生成的.exe 文件,双击即可运行,程序封装完成。

1.4 系统测试

本部分包括训练困惑度、测试效果和模型应用。

1.4.1 训练困惑度

在 Seq2Seq 模型中,使用困惑度评估最终效果,值越小则代表语言模型效果越好。本项目使用大网络进行训练,共计 48 000 步,在训练过程进行一段时间后,损失不再减少。开始执行训练与构建网络,如图 1-12 所示。

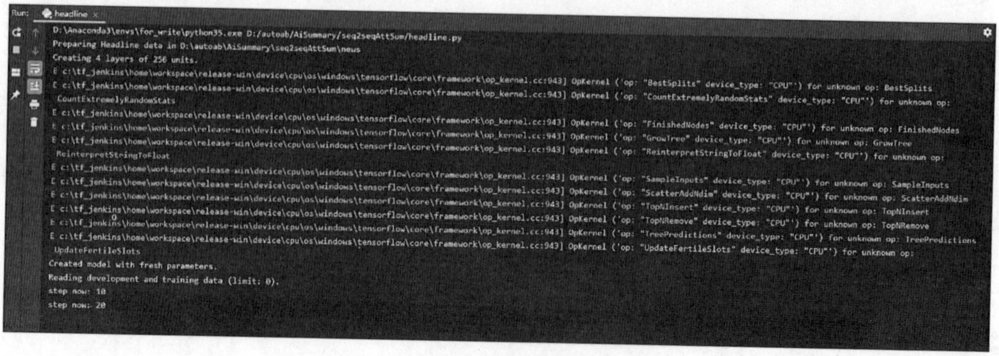

图 1-12 构建网络开始训练

在训练过程中,模型困惑度的值呈下降趋势,即语言的歧义性随着训练的进行而逐渐减小,模型效果逐渐变好。当模型运行到 30 000 步时,其下降趋势已趋于平缓,困惑度基本不再减小;到 47 000 步时,模型的 perplexity 值在小范围波动,最终,其困惑度最低点降至 232.62,如图 1-13 所示。

图 1-13 模型迭代训练结果

1.4.2 测试效果

载入训练好的模型,输入相关文本进行测试。使用 Seq2Seq 进行标题部分的输出,如图 1-14 所示。

从输出结果可以看到,模型的标题生成能力仍有所欠缺,仅能对简单的内容进行效果较好的标题实现,在处理难度较大的文本时,准确性还有待提高。

摘要提取与关键字提取部分,使用 TextRank 算法训练模型,用训练好的模型对给定的文本进行输出,经过多次测试,均得到了较好的结果,如图 1-15 所示。

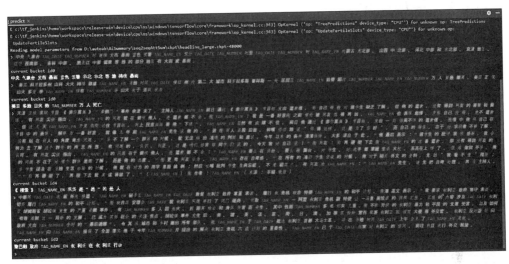

图 1-14 模型训练效果

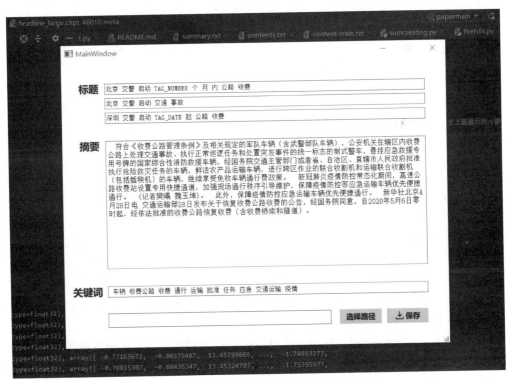

图 1-15 模型生成的标题、摘要和关键词

1.4.3 模型应用

由于程序已打包为可执行文件,故将.exe文件下载到计算机中,双击即可运行,应用初始界面如图1-16所示。

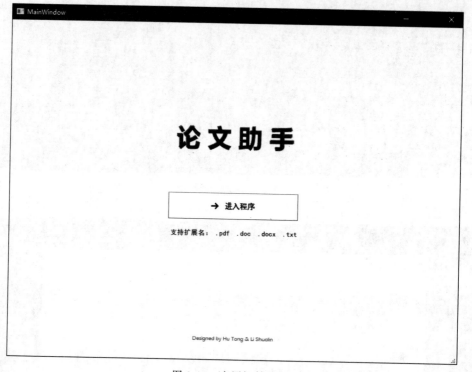

图1-16 应用初始界面

首页为项目名称、"进入程序"按钮及对支持处理文档格式的说明,单击"进入程序"按钮即可进入文件读取页。

在文件读取页中,分别通过单击打开地址、保存地址对应的浏览和需要处理的文档,并设置修改结果的保存路径及文件名,如图1-17所示。

读入文件后,预览及编辑页如图1-18所示。在此处对读入的内容进行预览及修改,单击"保存"按钮暂存修改无误的文档后,单击"生成"按钮,模型开始处理文本内容,进入等待页,如图1-19所示。

处理模型后,关闭当前窗口,程序会自动跳转至结果展示页。页面由上到下依次展示了三个不同的标题方案、论文摘要及关键词,用户可以在界面中直接复制,或在页面下方选择路径,将全部结果保存到本地机,如图1-20所示。

选择保存路径并单击"保存"按钮后,程序会将所有结果保存至指定路径下,并跳转至下载成功页,如图1-21所示。

图 1-17 保存路径及文件

图 1-18 预览及编辑页

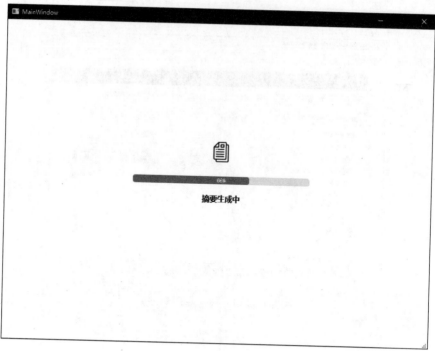

图 1-19 处理等待页

图 1-20 结果展示页图

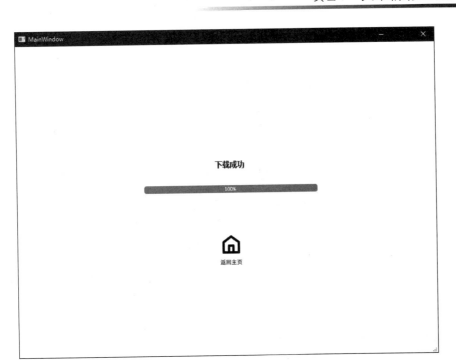

图 1-21　下载成功页图

处理完毕,用户直接关闭程序或单击"返回主页"按钮,跳转回首页处理其他文件。在 PC 端对程序进行测试,输入文件内容、输出内容图及输出结果文件分别如图 1-22～图 1-24 所示。

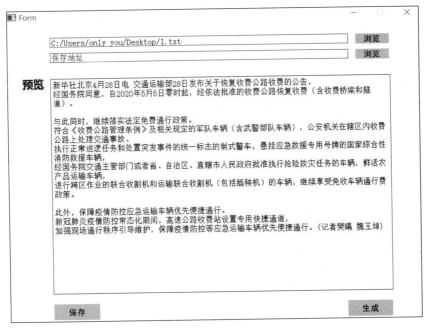

图 1-22　输入文件内容图

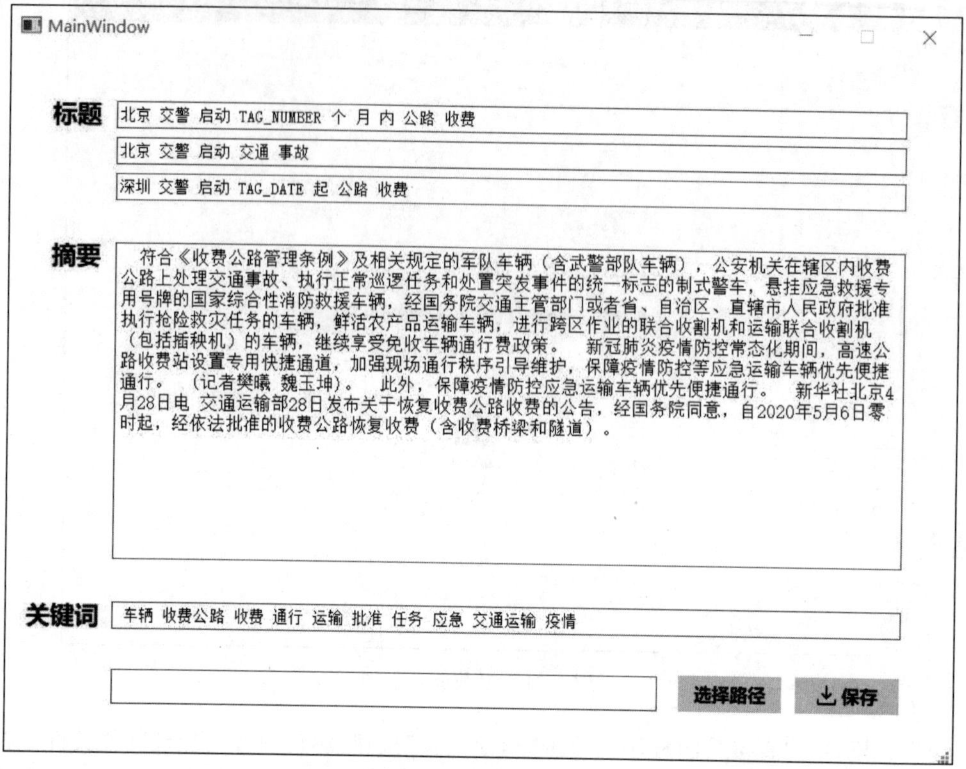

图 1-23 输出内容图

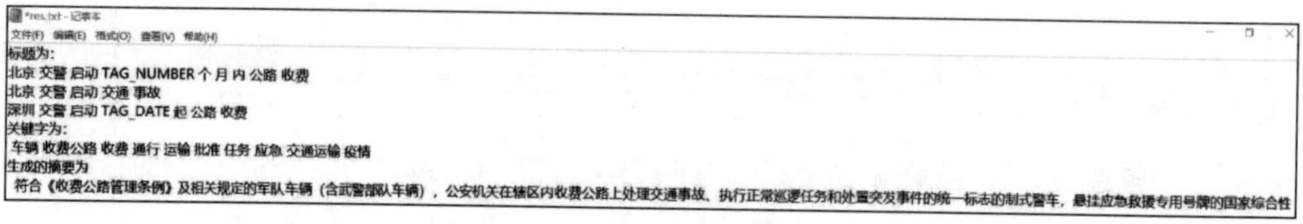

图 1-24 输出结果文件图

项目 2 Trump 推特的情感分析

PROJECT 2

本项目基于 LSTM(Long Short-Term Memory,长短期记忆网络)对 Trump 推特的情感色彩进行分类,通过 Tkinter 界面操纵,实现词频分析、模糊搜索等功能。

2.1 总体设计

本部分包括系统整体结构图和系统流程图。

2.1.1 系统整体结构图

系统整体结构如图 2-1 所示。

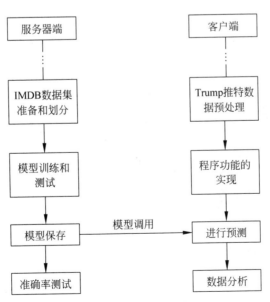

图 2-1 系统整体结构图

2.1.2 系统流程图

系统流程如图 2-2 所示。

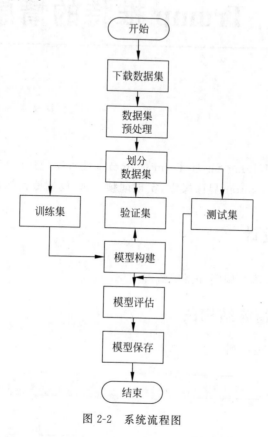

图 2-2 系统流程图

2.2 运行环境

本部分包括 Python 环境、TensorFlow 环境及工具包。

2.2.1 Python 环境

需要 Python 3.6 及以上配置，下载地址为 https://www.anaconda.com/，也可以下载虚拟机在 Linux 环境下运行代码。

2.2.2 TensorFlow 环境

打开 Anaconda Prompt，输入清华仓库镜像，输入命令：

```
conda config -- add channels https://mirrors.tuna.tsinghua.edu.cn/anaconda/pkgs/free/
conda config - set show_channel_urls yes
```

创建 Python 3.7 的环境，名称为 TensorFlow，此时 Python 版本与后面 TensorFlow 的版本有匹配问题，此步选择 Python 3.x，输入命令：

```
conda create - n tensorflow python = 3.5
```

在需要确认的地方，都输入 y。

在 Anaconda Prompt 中激活 TensorFlow 环境，输入命令：

```
activate tensorflow
```

安装 CPU 版本的 TensorFlow，输入命令：

```
pip install - upgrade -- ignore - installed tensorflow
```

至此，安装完毕。

2.2.3 工具包

添加镜像为清华大学后，用 conda install TensorFlow 即可安装。采用 Anaconda 自行安装后，其他的包都将配置完毕，如果采用 pip 命令，需要额外安装 matplotlib 和 PIL 库，分别进行图表绘制和图片处理，采用 pip install x(x 为安装包)指令在 cmd 中运行即可。

2.3 模块实现

本项目包括 4 个模块：准备数据、数据预处理、模型构建和模型测试。下面分别给出各模块的功能介绍及相关代码。

2.3.1 准备数据

数据集下载地址为 http://ai.stanford.edu/~amaas/data/sentiment/，包含 50 000 条偏向明显的评论，其中 25 000 条作为训练集，25 000 条作为测试集，label 为 pos(正向)和 neg(负向)。下载数据集后，用如下代码将数据集解压。

```
import tarfile
import os
def untar(fname, dirs):
    t = tarfile.open(fname)        #打开目标文件
    t.extractall(path = dirs)      #解压文件路径
if __name__ == "__main__":
    untar('aclImdb_v1.tar.gz', ".")
```

2.3.2 数据预处理

数据预处理的相关操作如下：

```python
# 读取下载的数据集
from keras.preprocessing import sequence
from keras.preprocessing.text import Tokenizer
import re
import os
def remove_html(text):                                      # 用正则表达式去除 HTML 标签
    r = re.compile(r'<[^>]+>')
    return r.sub('', text)
def read_file(filetype):                                    # 读取数据集内容和标签
    path = './aclImdb/'
    file_list = []
    positive = path + filetype + '/pos/'
    for f in os.listdir(positive):
        file_list += [positive + f]
    negative = path + filetype + '/neg/'
    for f in os.listdir(negative):
        file_list += [negative + f]
    print('filetype:', filetype, 'file_length:', len(file_list))
    label = ([1] * 12500 + [0] * 12500)    # 训练数据和测试数据，pos 和 neg 都是 12 500 条
    text = []
    for f_ in file_list:
        with open(f_, encoding = 'utf8') as f:
            text += [remove_html(''.join(f.readlines()))]   # 清除 HTML 标签
    return label, text
# 读取测试集和训练集
x_train, y_train = read_file('train')
x_test, y_test = read_file('test')
```

因为不能直接将字符串输入模型，所以建立一个字典，将一句话中的每个词转化为一个向量，称为词向量。它具有空间意义，并不是简单的映射——即意思相近的词向量在空间中的距离比较近，为方便训练，将数字列表的长度调成相等。

```python
token = Tokenizer(num_words = 3500)  # 建立一个有 3500 个单词的字典
token.fit_on_texts(y_train)
# 读取所有的训练数据评论，按照单词在评论中出现的次数进行排序，前 3500 个会列入字典
# 将评论数据转化为数字列表
train_seq = token.texts_to_sequences(y_train)
test_seq = token.texts_to_sequences(y_test)
prin(train_seq[0]) # 打印转化后的词向量
# 截长补短，让每一个数字列表长度都为 100
_train = sequence.pad_sequences(train_seq, maxlen = 100)
_test = sequence.pad_sequences(test_seq, maxlen = 100)
```

2.3.3 模型构建

数据加载进模型进行定义结构、模型及字典保存、预测结果展示。

1. 定义结构

定义模型结构相关操作如下。

```python
# 导入所需要的模块
from keras.models import Sequential
from keras.layers.core import Dense, Dropout, Activation
from keras.layers.embeddings import Embedding
from keras.layers.recurrent import LSTM
model_lstm = Sequential()
model_lstm.add(Embedding(output_dim = 32,  # 将数字列表转换为32维向量
                input_dim = 2500,  # 输入数据的维度是2500,因为之前建立的字典有2500个单词
                input_length = 100))  # 数字列表的长度为100
model_lstm.add(Dropout(0.25))
model_lstm.add(LSTM(32))
model_lstm.add(Dense(units = 256, activation = 'relu'))
model_lstm.add(Dropout(0.25))
# 输出层只有一个神经元,输出1表示正面评价,输出0表示负面评价
model_lstm.add(Dense(units = 1, activation = 'sigmoid'))
model_lstm.summary()
```

模型摘要如图 2-3 所示。

```
Model: "sequential_1"
Layer (type)                 Output Shape              Param #
embedding_1 (Embedding)      (None, 100, 32)           112000
dropout_1 (Dropout)          (None, 100, 32)           0
lstm_1 (LSTM)                (None, 32)                8320
dense_1 (Dense)              (None, 256)               8448
dropout_2 (Dropout)          (None, 256)               0
dense_2 (Dense)              (None, 1)                 257
Total params: 129 025
Trainable params: 129 025
Non-trainable params: 0
```

图 2-3　模型摘要

最后选择损失函数和优化器。由于面对的是一个二分类问题,网络输出是一个概率值,使用 binary_crossentropy(二元交叉熵),对于输出概率值的模型,交叉熵是最好的选择。

```python
model_lstm.compile(loss = 'binary_crossentropy',
    optimizer = 'adam', metrics = ['accuracy'])         # 选择损失函数和优化器
```

```python
# 开始训练
train_history = model_lstm.fit(_train, x_train, batch_size = 100,
                               epochs = 8, verbose = 2,
                               validation_split = 0.2)      # 设置参数
```

训练过程如图 2-4 所示。

```python
# 评估模型准确率
scores = model_lstm.evaluate(_test, x_test) # 第一个参数为 feature, 第二个参数为 label
```

```
Epoch 1/8
 - 12s - loss: 0.4835 - accuracy: 0.7596 - val_loss: 0.4531 - val_accuracy: 0.7876
Epoch 2/8
 - 12s - loss: 0.3222 - accuracy: 0.8607 - val_loss: 0.3129 - val_accuracy: 0.8488
Epoch 3/8
 - 10s - loss: 0.2993 - accuracy: 0.8756 - val_loss: 0.4574 - val_accuracy: 0.7742
Epoch 4/8
 - 10s - loss: 0.2857 - accuracy: 0.8816 - val_loss: 0.6443 - val_accuracy: 0.7294
Epoch 5/8
 - 10s - loss: 0.2735 - accuracy: 0.8875 - val_loss: 0.5849 - val_accuracy: 0.7200
Epoch 6/8
 - 9s - loss: 0.2562 - accuracy: 0.8939 - val_loss: 0.5938 - val_accuracy: 0.7574
Epoch 7/8
 - 10s - loss: 0.2452 - accuracy: 0.8985 - val_loss: 0.3939 - val_accuracy: 0.8348
Epoch 8/8
 - 10s - loss: 0.2290 - accuracy: 0.9072 - val_loss: 0.4449 - val_accuracy: 0.8092
```

图 2-4 训练过程

通过观察训练集和测试集的损失函数、准确率评估模型的训练程度，进行模型训练的进一步决策。模型训练的最佳状态为：训练集和测试集的损失函数（或准确率）不变且基本相等。

2. 模型及字典保存

为了能够被程序读取，需要将模型文件保存为 .h5 格式。Model 对象提供了 save() 和 save_wights() 两个方法。

save() 方法保存了模型结构、模型参数和优化器参数，使用 HDF5 文件保存模型。

```python
model_lstm.save('my_model.h5')      # 保存为 .h5 格式
```

模型保存后可以被重用，也可以移植到其他环境中使用。通过数据集制作一个词典，将预测的数据转化为词向量序列，在主程序中，假如重新预处理数据以制作词典会耗费时间。因此，为了优化系统，将预先处理好数据储存、优化运行时间、数据保存为 .json 格式的文件。

```python
import json
filename = 'train.json'
with open(filename, 'w') as file_obj:
    json.dump(y_train, file_obj)
```

3. 预测结果展示

通过比对函数检测预测结果是否正确。

```
_dict = {1:'正面的评论',0:'负面的评论'}
def display(i):
    print(y_test[i])
    print('label 真实值为:',_dict[x_test[i]],
          '预测结果为:',_dict[predict[i]])
```

2.3.4 模型测试

该模块主要包括：数据库模型导入及调用、功能类函数和可视化模块。

1. 数据库模型导入及调用

把数据库文件放入文件目录并调用。

```
import csv
import numpy as np
csv_reader = csv.reader(open("newtrumptweets.csv"))  #打开文件
text = []
for row in csv_reader:  #读入数据
text.append(row)
text1 = np.array(text)
content = text1[:,0]
date = text1[:,1]
retweet = text1[:,2]
likes = text1[:,3]
```

加载 Keras 库，调用 .json 文件中预处理好的数据建立词典，利用词典将调用数据库中的文本数据转化为词向量。

```
#导入所需的模块
from keras.preprocessing import sequence
from keras.preprocessing.text import Tokenizer
from keras.models import Sequential
from keras.layers.core import Dense,Dropout,Activation
from keras.layers.embeddings import Embedding
from keras.layers.recurrent import LSTM
from keras.models import load_model
#载入预处理好的数据
import json
filename = 'train.json'
with open(filename) as file_obj:
    train = json.load(file_obj)
token = Tokenizer(num_words = 3500)      #建立一个有3500个单词的字典
token.fit_on_texts(train)
#读取所有的训练数据评论,按照单词在评论中出现的次数进行排序,前3500个会列入字典
#载入保存好的模型
model = load_model('my_model.h5')
```

```python
# 调用模型进行预测
def plotshow(s1):                    # 预测及可视化图像展示
    if s1:
        q1 = []
        q2 = []
        q3 = []
        m = 0
        n = 0
        r1 = []
        for q in s1:
            q1.append(content[s1[m]])
            q2.append(likes[s1[m]])
            q3.append(retweet[s1[m]])
            m = m + 1
        content_seq = token.texts_to_sequences(q1)
        # 利用词典将文本数据转化为词向量序列
        _content = sequence.pad_sequences(content_seq, maxlen = 100)
        # 截长补短,将词向量序列长度调整一致,方便预测
        predict = model.predict_classes(_content)
        # 进行预测,返回类别索引,即该样本所属的类别标签
        predict = predict.reshape(-1)    # 转换成一维数组
        for q in s1:
            r1.append(display(n))
            r1.append('\n分析结果:')
            r1.append(_dict[predict[n]])
            n = n + 1
        # 画图可视化数据
        fig = plt.figure()
        ax1 = fig.add_subplot(211)
        plt.xlabel('likes')
        ax1.scatter(q2, predict)
        ax2 = fig.add_subplot(212)
        plt.xlabel('retweet')
        ax2.scatter(q3, predict)
        plt.tight_layout()
        plt.show()
    else:
        r1 = ["无可分析内容"]
    return r1  # 返回预测结果
```

2. 功能类函数

本部分包括词频统计模块、词云图制作模块和搜索模块。

1) 词频统计模块

```python
def order_dict(dicts, n):            # 排序字典
    result = []
```

```
        result1 = []
        p = sorted([(k, v) for k, v in dicts.items()], reverse = True)
        s = set()
        for i in p:
            s.add(i[1])
        for i in sorted(s, reverse = True)[:n]:
            for j in p:
                if j[1] == i:
                    result.append(j)
        for r in result:
            result1.append(r[0])
        return result1
def order_dict1(dicts,n):            #截取排序结果
        list1 = sorted(dicts.items(),key = lambda x:x[1])
        return list1[-1:-(n+1):-1]
        #return list1[-2:-(n+2):-1]去除统计结果为" "的情况
if __name__ == "__main__":
        str1 = ','.join(content)
        #划分单词
        import re
        array = re.split('[ ,.\n]',str1)
        #print('分词结果',array)
        #词频统计
        a = len(array) - 1
        for i in range(0,a):              #清洗单词
            if len(array[i])<= 4:
                array[i] = '#'
        dic = {}
        for i in array:
            if i not in dic:
                dic[i] = 1
            else:
                dic[i] += 1
        #数据清洗
        del[dic['twitter']]
        del[dic['#']]
        del[dic['https://www']]
        del[dic['don't']]
        del[dic['because']]
        del[dic['would']]
        del[dic['should']]
        del[dic['there']]
        del[dic['their']]
        del[dic['which']]
        del[dic['before']]
        del[dic['after']]
```

2）词云图制作模块

```python
def wd(pw):
    pw1 = np.array(pw)
    wl = pw1[:,0]
    wl = ','.join(wl)
    img = np.array(Image.open('wordcloud.jpg'))      # 载入背景图片
    # 生成一个词云对象
    wordcloud = WordCloud(
        background_color = "white",                   # 设定背景颜色
        scale = 10,                                   # 图像清晰度
        mask = img).generate(wl)                      # 设定词语图形状
    # 绘制图片
    plt.imshow(wordcloud)
    # 消除坐标轴
    plt.axis("off")
    plt.savefig('test1.jpg')
    plt.show()
```

3）搜索模块

```python
def fuzzyfinder(user_input, str1):
    if user_input:
        suggestions = []
        pattern = '.*?'.join(user_input)              # 去除符号
        regex = re.compile(pattern)                   # 编译正则表达式
        p1 = []
        s1 = []
        i = 0
        for item in content:
            match = regex.search(item)                # 检查当前项是否与 regex 匹配
            if match:
                p1.append(display(i))
                s1.append(i)
            i = i + 1
    else:
        p1 = ["请输入要搜索的关键词"]
        s1 = []
    return s1,p1                                      # 返回搜索结果
```

3. 可视化模块

可视化相关代码如下：

```python
from PIL import Image
from PIL import ImageTk
import tkinter as tk                                  # 使用 Tkinter 前需要先导入
import tkinter.messagebox
```

```python
import pickle
def create():                                    # 功能窗口生成
    window.destroy()                             # 销毁登录界面
    # 定义一个新的窗口
    window1 = tk.Tk()
    # 给窗口的可视化起名
    window1.title('Trump Observer')
    # 设定窗口的大小(长 * 宽)
    window1.geometry('800x600')                  # 这里的乘是小写 x
    canvas = tk.Canvas(window1, width = 300, height = 168, bg = 'green')
    image_file = ImageTk.PhotoImage(file = 'trump22.jpg')
    image = canvas.create_image(150, 0, anchor = 'n', image = image_file)
    canvas.pack(side = 'top')
    tk.Label(window1, text = 'Wellcome',font = ('Arial', 16)).pack()
# 在图形界面上设定输入框控件 entry 框并放置
e = tk.Entry(window1, show = None)   # 显示成明文形式
e.pack()
# 定义三个触发事件时的函数
def point1():   # 词频分析
    t.delete(0.0, 'end')
    t.insert('end',order_dict1(dic,50))
    wd(order_dict1(dic,50))
def point2():   # 模糊搜索
    t.delete(0.0, 'end')
    var = e.get()
    t.insert('end',fuzzyfinder(var, str1)[1])
    def point3():   # 情感分析
        t.delete(0.0, 'end')
        var = e.get()
        t.insert('end',plotshow(fuzzyfinder(var, str1)[0]))
        tkinter.messagebox.showinfo('0:负面评论;1:正面评论')
# 创建并放置三个按钮分别触发三种情况
b1 = tk.Button(window1, text = '词频统计', width = 10,
        height = 2, command = point1)
b1.pack()
b2 = tk.Button(window1, text = '搜索', width = 10,
        height = 2, command = point2)
b2.pack()
b3 = tk.Button(window1, text = '分析', width = 10,
        height = 2, command = point3)
b3.pack()
# 创建并放置一个多行文本框 text 用以显示,指定 height = 10 为文本框是三个字符高度
t = tk.Text(window1, height = 10)
t.pack()
window1.mainloop()
window = tk.Tk()
# 给窗口的可视化起名
```

```python
window.title('Wellcome to Trump Observer')
    # 设定窗口的大小(长 * 宽)
window.geometry('600x500')  # 这里的乘是小写 x
    # 加载欢迎图片
canvas = tk.Canvas(window, width = 280, height = 210, bg = 'green')
image_file = ImageTk.PhotoImage(file = 'trump11.jpg')
image = canvas.create_image(130, 0, anchor = 'n', image = image_file)
canvas.pack(side = 'top')
tk.Label(window, text = 'Wellcome', font = ('Arial', 16)).pack()
    # 用户信息
tk.Label(window, text = '用户名:', font = ('Arial', 14)).place(x = 20, y = 310)
tk.Label(window, text = '密码:', font = ('Arial', 14)).place(x = 20, y = 390)
    # 用户登录输入框
    # 用户名
var_usr_name = tk.StringVar()
entry_usr_name = tk.Entry(window, textvariable = var_usr_name, font = ('Arial', 14))
entry_usr_name.place(x = 200, y = 320)
    # 用户密码
var_usr_pwd = tk.StringVar()
entry_usr_pwd = tk.Entry(window, textvariable = var_usr_pwd, font = ('Arial', 14), show = '*')
entry_usr_pwd.place(x = 200, y = 400)
# 定义用户登录功能
def usr_login():
    # 这两行代码是获取输入的用户名和密码
    usr_name = var_usr_name.get()
    usr_pwd = var_usr_pwd.get()
    # 设置异常捕获,当第一次访问用户信息文件时是不存在的
    # 中间的两行即程序将输入信息和文件中的信息匹配
    try:
        with open('usrs_info.pickle', 'rb') as usr_file:
            usrs_info = pickle.load(usr_file)
    except FileNotFoundError:
# 没有"usr_file",则创建,并将管理员的用户名和密码写入,即用户名为"admin",密码为"admin"
        with open('usrs_info.pickle', 'wb') as usr_file:
            usrs_info = {'admin': 'admin'}
            pickle.dump(usrs_info, usr_file)
            usr_file.close()
    if usr_name in usrs_info:
        if usr_pwd == usrs_info[usr_name]:
# 如果用户名和密码与文件匹配成功,则会成功登录,并跳出弹窗 Hello 加上用户名
tkinter.messagebox.showinfo(title = 'Welcome', message = 'Hello,' + usr_name)
            create()
# 如果用户名匹配成功,而密码输入错误,则会弹出"密码错误"
        else:
            tkinter.messagebox.showerror(message = '密码错误')
else:  # 如果发现用户名不存在
    tkinter.messagebox.showerror(message = '用户名不存在')
```

```python
#定义用户注册功能
def usr_sign_up():
    def sign_to_Hongwei_Website():
        #以下三行是获取注册时所输入的信息
        np = new_pwd.get()
        npf = new_pwd_confirm.get()
        nn = new_name.get()
        #打开记录数据的文件,将注册信息读出
        with open('usrs_info.pickle', 'rb') as usr_file:
            exist_usr_info = pickle.load(usr_file)
        #如果两次密码输入不一致,则提示两次输入密码必须一致
        if np != npf:
            tkinter.messagebox.showerror('两次输入密码必须一致')
        #如果用户名已经在数据文件中,则提示错误,用户名已存在
        elif nn in exist_usr_info:
            tkinter.messagebox.showerror('错误', '用户名已存在')
        #如果输入无以上错误,则将注册输入的信息记录到文件中,并提示注册成功,然后销毁窗口
        else:
            exist_usr_info[nn] = np
            with open('usrs_info.pickle', 'wb') as usr_file:
                pickle.dump(exist_usr_info, usr_file)
            tkinter.messagebox.showinfo('注册成功')
            #销毁窗口
            window_sign_up.destroy()
    #定义长在窗口上的窗口
    window_sign_up = tk.Toplevel(window)
    window_sign_up.geometry('400x240')
    window_sign_up.title('Sign up window')
    new_name = tk.StringVar() #将输入的注册名赋值给变量
    tk.Label(window_sign_up, text = '用户名: ').place(x = 10, y = 10)
    #将用户名放置在坐标(10,10)
    entry_new_name = tk.Entry(window_sign_up, textvariable = new_name)
    #创建一个注册名的"entry",变量为"new_name"
    entry_new_name.place(x = 130, y = 10)  #"entry"放置在坐标(150,10)
    new_pwd = tk.StringVar()
    tk.Label(window_sign_up, text = '密码: ').place(x = 10, y = 50)
    entry_usr_pwd = tk.Entry(window_sign_up, textvariable = new_pwd, show = '*')
    entry_usr_pwd.place(x = 130, y = 50)
    new_pwd_confirm = tk.StringVar()
    tk.Label(window_sign_up, text = '再次确认密码: ').place(x = 10, y = 90)
    entry_usr_pwd_confirm = tk.Entry(window_sign_up, textvariable = new_pwd_confirm, show = '*')
    entry_usr_pwd_confirm.place(x = 130, y = 90)
    #下面是具体实例
    btn_comfirm_sign_up = tk.Button(window_sign_up, text = 'Sign up', command = sign_to_Hongwei_Website)
    btn_comfirm_sign_up.place(x = 180, y = 120)
#放置登录和注册按钮并设置触发情况
```

```
btn_login = tk.Button(window, text = '登录', command = usr_login, width = 15, height = 2)
btn_login.place(x = 140, y = 440)
btn_sign_up = tk.Button(window, text = '注册', command = usr_sign_up, width = 15, height = 2)
btn_sign_up.place(x = 340, y = 440)
#主窗口循环显示
window.mainloop()
```

2.4 系统测试

本部分包括模型效果和模型应用。

2.4.1 模型效果

模型训练效果如图 2-5 所示。

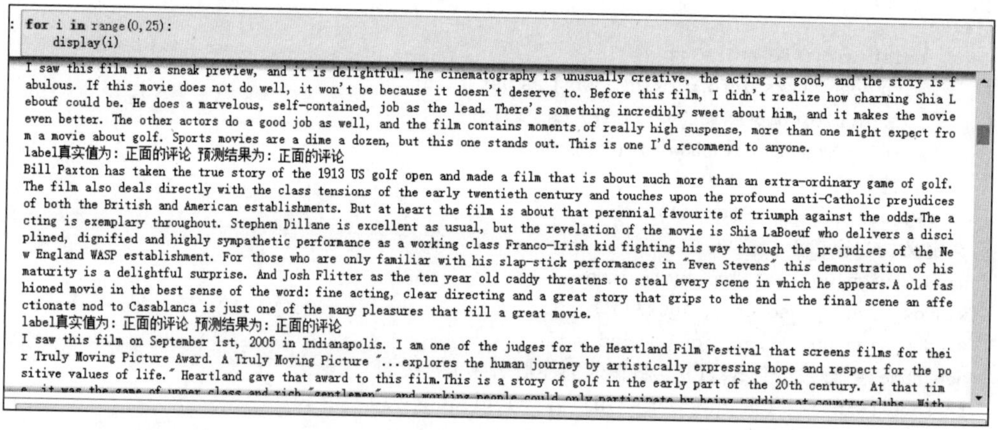

图 2-5　模型训练效果

2.4.2 模型应用

对格式为.ipynb 的文件编译成功后,运行文件生成可视化操作界面,初始登录界面如图 2-6 所示。

输入用户名及密码并登录成功后,主程序界面如图 2-7 所示。

界面从上至下分别是 1 张图片、文本输入框、3 个按钮和 1 个文本框显示结果。第一个按钮为数据库中文本内容的词频统计及词云图展示,如图 2-8 所示。

在文本框内输入要搜索的内容,搜索结果如图 2-9 所示。

单击"分析"按钮可将搜索结果进行情感分类,并根据点赞数和转发数绘制可视化图形,如图 2-10 所示。

图 2-6　初始登录界面

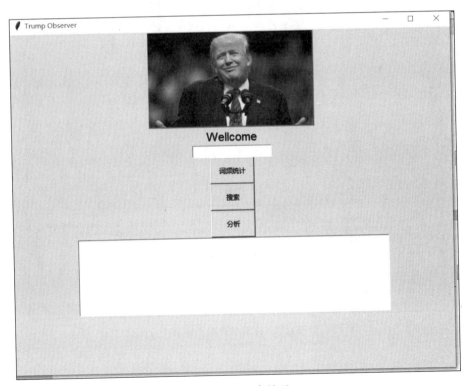

图 2-7　主程序界面

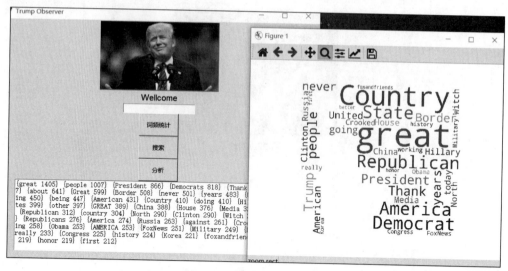

图 2-8　词频统计显示画面

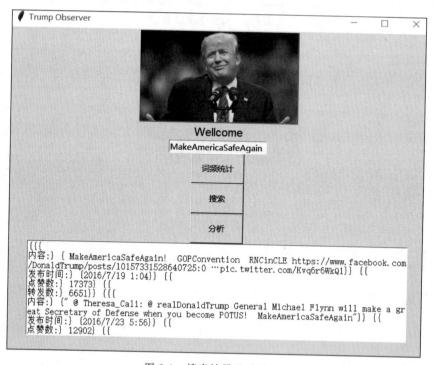

图 2-9　搜索结果显示画面

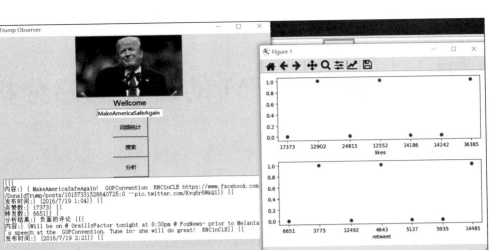

图 2-10 情感分析结果展示

项目 3　基于 LSTM 的影评情感分析

PROJECT 3

本项目基于 LSTM，使用分类数据集（Large Movie Review Dataset，LMRD）训练情感分析模型，实现移动端的文本情感推断设计。

3.1　总体设计

本部分包括系统整体结构图和系统前后端流程图。

3.1.1　系统整体结构图

系统整体结构如图 3-1 所示。

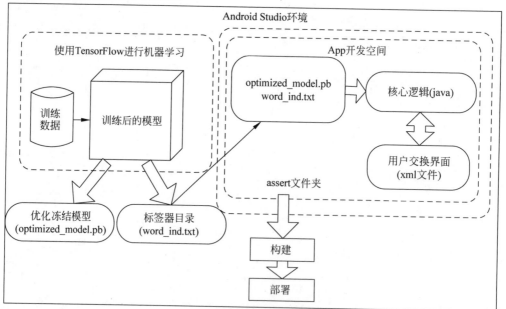

图 3-1　系统整体结构图

3.1.2 系统前后端流程图

系统前端流程如图 3-2 所示，系统后端流程如图 3-3 所示。

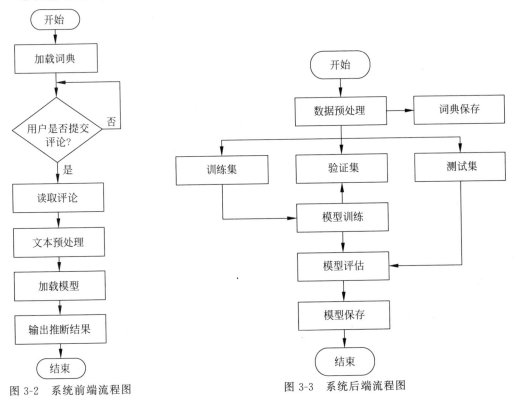

图 3-2　系统前端流程图　　　　图 3-3　系统后端流程图

3.2 运行环境

本部分包括 Python 环境、TensorFlow 环境和 Android 环境。

3.2.1 Python 环境

需要 Python 3.6 及以上配置，用 Anaconda 创建虚拟环境 MRSA（全标为 Movie Review Sentiment Analysis），完成所需 Python 环境的配置。

打开 Anaconda Prompt，输入命令：

```
conda create -n MRSA python=3.6
```

创建 MRSA 虚拟环境。

3.2.2 TensorFlow 环境

打开 Anaconda Prompt,激活所创建的 MRSA 虚拟环境,输入命令:

```
activate MRSA
```

安装 CPU 版本的 TensorFlow,输入命令:

```
pip install - upgrade -- ignore - installed tensorflow
```

安装完毕。

其他相关依赖包,包括 Keras、Re、Pickle、Fire、Pandas、Numpy,其安装方式和 TensorFlow 类似,直接在虚拟环境中 pip install package_name 或者 conda install package_name 即可完成。

3.2.3 Android 环境

安装 Android Studio 新建 Android 项目,打开 Android Studio,依次选择 File→New→New Project→Empty Activity→Next。

Name 定义 Movie Review Analysis,Save location 为项目保存的地址,可自行定义,Minimum API 为该项目能够兼容 Android 手机的最低版本,选择 16。单击 Finish 按钮,新建项目完成。App/build.gradle 里的内容有任何改动后,Android Studio 都会弹出信息提示。单击 Sync Now 按钮或 图标,同步该配置,"成功"表示配置完成。

3.3 模块实现

本项目包括 5 个模块:数据预处理、模型构建及训练、模型保存、词典保存和模型测试。下面分别给出各模块的功能介绍及相关代码。

3.3.1 数据预处理

本部分包括数据集合并、数据清洗、文本数值化和数据集划分。

1. 数据集合并

数据集下载地址为 http://ai.stanford.edu/~amaas/data/sentiment/。斯坦福大学提供的情感分类数据集中了 25 000 条电影评论用于训练,25 000 条用于测试。先将这 50 000 条数据合并,并保存为.csv 文件格式,相关代码如下:

```
# 导入原始数据
train_review_files_pos = os.listdir(path + 'train/pos/')
review_dest.append(path + 'train/pos/')
```

```python
train_review_files_neg = os.listdir(path + 'train/neg/')
review_dest.append(path + 'train/neg/')
test_review_files_pos = os.listdir(path + 'test/pos/')
review_dest.append(path + 'test/pos/')
test_review_files_neg = os.listdir(path + 'test/neg/')
review_dest.append(path + 'test/neg/')
#将标签合并
sentiment_label = [1] * len(train_review_files_pos) + \
                  [0] * len(train_review_files_neg) + \
                  [1] * len(test_review_files_pos) + \
                  [0] * len(test_review_files_neg)
#将所有评论合并
review_train_test = ['train'] * len(train_review_files_pos) + \
                    ['train'] * len(train_review_files_neg) + \
                    ['test'] * len(test_review_files_pos) + \
                    ['test'] * len(test_review_files_neg)
#将合并后的数据保存为.csv格式
df = pd.DataFrame()
df['Train_test_ind'] = review_train_test
df['review'] = reviews
df['sentiment_label'] = sentiment_label
df.to_csv(path + 'processed_file.csv', index = False)
```

合并后的.csv文件如图3-4所示。

```
data = pd.read_csv(path + 'processed_file.csv')
print('数据大小为: ',data.shape)
data.head()
```

数据大小为: (50000, 3)

	Train_test_ind	review	sentiment_label
0	train	bromwell high is a cartoon comedy it ran at th...	1
1	train	homelessness or houselessness as george carlin...	1
2	train	brilliant over acting by lesley ann warren bes...	1
3	train	this is easily the most underrated film inn th...	1
4	train	this is not the typical mel brooks film it was...	1

图3-4 合并后的.csv文件

2. 数据清洗

文本中一些非相干因素会影响最后模型的精度,采用正则表达式将所有标点符号去除,将大写字母转换成小写字母,相关代码如下:

```
def text_clean(text):
    #将所有大写字母转换成小写字母，并去除标点符号
    letters = re.sub("[^a-zA-z0-9\s]", " ",text)
    words = letters.lower().split()
    text = " ".join(words)
    return text
```

数据清洗结果如图 3-5 所示。

```
Bromwell High is a cartoon comedy. It ran at the same
time as some other programs about school life, such as
"Teachers". My 35 years in the teaching profession
lead me to believe that Bromwell High's satire is much
closer to reality than is "Teachers". The scramble to
survive financially, the insightful students who can
see right through their pathetic teachers' pomp, the
pettiness of the whole situation, all remind me of the
schools I knew and their students. When I saw the
episode in which a student repeatedly tried to burn
down the school, I immediately recalled ......... at
.......... High. A classic line: INSPECTOR: I'm here
to sack one of your teachers. STUDENT: Welcome to
Bromwell High. I expect that many adults of my age
think that Bromwell High is far fetched. What a pity
that it isn't!
```

(a) 原始文本

```
bromwell high is a cartoon comedy it ran at the same
time as some other programs about school life such as
teachers my 35 years in the teaching profession lead
me to believe that bromwell high s satire is much
closer to reality than is teachers the scramble to
survive financially the insightful students who can
see right through their pathetic teachers pomp the
pettiness of the whole situation all remind me of the
schools i knew and their students when i saw the
episode in which a student repeatedly tried to burn
down the school i immediately recalled at high a
classic line inspector i m here to sack one of your
teachers student welcome to bromwell high i expect
that many adults of my age think that bromwell high is
far fetched what a pity that it isn t
```

(b) 处理后的文本

图 3-5 数据清洗结果

3. 文本数值化

文本中每个单词对应唯一的索引(token)，依据索引将文本数值化。Keras tokenizer 通过采集前 50 000 个常用词，转换为数字索引或标记。为了处理方便，对于文本长度大于 1000 的评论，只取前 1000 个单词；若评论长度不足 1000，则在评论开始使用 0 填充。相关代码如下：

```
#采集前 50000 个常用词，把单词转换为数字索引或标记
max_features = 50000
tokenizer = Tokenizer(num_words = max_features, split = ' ')
tokenizer.fit_on_texts(df['review'].values)
X = tokenizer.texts_to_sequences(df['review'].values)
X_ = []
for x in X:
    x = x[:1000]
    X_.append(x)
X_ = pad_sequences(X_)
```

数值化结果如图 3-6 所示。

4. 数据集划分

将数据集划分为训练集、验证集及测试集，比例分别为 70%、15% 和 15%。相关代码如下：

```
y = df['sentiment_label'].values
```

```
records_processed 50000
[[    0    0    0 ...    4   11   16]
 [    0    0    0 ... 1173 22081   75]
 [    0    0    0 ...    9    1 1912]
 ...
 [    0    0    0 ...  167   32  363]
 [    0    0    0 ...  681    1 9109]
 [    0    0    0 ...   34  318   11]]
```

图 3-6 数值化结果

```
index = list(range(X_.shape[0]))
np.random.shuffle(index)
train_record_count = int(len(index) * 0.7)
validation_record_count = int(len(index) * 0.15)
train_indices = index[:train_record_count]
validation_indices = index[train_record_count:train_record_count +
                     validation_record_count]
test_indices = index[train_record_count + validation_record_count:]
X_train, y_train = X_[train_indices], y[train_indices]
X_val, y_val = X_[validation_indices], y[validation_indices]
X_test, y_test = X_[test_indices], y[test_indices]
```

划分后的数据集如图 3-7 所示。

```
x_train = np.load(path + 'X_train.npy')
x_val = np.load(path + 'X_val.npy')
x_test = np.load(path + 'X_test.npy')
print('训练数据集大小：', x_train.shape)
print('验证数据集大小：', x_val.shape)
print('测试数据集大小：', x_test.shape)

训练数据集大小：   (35000, 1000)
验证数据集大小：   (7500, 1000)
测试数据集大小：   (7500, 1000)
```

图 3-7 划分后的数据集

3.3.2 模型构建及训练

将数据加载进模型之后，需要定义模型结构、优化损失函数和性能指标。这里定义了两种结构进行训练，一是基于 BasicLSTM 的网络；二是基于 MultiRNN 的网络。

1. 定义模型结构

首先，构建一个简单的 LSTM 版本递归神经网络(BasicLSTM)，并在输入层后面放一个嵌入层。嵌入层的单词向量使用预先训练好的 100 维 Glove 向量初始化，该图层被定义为 trainable(可训练的)，这样，该单词向量嵌入层就可以根据训练数据自行更新。隐藏状态的维度和单元状态的维度也是 100。

其次，为获得文本中更多正确信息，进一步定义多层递归神经网络(MultiRNN)，共有三层，每层单元状态的维度分别是 100、200、100。定义嵌入层的相关代码如下：

```python
# 定义嵌入层
with tf.variable_scope('embedding'):
    self.emb_W = tf.get_variable('word_embeddings', [self.n_words, self.embedding_dim],
        initializer = tf.random_uniform_initializer(-1, 1, 0), trainable = True,
                    dtype = tf.float32)
    self.assign_ops = tf.assign(self.emb_W, self.emd_placeholder)
    self.embedding_input = tf.nn.embedding_lookup(self.emb_W, self.X, "embedding_input")
    print(self.embedding_input)
    self.embedding_input = tf.unstack(self.embedding_input, self.sentence_length, 1)
# 定义网络结构
with tf.variable_scope('LSTM_cell'):
    # 定义 BasicLSTM
    self.cell = tf.nn.rnn_cell.BasicLSTMCell(self.hidden_states)
    # 定义 MultiRNN
    # num_units = [100, 200, 100]
    # self.cells = [tf.nn.rnn_cell.BasicLSTMCell(num_unit) for num_unit in num_units]
    # self.cell = tf.nn.rnn_cell.MultiRNNCell(self.cells)
```

2. 优化损失函数

使用二进制交叉熵损失训练模型，并在损失函数中加入正则化以防止出现过拟合，同时使用 Adam（Adaptivemoment estimation）优化器训练模型，用精确度作为性能指标。相关代码如下：

```python
self.l2_loss = tf.nn.l2_loss(self.w, name = "l2_loss")
self.scores = tf.nn.xw_plus_b(self.output[-1], self.w, self.b, name = "logits")
self.prediction_probability = tf.nn.sigmoid(self.scores, name = 'positive_sentiment_probability')    # 计算属于1类的概率
self.predictions = tf.round(self.prediction_probability, name = 'final_prediction')
self.losses = tf.nn.sigmoid_cross_entropy_with_logits(logits = self.scores, labels = self.y)
            # 损失函数
self.loss = tf.reduce_mean(self.losses) + self.lambda1 * self.l2_loss
tf.summary.scalar('loss', self.loss)
self.optimizer = tf.train.AdamOptimizer(self.learning_rate).minimize(self.losses)
            # 优化器
self.correct_predictions = tf.equal(self.predictions, tf.round(self.y))
self.accuracy = tf.reduce_mean(tf.cast(self.correct_predictions, "float"), name = "accuracy")
tf.summary.scalar('accuracy', self.accuracy)
```

3. 模型实现

使用 tf.train.write_graph() 函数将模型图定义保存到 model.pbtxt 文件中，训练完成后，使用 tf.train.Saver() 函数将权重保存在 model_ckpt 中。model.pbtxt 和 model_ckpt 文件将被用于创建 protobuf 格式的 TensorFlow 模型优化版本，以便与 Android 应用集成，相关代码如下：

```
for epoch in range(self.epochs):  # 轮次
    gen_batch = self.batch_gen(self.X_train, self.y_train, self.batch_size)
    gen_batch_val = self.batch_gen(self.X_val, self.y_val, self.batch_size_val)
    for batch in range(self.num_batches):  # 批次
        X_batch, y_batch = next(gen_batch)
        X_batch_val, y_batch_val = next(gen_batch_val)
        sess.run(self.optimizer, feed_dict={self.X: X_batch, self.y: y_batch})
        if (batch + 1) % 10 == 0:
            c, a = sess.run([self.loss, self.accuracy], feed_dict={self.X: X_batch, self.y: y_batch})
            print(" Epoch = ", epoch + 1, " Batch = ", batch + 1, " Training Loss: ", "{:.9f}".format(c),
                  " Training Accuracy = ", "{:.9f}".format(a))
# 模型权值保存相关代码
builder = tf.saved_model.builder.SavedModelBuilder(saved_model_dir)
builder.add_meta_graph_and_variables(sess, [tf.saved_model.tag_constants.SERVING],
            signature_def_map = {
                tf.saved_model.signature_constants.DEFAULT_SERVING_SIGNATURE_DEF_KEY: signature},
            legacy_init_op = legacy_init_op)
builder.save()
tflite_model = tf.contrib.lite.toco_convert(sess.graph_def, [self.X[0]], [self.prediction_
 probability[0]], inference_type = 1, input_format = 1, output_format = 2, quantized_input_stats
 = None, drop_control_dependency = True)
open(self.path + "converted_model.tflite", "wb").write(tflite_model)
```

在train()函数中，根据传入批量大小使用生成器生成随机批次，生成器函数的定义如下：

```
def batch_gen(self, X, y, batch_size):
    index = list(range(X.shape[0]))
    np.random.shuffle(index)
    batches = int(X.shape[0] // batch_size)
    for b in range(batches):
        X_train, y_train = X[index[b * batch_size: (b + 1) * batch_size], :], y[
            index[b * batch_size: (b + 1) * batch_size]]
        yield X_train, y_train
```

通过合适的参数调用函数，创建批量的迭代器对象。使用next()函数，提取批量对象的下一个对象。在每个轮次开始时调用生成器函数，以保证每个轮次中的批量都是随机的。

3.3.3 模型保存

在model.pbtxt和model_ckpt的文件中保存训练好的模型并不能直接被Android应用程序使用。需要将其转换为protobuf格式（扩展名为.pb文件），与Android应用集成。优化的protobuf格式小于model.pbtxt和model_ckpt文件的大小。

首先，定义输入张量和输出张量的名称；其次，通过tensorflow.python.tools中的freeze_graph函数，使用这些输入和输出张量以及model.pbtxt和model_ckpt文件，将模型冻结；最后，被冻结的模型通过tensorflow.python.tools中的optimize_for_inference_lib

函数进一步优化，创建 protobuf 模型（即 optimized_model.pb），相关代码如下：

```
freeze_graph.freeze_graph(input_graph_path, input_saver_def_path,
                input_binary, checkpoint_path, output_node_names,
                restore_op_name, filename_tensor_name,
                output_frozen_graph_name, clear_devices, "")
input_graph_def = tf.GraphDef()
with tf.gfile.Open(output_frozen_graph_name, "rb") as f:
    data = f.read()
    input_graph_def.ParseFromString(data)
output_graph_def = optimize_for_inference_lib.optimize_for_inference(
        input_graph_def,
        ["inputs/X"],                    #输入节点构成的数组
        ["positive_sentiment_probability"],
        tf.int32.as_datatype_enum        #输出节点构成的数组
        )
#保存优化后的模型图
f = tf.gfile.FastGFile(output_optimized_graph_name, "w")
f.write(output_graph_def.SerializeToString())
```

3.3.4 词典保存

在预处理期间，训练 Keras tokenizer，将单词替换为数字索引，处理后的电影评论提供给 LSTM 模型进行训练。保留频率最高的前 50 000 个单词，并将电影评论序列的最大长度限制为 1000。尽管训练后的 Keras tokenizer 被保存并用于推断，但不能直接被 Android 应用程序使用。

将 Keras tokenizer 还原，50 000 个单词及其相应的单词索引保存在文本文件中。此文本文件可以在 Android 应用程序中使用，以构建单词到索引的词典，用来转换电影评论的文本。单词到索引映射可以通过 tokenizer.word_index 从加载的 Keras tokenizer 对象进行检索。相关代码如下：

```
def tokenize(path, path_out):
    #保存词典
    with open(path, 'rb') as handle:
        tokenizer = pickle.load(handle)
    dict_ = tokenizer.word_index
    keys = list(dict_.keys())[:50000]
    values = list(dict_.values())[:50000]
    total_words = len(keys)
    f = open(path_out,'w')
    for i in range(total_words):
        line = str(keys[i]) + ',' + str(values[i]) + '\n'
        f.write(line)
    f.close()
```

3.3.5 模型测试

完成模型训练后,移植到移动端,在设计移动应用程序时包括交互界面设计及核心逻辑设计。

1. 交互界面设计

移动应用程序界面设计的相应代码采用 XML 文件格式。应用程序包含一个简单的电影评论文本框,用户在其中输入他们对于电影的评论,完成后单击 SUBMIT 按钮,电影评论将被传递给应用程序的核心逻辑模块,该模块处理电影评论文本,并将其传递给 TensorFlow 优化模型进行推断,针对电影评论的情感打分,该分数会转换为相应的星级,并显示在移动应用程序中。

用于帮助用户和移动应用程序核心逻辑进行彼此交互的变量是在 XML 文件中通过 android:id 选项声明的。例如,用户提供的电影评论可以使用 Review 变量进行处理,对应 XML 文件中的定义为:

```
android:id = "@ + id/submit"
```

相关代码如下:

```xml
res/layout/activity_main.xml
<?xml version = "1.0" encoding = "utf-8"?>
<android.support.constraint.ConstraintLayout xmlns:android = "http://schemas.android.com/apk/res/android"
    xmlns:app = "http://schemas.android.com/apk/res-auto"
    xmlns:tools = "http://schemas.android.com/tools"
    android:layout_width = "match_parent"
    android:layout_height = "match_parent"
    tools:context = ".MainActivity"
    tools:layout_editor_absoluteY = "81dp">
    <TextView
        android:id = "@ + id/desc"
        android:layout_width = "100dp"
        android:layout_height = "26dp"
        android:layout_marginEnd = "8dp"
        android:layout_marginLeft = "44dp"
        android:layout_marginRight = "8dp"
        android:layout_marginStart = "44dp"
        android:layout_marginTop = "36dp"
        android:text = "Movie Review"
        app:layout_constraintEnd_toEndOf = "parent"
        app:layout_constraintHorizontal_bias = "0.254"
        app:layout_constraintStart_toStartOf = "parent"
        app:layout_constraintTop_toTopOf = "parent"
        tools:ignore = "HardcodedText" />
```

```xml
<EditText
    android:id = "@+id/Review"
    android:layout_width = "319dp"
    android:layout_height = "191dp"
    android:layout_marginEnd = "8dp"
    android:layout_marginLeft = "8dp"
    android:layout_marginRight = "8dp"
    android:layout_marginStart = "8dp"
    android:layout_marginTop = "24dp"
    app:layout_constraintEnd_toEndOf = "parent"
    app:layout_constraintStart_toStartOf = "parent"
    app:layout_constraintTop_toBottomOf = "@+id/desc" />
<RatingBar
    android:id = "@+id/ratingBar"
    android:layout_width = "240dp"
    android:layout_height = "49dp"
    android:layout_marginEnd = "8dp"
    android:layout_marginLeft = "52dp"
    android:layout_marginRight = "8dp"
    android:layout_marginStart = "52dp"
    android:layout_marginTop = "28dp"
    app:layout_constraintEnd_toEndOf = "parent"
    app:layout_constraintHorizontal_bias = "0.238"
    app:layout_constraintStart_toStartOf = "parent"
    app:layout_constraintTop_toBottomOf = "@+id/score"
    tools:ignore = "MissingConstraints" />
<TextView
    android:id = "@+id/score"
    android:layout_width = "125dp"
    android:layout_height = "39dp"
    android:layout_marginEnd = "8dp"
    android:layout_marginLeft = "96dp"
    android:layout_marginRight = "8dp"
    android:layout_marginStart = "96dp"
    android:layout_marginTop = "32dp"
    android:ems = "10"
    android:inputType = "numberDecimal"
    app:layout_constraintEnd_toEndOf = "parent"
    app:layout_constraintHorizontal_bias = "0.135"
    app:layout_constraintStart_toStartOf = "parent"
    app:layout_constraintTop_toBottomOf = "@+id/submit" />
<Button
    android:id = "@+id/submit"
    android:layout_width = "wrap_content"
    android:layout_height = "35dp"
    android:layout_marginEnd = "8dp"
    android:layout_marginLeft = "136dp"
```

```
                android:layout_marginRight = "8dp"
                android:layout_marginStart = "136dp"
                android:layout_marginTop = "24dp"
                android:text = "SUBMIT"
                app:layout_constraintEnd_toEndOf = "parent"
                app:layout_constraintHorizontal_bias = "0.0"
                app:layout_constraintStart_toStartOf = "parent"
                app:layout_constraintTop_toBottomOf = "@ + id/Review" />
    </android.support.constraint.ConstraintLayout>
```

该文件提供了 5 个控件。其中:1 个 Button,用于提交电影评论;2 个 TextView,分别显示影评和预测电影评论为正面的概率;1 个 RatingBar,显示星级评分;1 个 EditText,获取用户输入的评论。

2. 核心逻辑设计

Android 应用程序的核心逻辑是处理用户请求以及传递的数据,将结果返回给用户。作为应用程序的一部分,核心逻辑将接收用户提供的电影评论,并处理原始数据,将其转换为可以被训练好的 LSTM 模型进行推断的格式。

Java 中的 OnClickListener()函数用于监视用户是否已提交处理请求。在可以将数据输入经过优化训练好的 LSTM 模型进行推断之前,用户提供电影评论中的每个单词都需要被转化为索引。因此,除了优化 protobuf 模型,单词字典及其对应的索引也需要预先存储在设备上。使用 TensorFlowInferenceInterface()方法通过训练好的模型来运行推断。经过优化的 protobuf 模型和单词字典及其相应的索引存储在 assets 文件夹中。

应用程序核心逻辑需要完成的任务如下:

(1) 将单词到索引的字典加载到 WordToInd HashMap 中。单词到索引字典是在训练模型之前预处理文本时从 tokenizer 派生而来的。相关代码如下:

```
final Map < String, Integer > WordToInd = new HashMap < String, Integer >();
        BufferedReader reader = null;
        try { //单词到索引的字典加载
            reader = new BufferedReader(
                new InputStreamReader(getAssets().open("word_ind.txt")));
            String line;
            while ((line = reader.readLine()) != null)
            {//读入
                String[] parts = line.split("\n")[0].split(",",2);
                if (parts.length >= 2)
                {
                    String key = parts[0];
                    int value = Integer.parseInt(parts[1]);
                    WordToInd.put(key,value);
                } else
                {
                }
```

```java
            }
        } catch (IOException e) { //捕捉异常
        } finally {
            if (reader != null) {
                try {
                    reader.close();
                } catch (IOException e) {
                }
            }
        }
```

(2) 通过监听 OnClickListener()方法判断用户是否已提交电影评论进行推断。

(3) 如果已提交,则从 XML 绑定的 Review 对象中读取。

首先,通过删除标点符号等操作清理评论文本;其次,进行单词分词。每个单词都使用 WordToInd HashMap 转换为相应的索引。这些索引构成输入 TensorFlow 模型并用于推断的 InputVec 向量,向量的长度为 1000。因此,如果评论少于 1000 个单词,则用 0 在向量开头进行填充。相关代码如下:

```java
final Map<String, Integer> WordToInd = new HashMap<String, Integer>();
    BufferedReader reader = null;
    try {   //读入缓存
        reader = new BufferedReader(
            new InputStreamReader(getAssets().open("word_ind.txt")));
        String line;
        while((line = reader.readLine()) != null)
        {
            String[] parts = line.split("\n")[0].split(",",2);
            if(parts.length >= 2)
            {
                String key = parts[0];
                int value = Integer.parseInt(parts[1]);
                WordToInd.put(key,value);
            } else
            {
            }
        }
    } catch(IOException e) {   //捕捉异常
    } finally {
        if(reader != null) {
            try {
                reader.close();
            } catch(IOException e) {
            }
        }
    }
```

(4) 从 assets 文件夹将经过优化的 protobuf 模型（扩展名为.pb）载入内存，使用 TensorFlowInferenceInterface 功能创建 mInferenceInterface 对象，与原始模型一样，需要定义输入/输出节点，相关代码如下：

```
private TensorFlowInferenceInterface mInferenceInterface;
private static final String MODEL_FILE = "file:///android_asset/optimized_model.pb";
//模型存放路径
private static final String INPUT_NODE = "inputs/X";
private static final String OUTPUT_NODE = "positive_sentiment_probability";
```

对于模型，它们被定义为 INPUT_NODE 和 OUTPUT_NODE，分别包含 TensorFlow 输入占位符的名称和输出的评分概率操作。mInferenceInterface 对象的 feed() 方法用于将 InputVec 赋值给模型的 INPUT_NODE，而 mInferenceInterface 的 run() 方法用于执行 OUTPUT_NODE。最后，调用 mInferenceInterface 的 fetch() 得到用浮点变量 value_ 表示推断结果。相关代码如下：

```
mInferenceInterface.feed(INPUT_NODE, InputVec, 1, 1000);
mInferenceInterface.run(new String[] {OUTPUT_NODE}, false);
System.out.println(Float.toString(value_[0]));
mInferenceInterface.fetch(OUTPUT_NODE, value_);
   System.out.println(Float.toString(value_[0]));
```

(5) 首先，将 value_ 乘以 5 得到情感得分（评论为正面评论的概率）；其次，提供给 Android 应用程序的交互对象 ratingBar 变量。相关代码如下：

```
  double scoreIn;
scoreIn = value_[0] * 5;
double ratingIn = scoreIn;
String stringDouble = Double.toString(scoreIn);
score.setText(stringDouble);
   ratingBar.setRating((float) ratingIn);
```

此外，还需要编辑应用程序的 build.gradle 文件，将需要的包添为依赖项。

3.4 系统测试

本部分包括数据处理、模型训练、词典保存及模型效果。

3.4.1 数据处理

在 PyCharm 终端输入 python preprocess.py --path E:/MRSA/aclImdb/，输出结果如图 3-8 所示。

```
(tensorflow) C:\Users\dy-d\PycharmProjects\MRSA>python preprocess.py --path E:/MRSA/aclImdb/
Using TensorFlow backend.
records_processed 50000
5.458 min: Process
```

图 3-8　数据处理输出结果

3.4.2　模型训练

在 PyCharm 终端输入如下命令：

python movie_review_model_train.py process_main -- path E:/MRSA/ -- epochs 10

开始训练，模型经过 10 个轮次的适度训练，避免过拟合。优化器的学习率为 0.001，训练和验证的批量大小分别设置为 250 和 50。将训练输出结果保存在 .txt 文件中。BasicLSTM 的训练结果如图 3-9 所示，MultiLSTM 训练结果如图 3-10 所示。

```
250 50
Epoch= 1   Validation Loss:   0.616187274   Validation Accuracy= 0.680000007
Epoch= 2   Validation Loss:   0.524188161   Validation Accuracy= 0.740000010
Epoch= 3   Validation Loss:   0.436133772   Validation Accuracy= 0.819999993
Epoch= 4   Validation Loss:   0.390949339   Validation Accuracy= 0.819999993
Epoch= 5   Validation Loss:   0.515198827   Validation Accuracy= 0.759999990
Epoch= 6   Validation Loss:   0.343094289   Validation Accuracy= 0.860000014
Epoch= 7   Validation Loss:   0.261696249   Validation Accuracy= 0.939999998
Epoch= 8   Validation Loss:   0.244921491   Validation Accuracy= 0.899999976
Epoch= 9   Validation Loss:   0.411403835   Validation Accuracy= 0.839999974
Epoch= 10  Validation Loss:   0.355748057   Validation Accuracy= 0.879999995
Test Loss:   0.295403659   Test Accuracy= 0.892000020
4.527 hrs: Model train
```

图 3-9　BasicLSTM 训练结果

```
250 50
Epoch= 1   Validation Loss:   0.538058639   Validation Accuracy= 0.759999990
Epoch= 2   Validation Loss:   0.555779696   Validation Accuracy= 0.759999990
Epoch= 3   Validation Loss:   0.495759934   Validation Accuracy= 0.779999971
Epoch= 4   Validation Loss:   0.456312269   Validation Accuracy= 0.839999974
Epoch= 5   Validation Loss:   0.342410803   Validation Accuracy= 0.839999974
Epoch= 6   Validation Loss:   0.361103296   Validation Accuracy= 0.860000014
Epoch= 7   Validation Loss:   0.477218360   Validation Accuracy= 0.779999971
Epoch= 8   Validation Loss:   0.324074388   Validation Accuracy= 0.839999974
Epoch= 9   Validation Loss:   0.373119235   Validation Accuracy= 0.800000012
Epoch= 10  Validation Loss:   0.378401011   Validation Accuracy= 0.839999974
Test Loss:   0.362354994   Test Accuracy= 0.856000006
208912.500 s: Model train
```

图 3-10　MultiLSTM 训练结果

通过对比，MultiLSTM 模型训练集的准确率达到 94%，在验证集、测试集上的准确度均随着训练的进展而减少，发生了过拟合现象。所以最终移植到 Android 端时，采用 BasicLSTM 模型。

3.4.3　词典保存

在 Pycharm 终端输入 python freeze_code.py -- path E:/MRSA/ -- MODEL_NAME model,将模型冻结为 protobuf 格式,终端输出运行时间为 1.177min。

在 Pycharm 终端输入 python tokenizer_2_txt.py -- path 'E:/MRSA//aclImdb/tokenizer.pickle' --path_out 'E:/MRSA/word_ind.txt',即可保存词典。

保存的 optimized_model.pb 和 word_ind.txt 文件会移植到移动端。

3.4.4 模型效果

本部分包括程序下载运行和应用使用说明。

1. 程序下载运行

Android 项目编译成功后,在真机上进行测试,模拟器运行较慢,不建议使用。运行到真机的方法如下:

(1) 将手机数据线连接到计算机,开启开发者模式,打开 USB 调试,单击 Android 项目的"运行"按钮,将出现连接手机选项,单击即可。

(2) Android Studio 生成 .apk 文件,发送到手机,在手机上下载 .apk 文件,安装即可。

2. 应用使用说明

打开 App,应用初始界面如图 3-11 所示。

界面从上至下分别是文本框显示 Movie Review、文本编辑框、按钮、文本框显示概率,RatingBar 显示评分值。

此时在文本框内输入有关电影 *The Shawshank Redemption* 的评论,如图 3-12 所示。

图 3-11　应用初始界面

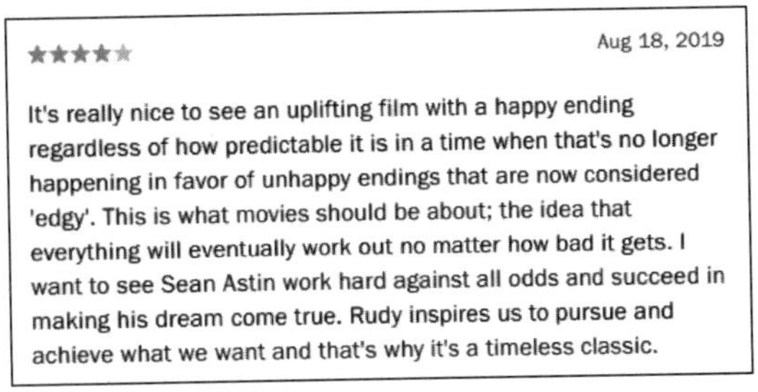

图 3-12　*The Shawshank Redemption* 电影评论

单击 SUBMIT 按钮，显示文本框内输出预测概率为 4.293/5，如图 3-13 所示。而评论员给这部电影的评分为 4/5，预测的评分更加精细化。

单击 SUBMIT 按钮，显示文本框内输出预测概率为 3.246/5，如图 3-14 所示。而 Rotten Tomatoes 对这部电影的平均评分为 3.5/5。

图 3-13　移动应用预测结果 1

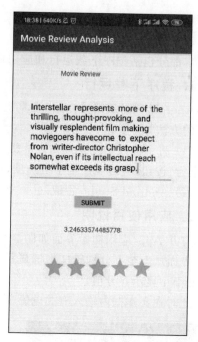

图 3-14　移动应用预测结果 2

从对比结果可以看出，应用程序能够为电影评论提供更加精细化的评分，用于电影评分的修正。以上各电影评分来自如下两个链接：

（1）https://www.rottentomatoes.com/m/shawshank_redemption/reviews?type=user；

（2）https://www.rottentomatoes.com/m/interstellar_2014/。

项目 4　Image2Poem——根据图像生成古体诗句

PROJECT 4

本项目使用 MS COCO 数据集,通过对图像进行简单描述,基于 CNN 和 LSTM 神经网络训练模型,进行诗歌创作,并以网站的形式展示,方便用户进行交互。

4.1　总体设计

本部分包括系统整体结构图和系统流程图。

4.1.1　系统整体结构图

系统整体结构如图 4-1 所示。

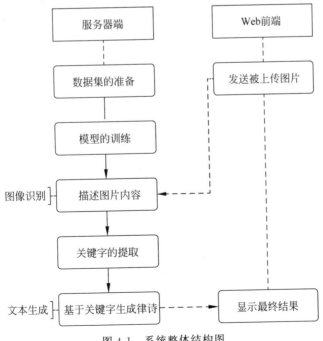

图 4-1　系统整体结构图

4.1.2 系统流程图

系统流程如图 4-2 所示。

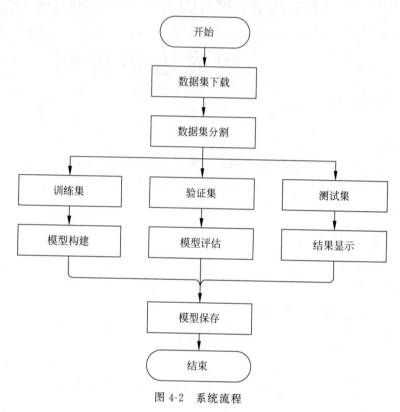

图 4-2 系统流程

4.2 运行环境

本项目使用的 Python 模块名及其版本如表 4-1 所示。

表 4-1 Python 模块名及其版本

模 块 名	版 本
Python	3.6.2
TensorFlow	1.12
Gensim	3.8.1
jieba	0.39
Numpy	1.18.2
Pandas	1.0.1

续表

模 块 名	版 本
joblib	0.14.1
Tqdm	4.43.0
Matplotlib	3.2.0
Scikit-image	0.16.2
Nltk	3.4.5
Django	3.0

4.2.1 Python 环境

基于 Python 3.6 版本,在 Windows 环境下推荐下载 Anaconda 完成 Python 所需的配置,下载地址为:https://www.anaconda.com/。

4.2.2 TensorFlow 安装

由于模型训练量比较大,需要使用 TensorFlow-GPU 版本。安装 TensorFlow-GPU,需要对应版本的 GPU 驱动、CUDA、cuDNN 软件。运行机器的 GPU 驱动只支持 CUDA 9.2 以下版本,如图 4-3 所示。

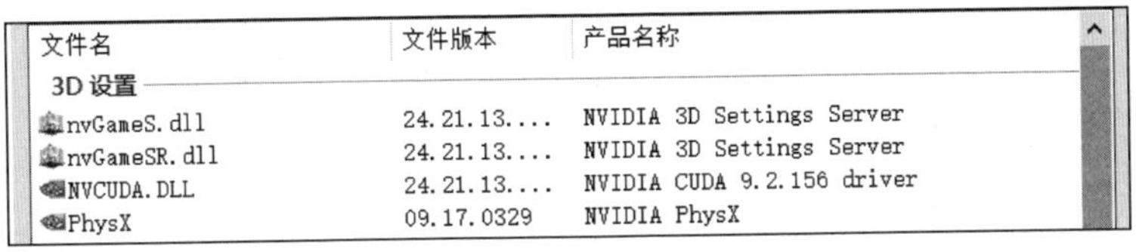

图 4-3 GPU 驱动支持

下载地址为 https://developer.nvidia.com/cuda-toolkit-archive,本项目选择 CUDA 9.0 版本。通过如下地址下载对应版本的 cuDNN 软件:https://developer.nvidia.com/rdp/cudnn-archive。安装 TensorFlow-GPU 后,在 Anaconda Prompt 中创建并激活 TensorFlow-GPU 环境,输入命令:

```
> conda create -n tensorflow-gpu python=3.6
> activate tensorflow-gpu
```

pip 指令安装 TensorFlow-GPU,注意版本要满足环境要求,输入命令:

```
> pip install tensorflow_gpu==1.12.0
```

安装完毕。

4.2.3 其他 Python 模块的安装

分别安装 gensim、jieba、numpy 和 django，输入如下命令：

```
> pip install gensim == 3.8.1
> pip install jieba == 0.39
> pip install numpy == 1.18.2
> pip install django == 3.0
```

安装完毕。

4.2.4 百度通用翻译 API 开通及使用

开放平台地址为 http://api.fanyi.baidu.com/。注册百度账号并完成通用翻译 API 的申请后获得 APP ID 和密钥，如图 4-4 所示。

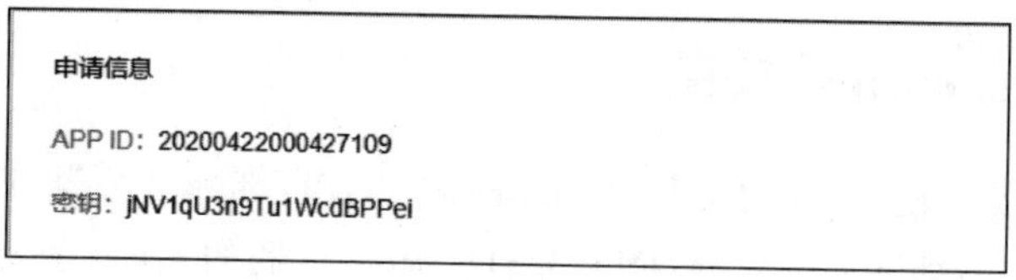

图 4-4　百度翻译用户申请信息

4.3 模块实现

本项目包括 7 个模块：数据准备、Web 后端准备、百度通用翻译、全局变量声明、创建模型、模型训练及保存、模型调用。下面分别给出各模块的功能介绍及相关代码。

4.3.1 数据准备

本部分包括数据集的获取、数据获取、预处理与加载、中文语料预处理。

1. 数据集的获取

数据集官网链接为 http://cocodataset.org/。
下载链接为：

http://images.cocodataset.org/zips/train2014.zip；
http://images.cocodataset.org/zips/val2014.zip；
http://images.cocodataset.org/zips/test2014.zip；
http://images.cocodataset.org/annotations/annotations_trainval2014.zip。

只需完成图像标题任务，使用标题标注即可，共有 5 个关键字：

```
{
info{ #基本信息
    "year" : int,
    "version" : str,
    "description" : str,
    "contributor" : str,
    "url" : str,
    "date_created" : datetime,
    }
image{ #图像
    "id" : int,
    "width" : int,
    "height" : int,
    "file_name" : str,
    "license" : int,
    "flickr_url" : str,
    "coco_url" : str,
    "date_captured" : datetime,
    }
license{ #授权
    "id" : int,
    "name" : str,
    "url" : str,
    }
categories{ #类别
            #caption 标注下,该项为空
            }
annotation{ #标注
    "id" : int,
    "image_id" : int,
    "caption" : str,
     }
}
```

2. 数据获取

(1) 古诗数据：包括元、明、清时期的古诗。

下载链接为 https://github.com/DevinZ1993/Chinese-Poetry-Generation/tree/master/raw。

(2) 通常使用中文维基百科（https://zh.wikipedia.org/wiki）和百度百科词条（https://baike.baidu.com/）。综合考虑收集难度、使用的便利性和内容质量，将中文维基百科语料作为古诗数据的额外补充，以解决模型不能理解现代词汇的问题。下载链接为 https://dumps.wikimedia.org/zhwiki/20200401/zhwiki-20200401-pages-articles1.xml-p1p162886.bz2。

由于下载的是压缩文件，需要用 WikiExtractor 进行提取。WikiExtractor 下载地址为 https://github.com/attardi/wikiextractor/blob/master/WikiExtractor.py。

在 Python 3.x 的环境下运行以下指令：

```
> python WikiExtractor.py -b 500M -o wiki_00 zhwiki-10200401-pages-articles1.xml-p1p2886.bz2
```

将维基百科的全部中文正文输出到 wiki_00 文件中。通过该方法生成的文本中，简体字与繁体字混合，不利于处理，因此，使用开源工具 OpenCC 进行转换。

3. 预处理与加载

MS COCO 数据集在 GitHub 网站中上传了各种版本的 API 调用接口，可以为训练数据的处理提供现成的模块调用，但是，部分代码使用的是 Python 2.7，编程时要略微修改。下载链接为 https://github.com/cocodataset/cocoapi/tree/master/PythonAPI 和 https://github.com/tylin/coco-caption。

进行词汇表 Vocabulary 的建立，以便 Decoder 翻译时使用。

```python
#调用 COCO API
from utils.coco.coco import COCO
#建立词汇表
from utils.vocabulary import Vocabulary
def prepare_train_data(config):
    #准备训练用的数据
    coco = COCO(config.train_caption_file)
    #按句子长度筛选数据集里的 caption
    coco.filter_by_cap_len(config.max_caption_length)
    #开始建立词汇表
    print("Building the vocabulary...")
    vocabulary = Vocabulary(config.vocabulary_size)
    if not os.path.exists(config.vocabulary_file):
        vocabulary.build(coco.all_captions())
        vocabulary.save(config.vocabulary_file)
    else:
        vocabulary.load(config.vocabulary_file)
    print("Vocabulary built.")
    print("Number of words = %d" %(vocabulary.size))
    coco.filter_by_words(set(vocabulary.words))
    #开始载入数据集中的信息
    print("Processing the captions...")
    if not os.path.exists(config.temp_annotation_file):
        captions = [coco.anns[ann_id]['caption'] for ann_id in coco.anns]
        image_ids = [coco.anns[ann_id]['image_id'] for ann_id in coco.anns]
        image_files = [os.path.join(config.train_image_dir,
                                    coco.imgs[image_id]['file_name'])
                       for image_id in image_ids]
        #图片信息的载入
        annotations = pd.DataFrame({'image_id': image_ids,
                                    'image_file': image_files,
                                    'caption': captions})
        annotations.to_csv(config.temp_annotation_file)
```

```python
    # 标题的载入
    else:
        annotations = pd.read_csv(config.temp_annotation_file)
        captions = annotations['caption'].values
        image_ids = annotations['image_id'].values
        image_files = annotations['image_file'].values
    if not os.path.exists(config.temp_data_file):
        word_idxs = []
        masks = []
        for caption in tqdm(captions):
            current_word_idxs_ = vocabulary.process_sentence(caption)
            current_num_words = len(current_word_idxs_)
            current_word_idxs = np.zeros(config.max_caption_length,
                                        dtype = np.int32)
            current_masks = np.zeros(config.max_caption_length)
            current_word_idxs[:current_num_words] = np.array(current_word_idxs_)
            current_masks[:current_num_words] = 1.0
            word_idxs.append(current_word_idxs)
            masks.append(current_masks)
        word_idxs = np.array(word_idxs)
        masks = np.array(masks)
        data = {'word_idxs': word_idxs, 'masks': masks}
        np.save(config.temp_data_file, data)
    else:
        data = np.load(config.temp_data_file).item()
        word_idxs = data['word_idxs']
        masks = data['masks']
    print("Captions processed.")
    print("Number of captions = %d" % (len(captions)))
    # 完成数据集载入到内存
    print("Building the dataset...")
    dataset = DataSet(image_ids,
                     image_files,
                     config.batch_size,
                     word_idxs,
                     masks,
                     True,
                     True)
    print("Dataset built.")
    return dataset
```

4. 中文语料预处理

在生成古诗主题词的计划阶段，需要找出与用户输入最接近的 4 个词语作为每一行诗句的主题。首先，根据规则从原始数据中提取有用的词语；其次，使用 TextRank 计算出这些词语的权重；最后，保存 RankedWords 模型。

```python
        _stopwords_path = os.path.join(raw_dir, 'stopwords.txt')
        _damp = 0.85
        def train_planner():
            #尝试另一个关键词拓展模型
            print("Training Word2Vec-based planner...")
            if not os.path.exists(save_dir):
                os.mkdir(save_dir)
            if not check_uptodate(plan_data_path):
                gen_train_data()
            word_lists = []
            with open(plan_data_path, 'r') as fin:
                for line in fin.readlines():
                    #各句之间用制表符隔开
                    word_lists.append(line.strip().split('\t'))
            #生成一个Word2Vec模型
            model = models.Word2Vec(word_lists, size = 512, min_count = 5)
            model.save(_plan_model_path)
    #从原始文件中读取所有停止词
    def _get_stopwords():
        stopwords = set()
        with open(_stopwords_path, 'r', encoding = 'UTF-8') as fin:
            for line in fin.readlines():
                stopwords.add(line.strip())
        return stopwords
    #使用另一个关键词提取算法
    class RankedWords(Singleton):
        def __init__(self): #初始化
            self.stopwords = _get_stopwords()
            if not check_uptodate(wordrank_path):
                self._do_text_rank()
            with open(wordrank_path, 'r') as fin:
                self.word_scores = json.load(fin)
            self.word2rank = dict((word_score[0], rank)
                    for rank, word_score in enumerate(self.word_scores))
        def _do_text_rank(self): #文本排序
            print("Do text ranking...")
            adjlists = self._get_adjlists()
            print("[TextRank] Total words: %d" % len(adjlists))
            #scores初始化,可以理解为scores越大,该词就越重要,越容易被选中
            scores = dict()
            for word in adjlists:
                scores[word] = [1.0, 1.0]
            #进行同步的数值迭代
            itr = 0
            while True:
                sys.stdout.write("[TextRank] Iteration %d..." % itr)
                sys.stdout.flush()
```

项目4 Image2Poem——根据图像生成古体诗句

```python
        # 通过遍历词关联表查看每一个词与其他词的关联次数和重要程度,据此计算该词的 score
        for word, adjlist in adjlists.items():
            # _damp 决定了每次进行更新时的步长
            scores[word][1] = (1.0 - _damp) + _damp * \
                sum(adjlists[other][word] * scores[other][0]
                    for other in adjlist)
        eps = 0
        for word in scores:
            eps = max(eps, abs(scores[word][0] - scores[word][1]))
            scores[word][0] = scores[word][1]
        print(" eps = %f" % eps)
        # 当精度足够高时,结束训练
        if eps <= 1e-6:
            break
        itr += 1
    # 基于字典的比较,以 TextRank 得分为准
    segmenter = Segmenter()
    def cmp_key(x):
        word, score = x
        return (0 if word in segmenter.sxhy_dict else 1, -score)
    words = sorted([(word, score[0]) for word, score in scores.items()],
            key = cmp_key)
    # 保存 ranked words 和对应的 scores
    with open(wordrank_path, 'w') as fout:
        json.dump(words, fout)
# 从原始数据中获得各词之间的关联程度
def _get_adjlists(self):
    print("[TextRank] Generating word graph ...")
    segmenter = Segmenter()
    poems = Poems()
    adjlists = dict()
    # 计数次数
    for poem in poems:
        for sentence in poem:
            words = []
            # 根据诗句含义分割出句子的技巧部分,例如,押韵用词等
            for word in segmenter.segment(sentence):
                if word not in self.stopwords:
                    words.append(word)
            for word in words:
                if word not in adjlists:
                    adjlists[word] = dict()
            for i in range(len(words)):
                # 如果两个词具有联系,那么增加它们的关联权重,否则初始化权重为 1.0
                for j in range(i + 1, len(words)):
                    if words[j] not in adjlists[words[i]]:
                        adjlists[words[i]][words[j]] = 1.0
```

```python
            else:
                adjlists[words[i]][words[j]] += 1.0
        if words[i] not in adjlists[words[j]]:
            adjlists[words[j]][words[i]] = 1.0
        else:
            adjlists[words[j]][words[i]] += 1.0
    # 归一化权重
    for a in adjlists:
        sum_w = sum(w for _, w in adjlists[a].items())
        for b in adjlists[a]:
            adjlists[a][b] /= sum_w
    return adjlists
def __getitem__(self, index):
    if index < 0 or index >= len(self.word_scores):
        return None
    return self.word_scores[index][0]
def __len__(self):
    return len(self.word_scores)
def __iter__(self):
    return map(lambda x: x[0], self.word_scores)
def __contains__(self, word):
    return word in self.word2rank
def get_rank(self, word):  # 获取排序
    if word not in self.word2rank:
        return len(self.word2rank)
    return self.word2rank[word]
```

4.3.2 Web 后端准备

本项目选用 Django 作为 Web 后端引擎，使用 django-admin 工具进行 Django 项目的初始化：

```
> django-admin startproject image2poem    # 新建项目
> django-admin startapp app1               # 新建应用
```

为了保存待识别图片的信息，需要在 SQLite 中建立数据库表，用于存储图片的路径及其他信息。在 app1 中的 models.py 文件中添加一个图片类。

```
from django.db import models
class IMG(models.Model):  # 创建模型
    img = models.ImageField(upload_to='img')
    name = models.CharField(max_length=20)
```

在命令行中输入下面的指令，完成 SQLite 数据库表的初始化。

```
python manage.py makemigrations app1
```

4.3.3 百度通用翻译

获得百度通用翻译 API 的 AppID 与 Key 后,编写专门用于翻译英文图片描述的类。在整合应用中,只需调用该类实例的方法即可。

```python
#!/usr/bin/python
# coding=utf-8
import requests
import hashlib
import random
import json
App_ID = '20200422000427109'
KEY = 'jNV1qU3n9Tu1WcdBPPei'
# HTTP 与 HTTPS 所使用的 URI 不同
HTTP_URL = 'http://api.fanyi.baidu.com/api/trans/vip/translate'
class Translator:
    def __init__(self):
        # 使用相同的 salt
        self.salt = random.randint(11111111, 99999999)
    def __sign__(self, query):
        sign = App_ID + query + str(self.salt) + KEY
        hl = hashlib.md5()
        hl.update(sign.encode(encoding="UTF-8"))
        return hl.hexdigest()
    def __request_url_generate(self, query):
        sign = self.__sign__(query)
        # 文本翻译
        return '{}?q={}&from=en&to=zh&appid={}&salt={}&sign={}'.format(HTTP_URL, query, App_ID, str(self.salt), sign)
    def baidu_general_translate(self, query):
        if query == '':
            return 'Input could not be empty'
        url = self.__request_url_generate(query)
        req = requests.get(url)
        # 如果响应的状态码不正确则提示错误
        if req.status_code == requests.codes.ok:
            text = req.text
            # 解析返回的 json
            data = json.loads(text)
            return data['trans_result'][0]['dst']
        else:
            return 'Fail to translate query, please check for the integrity of user info'
```

4.3.4 全局变量声明

为防止变量污染情况出现,提前声明要使用的全局变量。

```python
class Config(object):
    #各种(超)参数的包装类
    def __init__(self):
        #关于模型体系结构
        self.cnn = 'vgg16'                      #可选"VGG16"或者"resnet50"
        self.max_caption_length = 20
        self.dim_embedding = 512
        self.num_lstm_units = 512
        self.num_initalize_layers = 2           #1 或 2
        self.dim_initalize_layer = 512
        self.num_attend_layers = 2              #1 或 2
        self.dim_attend_layer = 512
        self.num_decode_layers = 2              #1 或 2
        self.dim_decode_layer = 1024
        #关于权重的初始化与正则化
        self.fc_kernel_initializer_scale = 0.08
        self.fc_kernel_regularizer_scale = 1e-4
        self.fc_activity_regularizer_scale = 0.0
        self.conv_kernel_regularizer_scale = 1e-4
        self.conv_activity_regularizer_scale = 0.0
        self.fc_drop_rate = 0.5
        self.lstm_drop_rate = 0.3
        self.attention_loss_factor = 0.01
        #关于优化参数
        self.num_epochs = 90                    #100 可选
        self.batch_size = 16                    #32 可选
        self.optimizer = 'Adam'  #"Adam"、"RMSProp"、"Momentum"或"SGD"
        self.initial_learning_rate = 0.0001
        self.learning_rate_decay_factor = 1.0
        self.num_steps_per_decay = 100000
        self.clip_gradients = 5.0
        self.momentum = 0.0
        self.use_nesterov = True
        self.decay = 0.9
        self.centered = True
        self.beta1 = 0.9
        self.beta2 = 0.999
        self.epsilon = 1e-6
        #关于模型保存
        self.save_period = 1000
        self.save_dir = './models/'
        self.summary_dir = './summary/'
        #关于词汇表建立
        self.vocabulary_file = './vocabulary.csv'
        self.vocabulary_size = 5000
        #关于模型训练
        self.train_image_dir = './train/images/train2014/'
```

```
        self.train_caption_file = './train/captions_train2014.json'
        self.temp_annotation_file = './train/anns.csv'
        self.temp_data_file = './train/data.npy'
        #关于模型评估
        self.eval_image_dir = './val/images/val2014/'
        self.eval_caption_file = './val/captions_val2014.json'
        self.eval_result_dir = './val/results/'
        self.eval_result_file = './val/results.json'
        self.save_eval_result_as_image = False
        #关于模型测试
        self.test_image_dir = './test/images/'
        self.test_result_dir = './test/results/'
        self.test_result_file = './test/results.csv'
```

4.3.5 创建模型

本部分包括 Image-Caption 的 CNN 模型、RNN 模型和 Poetry Generator。

1. Image-caption 的 CNN 模型

CNN 共有 13 个卷积层、3 个全连接层和 5 个池化层,图片先归一化成 $224 \times 224 \times 3$ 的规格后,经过卷积池化后得到长度 $L=14 \times 14$,维度 $D=512$ 的特征向量,完成编码工作。

```
def build_cnn(self):
    #构建CNN网络模型
    print("Building the CNN...")
    if self.config.cnn == 'vgg16':
        self.build_vgg16()
    else:
        print('can not build cnn! ')
    print("CNN built.")
def build_vgg16(self):
    #构建VGG16网络
    config = self.config
    #创建图片 placeholder
    images = tf.placeholder(
        dtype = tf.float32,
        shape = [config.batch_size] + self.image_shape)
    #第一次卷积、池化
    conv1_1_feats = self.nn.conv2d(images, 64, name = 'conv1_1')
    conv1_2_feats = self.nn.conv2d(conv1_1_feats, 64, name = 'conv1_2')
    pool1_feats = self.nn.max_pool2d(conv1_2_feats, name = 'pool1')
    #第二次卷积、池化
    conv2_1_feats = self.nn.conv2d(pool1_feats, 128, name = 'conv2_1')
    conv2_2_feats = self.nn.conv2d(conv2_1_feats, 128, name = 'conv2_2')
    pool2_feats = self.nn.max_pool2d(conv2_2_feats, name = 'pool2')
    #第三次卷积、池化
```

```python
        conv3_1_feats = self.nn.conv2d(pool2_feats, 256, name = 'conv3_1')
        conv3_2_feats = self.nn.conv2d(conv3_1_feats, 256, name = 'conv3_2')
        conv3_3_feats = self.nn.conv2d(conv3_2_feats, 256, name = 'conv3_3')
        pool3_feats = self.nn.max_pool2d(conv3_3_feats, name = 'pool3')
        #第四次卷积、池化
        conv4_1_feats = self.nn.conv2d(pool3_feats, 512, name = 'conv4_1')
        conv4_2_feats = self.nn.conv2d(conv4_1_feats, 512, name = 'conv4_2')
        conv4_3_feats = self.nn.conv2d(conv4_2_feats, 512, name = 'conv4_3')
        pool4_feats = self.nn.max_pool2d(conv4_3_feats, name = 'pool4')
        #第五次卷积
        conv5_1_feats = self.nn.conv2d(pool4_feats, 512, name = 'conv5_1')
        conv5_2_feats = self.nn.conv2d(conv5_1_feats, 512, name = 'conv5_2')
        conv5_3_feats = self.nn.conv2d(conv5_2_feats, 512, name = 'conv5_3')
        reshaped_conv5_3_feats = tf.reshape(conv5_3_feats,
                                    [config.batch_size, 196, 512])
        self.conv_feats = reshaped_conv5_3_feats
        self.num_ctx = 196
        self.dim_ctx = 512
        self.images = images
```

2. Image-caption 的 RNN 模型

把 CNN 网络生成的特征向量输入 RNN 模型中，RNN 网络使用 LSTM（Long Short-Term Memory，长短期记忆）模型。每层的输入不仅由当前层决定，而且还由上一层输出决定，并在上层输出引入注意力权重参数，实现带有注意力机制的翻译，最终根据权重值安排用词，生成完整句子。

```python
    def build_rnn(self):
        #构建 RNN
        print("Building the RNN...")
        config = self.config
        #创建卷积特征 placeholders
        if self.is_train:
            contexts = self.conv_feats
            sentences = tf.placeholder(
                dtype = tf.int32,
                shape = [config.batch_size, config.max_caption_length])
            masks = tf.placeholder(    #插入一个张量的占位符
                dtype = tf.float32,
                shape = [config.batch_size, config.max_caption_length])
        else:
            contexts = tf.placeholder(
                dtype = tf.float32,
                shape = [config.batch_size, self.num_ctx, self.dim_ctx])
            last_memory = tf.placeholder(    #插入一个张量的占位符
                dtype = tf.float32,
                shape = [config.batch_size, config.num_lstm_units])
```

```python
        last_output = tf.placeholder(          # 插入一个张量的占位符
            dtype = tf.float32,
            shape = [config.batch_size, config.num_lstm_units])
        last_word = tf.placeholder(            # 插入一个张量的占位符
            dtype = tf.int32,
            shape = [config.batch_size])
# 设置单词嵌入
with tf.variable_scope("word_embedding"):
    embedding_matrix = tf.get_variable(
        name = 'weights',
        shape = [config.vocabulary_size, config.dim_embedding],
        initializer = self.nn.fc_kernel_initializer,
        regularizer = self.nn.fc_kernel_regularizer,
        trainable = self.is_train)
# 设置 LSTM 网络模型
lstm = tf.nn.rnn_cell.LSTMCell(
    config.num_lstm_units,
    initializer = self.nn.fc_kernel_initializer)
if self.is_train:
    lstm = tf.nn.rnn_cell.DropoutWrapper(
        lstm,
        input_keep_prob = 1.0 - config.lstm_drop_rate,
        output_keep_prob = 1.0 - config.lstm_drop_rate,
        state_keep_prob = 1.0 - config.lstm_drop_rate)
# 使用平均值初始化 LSTM 模型
with tf.variable_scope("initialize"):
    context_mean = tf.reduce_mean(self.conv_feats, axis = 1)
    initial_memory, initial_output = self.initialize(context_mean)
    initial_state = initial_memory, initial_output
# 运行准备
predictions = []
if self.is_train:
    alphas = []
    cross_entropies = []
    predictions_correct = []
    num_steps = config.max_caption_length
    last_output = initial_output
    last_memory = initial_memory
    last_word = tf.zeros([config.batch_size], tf.int32)
else:
    num_steps = 1
last_state = last_memory, last_output
# 陆续生成单词
for idx in range(num_steps):
    # 注意力机制的引入
    with tf.variable_scope("attend"):
        alpha = self.attend(contexts, last_output)
```

```python
            context = tf.reduce_sum(contexts * tf.expand_dims(alpha, 2),
                                    axis = 1)
            if self.is_train:
                tiled_masks = tf.tile(tf.expand_dims(masks[:, idx], 1),
                                      [1, self.num_ctx])
                masked_alpha = alpha * tiled_masks
                alphas.append(tf.reshape(masked_alpha, [-1]))
        #嵌入最后一个单词
        with tf.variable_scope("word_embedding"):
            word_embed = tf.nn.embedding_lookup(embedding_matrix,
                                                last_word)
    #应用LSTM模型
        with tf.variable_scope("lstm"):
            current_input = tf.concat([context, word_embed], 1)
            output, state = lstm(current_input, last_state)
            memory, _ = state
    #将LSTM的扩展输出解码成一个单词
        with tf.variable_scope("decode"):
            expanded_output = tf.concat([output,
                                         context,
                                         word_embed],
                                        axis = 1)
            logits = self.decode(expanded_output)
            probs = tf.nn.softmax(logits)
            prediction = tf.argmax(logits, 1)
            predictions.append(prediction)
    #计算每一步训练的损失值
        if self.is_train:
            cross_entropy = tf.nn.sparse_softmax_cross_entropy_with_logits(
                    labels = sentences[:, idx],
                    logits = logits)
            masked_cross_entropy = cross_entropy * masks[:, idx]
            cross_entropies.append(masked_cross_entropy)    #交叉熵
            ground_truth = tf.cast(sentences[:, idx], tf.int64)
            prediction_correct = tf.where(
                tf.equal(prediction, ground_truth),
                tf.cast(masks[:, idx], tf.float32),
                tf.cast(tf.zeros_like(prediction), tf.float32))
            predictions_correct.append(prediction_correct)    #预测返回
            last_output = output
            last_memory = memory
            last_state = state
            last_word = sentences[:, idx]
        tf.get_variable_scope().reuse_variables()
    #计算最终的损失值
    if self.is_train:
        cross_entropies = tf.stack(cross_entropies, axis = 1)
```

```
        cross_entropy_loss = tf.reduce_sum(cross_entropies) \
                            / tf.reduce_sum(masks)
    alphas = tf.stack(alphas, axis = 1)
    alphas = tf.reshape(alphas, [config.batch_size, self.num_ctx, -1])
    attentions = tf.reduce_sum(alphas, axis = 2)
    diffs = tf.ones_like(attentions) - attentions
    attention_loss = config.attention_loss_factor \          #注意力损失
                    * tf.nn.l2_loss(diffs) \
                    / (config.batch_size * self.num_ctx)
    reg_loss = tf.losses.get_regularization_loss()           #正则损失
    total_loss = cross_entropy_loss + attention_loss + reg_loss
    predictions_correct = tf.stack(predictions_correct, axis = 1)
    accuracy = tf.reduce_sum(predictions_correct) \          #准确率
                / tf.reduce_sum(masks)
self.contexts = contexts
if self.is_train:                                            #输出参数的值
    self.sentences = sentences
    self.masks = masks
    self.total_loss = total_loss
    self.cross_entropy_loss = cross_entropy_loss
    self.attention_loss = attention_loss
    self.reg_loss = reg_loss
    self.accuracy = accuracy
    self.attentions = attentions
else:
    self.initial_memory = initial_memory
    self.initial_output = initial_output
    self.last_memory = last_memory
    self.last_output = last_output
    self.last_word = last_word
    self.memory = memory
    self.output = output
    self.probs = probs
print("RNN built.")
```

3. Poetry Generator

Poetry Generator 使用基于注意力模型的 RNN 网络。首先,把生成的关键词编码成向量,添加到上下文向量中;其次,生成第一句古诗向量,添加到上下文向量中;最后,每一句诗都根据上下文向量进行生成,并且添加到上下文向量中。进行 4 次循环后,投射器将上下文向量解码得到字符化的诗句,优化器使用 Adam Poetry Generator 模型的相关代码如下:

```
def _build_keyword_encoder(self):
    #将关键词编码为向量
    self.keyword = tf.placeholder(                           #关键词占位符
            shape = [_BATCH_SIZE, None, CHAR_VEC_DIM],
            dtype = tf.float32,
```

```python
                name = "keyword")
        self.keyword_length = tf.placeholder(              # 关键词长度占位符
                shape = [_BATCH_SIZE],
                dtype = tf.int32,
                name = "keyword_length")
        _, bi_states = tf.nn.bidirectional_dynamic_rnn(    # 双向动态 RNN
                cell_fw = tf.contrib.rnn.GRUCell(_NUM_UNITS / 2),
                cell_bw = tf.contrib.rnn.GRUCell(_NUM_UNITS / 2),
                inputs = self.keyword,
                sequence_length = self.keyword_length,
                dtype = tf.float32,
                time_major = False,
                scope = "keyword_encoder")
        self.keyword_state = tf.concat(bi_states, axis = 1)
        tf.TensorShape([_BATCH_SIZE, _NUM_UNITS]).\        # 返回表示维度的向量
                assert_same_rank(self.keyword_state.shape)
    # 上下文编码器是保证诗词语义连续性的关键
    # 建立上下文编码器
    def _build_context_encoder(self):
    # 将上下文编码为向量 list
        self.context = tf.placeholder(
                shape = [_BATCH_SIZE, None, CHAR_VEC_DIM],
                dtype = tf.float32,
                name = "context")
        self.context_length = tf.placeholder(              # 上下文长度占位符
                shape = [_BATCH_SIZE],
                dtype = tf.int32,
                name = "context_length")
        bi_outputs, _ = tf.nn.bidirectional_dynamic_rnn(   # 双向动态 RNN
                cell_fw = tf.contrib.rnn.GRUCell(_NUM_UNITS / 2),
                cell_bw = tf.contrib.rnn.GRUCell(_NUM_UNITS / 2),
                inputs = self.context,
                sequence_length = self.context_length,
                dtype = tf.float32,
                time_major = False,
                scope = "context_encoder")
        self.context_outputs = tf.concat(bi_outputs, axis = 2)
        tf.TensorShape([_BATCH_SIZE, None, _NUM_UNITS]).\  # 返回维度的向量
                assert_same_rank(self.context_outputs.shape)
    # 建立解码器模型
    def _build_decoder(self):
    # 将关键词上下文解码为向量序列
        attention = tf.contrib.seq2seq.BahdanauAttention(  # 注意力机制
                num_units = _NUM_UNITS,
                memory = self.context_outputs,
                memory_sequence_length = self.context_length)
        decoder_cell = tf.contrib.seq2seq.AttentionWrapper( # 解码
```

```python
            cell = tf.contrib.rnn.GRUCell(_NUM_UNITS),
            attention_mechanism = attention)
    self.decoder_init_state = decoder_cell.zero_state(        #初始化状态
            batch_size = _BATCH_SIZE, dtype = tf.float32).\
                clone(cell_state = self.keyword_state)
    self.decoder_inputs = tf.placeholder(                     #解码输入
            shape = [_BATCH_SIZE, None, CHAR_VEC_DIM],
            dtype = tf.float32,
            name = "decoder_inputs")
    self.decoder_input_length = tf.placeholder(               #解码输入长度
            shape = [_BATCH_SIZE],
            dtype = tf.int32,
            name = "decoder_input_length")
    self.decoder_outputs, self.decoder_final_state = tf.nn.dynamic_rnn(
            cell = decoder_cell,  #解码输出
            inputs = self.decoder_inputs,
            sequence_length = self.decoder_input_length,
            initial_state = self.decoder_init_state,
            dtype = tf.float32,
            time_major = False,
            scope = "training_decoder")
    tf.TensorShape([_BATCH_SIZE, None, _NUM_UNITS]).\         #返回维度的向量
            assert_same_rank(self.decoder_outputs.shape)
#建立投影模型
def _build_projector(self):
    #projector将解码器输出的向量投影到人类可读的字符空间中
    softmax_w = tf.Variable(
            tf.random_normal(shape = [_NUM_UNITS, len(self.char_dict)],
                mean = 0.0, stddev = 0.08),
            trainable = True)
    softmax_b = tf.Variable(
            tf.random_normal(shape = [len(self.char_dict)],
                mean = 0.0, stddev = 0.08),
            trainable = True)
    reshaped_outputs = self._reshape_decoder_outputs()        #变为指定的输出
    self.logits = tf.nn.bias_add(
            tf.matmul(reshaped_outputs, softmax_w),
            bias = softmax_b)
    self.probs = tf.nn.softmax(self.logits)
def _reshape_decoder_outputs(self):
    #将解码器的输出维度转化为 shape[?, _NUM_UNITS]
    def concat_output_slices(idx, val):
        output_slice = tf.slice(
                input_ = self.decoder_outputs,
                begin = [idx, 0, 0],
                size = [1, self.decoder_input_length[idx], _NUM_UNITS])
        return tf.add(idx, 1),\
```

```python
                    tf.concat([val, tf.squeeze(output_slice, axis = 0)],
                        axis = 0)
            tf_i = tf.constant(0)
            tf_v = tf.zeros(shape = [0, _NUM_UNITS], dtype = tf.float32)
            _, reshaped_outputs = tf.while_loop(                    # 通过循环实现
                cond = lambda i, v: i < _BATCH_SIZE,
                body = concat_output_slices,
                loop_vars = [tf_i, tf_v],
                shape_invariants = [tf.TensorShape([]),
                    tf.TensorShape([None, _NUM_UNITS])])
            tf.TensorShape([None, _NUM_UNITS]).\
                assert_same_rank(reshaped_outputs.shape)
            return reshaped_outputs
    def _build_optimizer(self):
        # 定义交叉熵以及为了降低交叉熵所使用的优化器
        self.targets = tf.placeholder(
            shape = [None],
            dtype = tf.int32,
            name = "targets")
        labels = tf.one_hot(self.targets, depth = len(self.char_dict))
        cross_entropy = tf.losses.softmax_cross_entropy(
            onehot_labels = labels,
            logits = self.logits)
        self.loss = tf.reduce_mean(cross_entropy)               # 计算损失
        self.learning_rate = tf.clip_by_value(                  # 学习率
            tf.multiply(1.6e-5, tf.pow(2.1, self.loss)),
            clip_value_min = 0.0002,
            clip_value_max = 0.02)
        self.opt_step = tf.train.AdamOptimizer(
            learning_rate = self.learning_rate).\
                minimize(loss = self.loss)
    def _build_graph(self):
        # 按照顺序建立各模型
        self._build_keyword_encoder()
        self._build_context_encoder()
        self._build_decoder()
        self._build_projector()
        self._build_optimizer()
    def __init__(self):
        self.g = tf.Graph()
        # 在TensorFlow的默认图中建立模型
        with self.g.as_default():
            self.char_dict = CharDict()
            self.char2vec = Char2Vec()
            self._build_graph()
            if not os.path.exists(save_dir):
                os.mkdir(save_dir)
```

```
            self.saver = tf.train.Saver(tf.global_variables())
            self.trained = False
            #保存已经建好的图
            self.merged = None
            self.summary = None
            self.summary_writer = None
            #读取已经建立的RankedWord模型,防止在infere的过程中多次载入
            self.generation_initialized = False
            def _initialize_session(self, session):
    checkpoint = tf.train.get_checkpoint_state(save_dir)
    #检查是否已经存在
    if not checkpoint or not checkpoint.model_checkpoint_path:
            init_op = tf.group(tf.global_variables_initializer(),
                        tf.local_variables_initializer())
            session.run(init_op)
    else:
            self.saver.restore(session, checkpoint.model_checkpoint_path)
            self.trained = True
```

4.3.6　模型训练及保存

定义模型架构和编译之后,使用训练集训练模型,其中图片描述、诗歌生成训练是分开进行的。进入项目路径后,使用如下指令开始训练:

```
> python image_caption/train_imagecaption.py -- phase = train
> python image_caption/train_poem.py - a #同时训练planning与generator
```

1. Image-caption

训练函数的实现过程如下:

```
def train(self, sess, train_data):
    #使用COCO train2014数据训练模型
    print("Training the model...")
    config = self.config
    #确保训练过程中的数据保存路径
    if not os.path.exists(config.summary_dir):
        os.mkdir(config.summary_dir)
    train_writer = tf.summary.FileWriter(config.summary_dir,
                                        sess.graph)
    for _ in tqdm(list(range(config.num_epochs)), desc = 'epoch'):    #轮次
        for _ in tqdm(list(range(train_data.num_batches)), desc = 'batch'):
            batch = train_data.next_batch()                           #批次
            image_files, sentences, masks = batch
            images = self.image_loader.load_images(image_files)
            feed_dict = {self.images: images,
                        self.sentences: sentences,
```

```python
                            self.masks: masks}
            _, summary, global_step = sess.run([self.opt_op,
                                                self.summary,
                                                self.global_step],
                                               feed_dict = feed_dict)
            if (global_step + 1) % config.save_period == 0:
                self.save()
        train_writer.add_summary(summary, global_step)
        train_data.reset()
    # 调用 save()函数保存模型
    self.save()
    train_writer.close()
    print("Training complete.")
def save(self):
    # 保存模型
    config = self.config
    data = {v.name: v.eval() for v in tf.global_variables()}
    save_path = os.path.join(config.save_dir, str(self.global_step.eval()))
    print((" Saving the model to %s..." % (save_path + ".npy")))
    np.save(save_path, data)
    info_file = open(os.path.join(config.save_dir, "config.pickle"), "wb")
    config_ = copy.copy(config)    # 使用 copy 模块赋值并使用新的内存空间存储
    config_.global_step = self.global_step.eval()
    joblib.dump(config_, info_file)
    info_file.close()
    print("Model saved.")
```

2. Poetry Generator

训练函数的实现过程如下:

```python
def train(self, n_epochs = 1000):
    print("Training RNN - based generator ...")
    with tf.Session() as session:
        # 初始化 session 后才可以进行训练
        self._initialize_session(session)
        self.merged = tf.summary.merge([tf.summary.scalar('loss', self.loss)])
        self.summary_writer = tf.summary.FileWriter('./event/', graph = session.graph)
        try:
            for epoch in range(n_epochs):                    # 轮次
                batch_no = 0
                merge_count = 0
                for keywords, contexts, sentences \
                        in batch_train_data(_BATCH_SIZE): # 批次
                    sys.stdout.write("[Seq2Seq Training] epoch = %d, " \
                        "line %d to %d..." %
                        (epoch, batch_no * _BATCH_SIZE,
```

```python
                            (batch_no + 1) * _BATCH_SIZE))
                    sys.stdout.flush()
                    self._train_a_batch(session, epoch,
                            keywords, contexts, sentences)
                    batch_no += 1
                    if 0 == batch_no % 64:
                        self.saver.save(session, _model_path)
                        self.summary_writer.add_summary(self.summary, merge_count)
                        merge_count += 1
                # 每个epoch都进行一次模型的保存
                self.saver.save(session, _model_path)
            print("Training is done.")
        except KeyboardInterrupt:
            print("Training is interrupted.")
    def _train_a_batch(self, session, epoch, keywords, contexts, sentences):
        keyword_data, keyword_length = self._fill_np_matrix(keywords)
        context_data, context_length = self._fill_np_matrix(contexts)
        decoder_inputs, decoder_input_length = self._fill_np_matrix(
                [start_of_sentence() + sentence[:-1] \
                    for sentence in sentences])        # 解码输入及长度
        targets = self._fill_targets(sentences)
        feed_dict = {                                  # 输入字典
                self.keyword : keyword_data,
                self.keyword_length : keyword_length,
                self.context : context_data,
                self.context_length : context_length,
                self.decoder_inputs : decoder_inputs,
                self.decoder_input_length : decoder_input_length,
                self.targets : targets
                }
        self.summary, loss, learning_rate, _ = session.run(   # 参数值输出
                [self.merged, self.loss, self.learning_rate, self.opt_step],
                feed_dict = feed_dict)
        print(" loss = %f, learning_rate = %f" % (loss, learning_rate))
    # 对各种向量进行初始化,填充numpy数组
    def _fill_np_matrix(self, texts):
        max_time = max(map(len, texts))
        matrix = np.zeros([_BATCH_SIZE, max_time, CHAR_VEC_DIM],
                dtype = np.float32)
        for i in range(_BATCH_SIZE):                   # 在批次和长度范围内
            for j in range(max_time):
                matrix[i, j, :] = self.char2vec.get_vect(end_of_sentence())
        for i, text in enumerate(texts):
            matrix[i, :len(text)] = self.char2vec.get_vects(text)
        seq_length = [len(texts[i]) if i < len(texts) else 0 \
                for i in range(_BATCH_SIZE)]
        return matrix, seq_length
```

```python
#将 sentence 中的字符转化为 char_dict 中的序号
def _fill_targets(self, sentences):
    targets = []
    for sentence in sentences:
        targets.extend(map(self.char_dict.char2int, sentence))
    return targets
```

4.3.7 模型调用

创建 Captioner 与 PortyGenerator 两类专门负责神经网络模型的载入、Web 用户输入的转发和返回神经网络模型的输出功能。首先,启动应用服务器的同时完成模型载入;其次,处理用户请求时调用 Captioner 与 PortyGenerator 的函数,完成整个识别与创作过程;最后,将结果返回 HTML 页面当中。

1. Captioner

Captioner 类的定义及相关操作如下。

```python
class Captioner:
    def __init__(self):
        #caption_generator 需要将该目录下的文件加载到当前目录
        #如果单独运行该模块,需要修改此处和下面函数末尾处
        os.chdir('./image_caption')
        FLAGS = tf.app.flags.FLAGS
        tf.flags.DEFINE_string('phase', 'test',
                        'The phase can be train, eval or test')   #用语句控制流程
        tf.flags.DEFINE_boolean('load', True,
                        'Turn on to load a pretrained model from either \
                        the latest checkpoint or a specified file')
        tf.flags.DEFINE_string('model_file', 'models//289999.npy',
                        'If sepcified, load a pretrained model from this file')
        #终止模型再次训练
        tf.flags.DEFINE_boolean('load_cnn', True,
                        'Turn on to load a pretrained CNN model')
        tf.flags.DEFINE_string('cnn_model_file', './cnn_models/vgg16_no_fc.npy',
                        'The file containing a pretrained CNN model')
        tf.flags.DEFINE_boolean('train_cnn', True,
                        'Turn on to train both CNN and RNN. \
                        Otherwise, only RNN is trained')
        tf.flags.DEFINE_integer('beam_size', 3,
                        'The size of beam search for caption generation')
        self.config = Config()    #集束搜索配置
        self.config.phase = FLAGS.phase
        self.config.train_cnn = FLAGS.train_cnn
        self.config.beam_size = FLAGS.beam_size
        sess = tf.Session()
```

```python
        self.model = CaptionGenerator(self.config)
        self.model.load(sess, FLAGS.model_file)
        #tf.get_default_graph().finalize()
        self.caption_sess = sess
        os.chdir('../')
    def generate_caption(self, img_path):
        # img_path:目标图片的相对路径
        os.chdir('./image_caption')
        #测试
        data, vocabulary = prepare_test_data(self.config, [img_path])
        caption = self.model.test(self.caption_sess, data, vocabulary)
        os.chdir('../')
        return caption
```

2. PortyGenerator

PortyGenerator 类的定义及相关操作如下。

```python
class PoetryGenerator:
    #加载模型
    def __init__(self):
        os.chdir('image_caption')
        self.planner = Planner()
        self.generator = Generator()
        os.chdir('../')
    #读取用户输入生成古诗
    def generate_poem(self, hints):
        os.chdir('image_caption')
        keywords = self.planner.plan(hints)
        print('生成的关键词:{}'.format(keywords))
        poem = self.generator.generate(keywords)
        os.chdir('../')
        return poem
```

3. Django 视图及模板创建

Django 视图创建及相关操作如下。

```python
#创建视图(view)处理 HTTP 请求
def uploadImg(request):
    #图片上传
    if request.method == 'POST':
        uploaded_img = IMG(
            img = request.FILES.get('img'),
            name = request.FILES.get('img').name
        )
        uploaded_img.save()
        #进行切片是为了生成器路径正确读取
```

```python
            print(uploaded_img.img.url[1:])
            captions = gl_setting.get_value('CAPTION_GENERATOR').generate_caption(uploaded_img.img.url[1:])
            print('生成的caption: {}'.format(captions[0]))
            chinese = gl_setting.get_value('TRANSLATOR').baidu_general_translate(captions[0])
            print('翻译结果: {}'.format(chinese))
            poem = gl_setting.get_value('POEM_GENERATOR').generate_poem(chinese)
            print('生成的古诗: ')
            for sentence in poem:
                print(sentence)
            content = {
                'img': uploaded_img,
                'poem': poem,
            }
            return render(request, 'app1/uploading.html', content)
        else:
            return render(request, 'app1/uploading.html')
#创建HTML文件模板,渲染视图输出
<!-- showing.html -->
<!DOCTYPE html>
<html lang="en">
<head>
    <meta charset="UTF-8">
    <title>Title</title>
</head>
<body>
    {% for img in imgs %}
        <img src="{{ img.img.url }}" />
    {% endfor %}
</body>
</html>
```

4.4 系统测试

本部分包括训练准确率、模型效果及整合应用。

4.4.1 训练准确率

本部分包括Image-caption和Poetry Generator模型。

1. Image-caption 模型

MS COCO 的 API 接口中提供了多样模型评判工具,只要在函数中调用即可。本项目使用 BLEU、ROUGE、CIDEr 三种指标验证准确率。

（1）BLEU（Bilingual Evaluation Understudy），是双语评估替补，取值为0～1，用生成的句子与参考句子进行匹配，能匹配上的单词越多，其分值越高。

（2）ROUGE（Recall-Oriented Understudy for Gisting Evaluation）是一种与BLEU类似的评判标准，基于召回率的相似度度量方法。但它评判的是最长公共子句，即生成的句子从第一个与参考句子匹配上的单词计数，能匹配上的单词所组成的子句越长，分值越高。所以ROUGE是连续的短句匹配，BLEU是离散的单词匹配。

（3）CIDEr（Consensus-based Image Description Evaluation）把每个句子进行TF-IDF向量化处理，用余弦相似度的方法将生成句子和参考句子的空间距离进行比对，如图4-5所示。

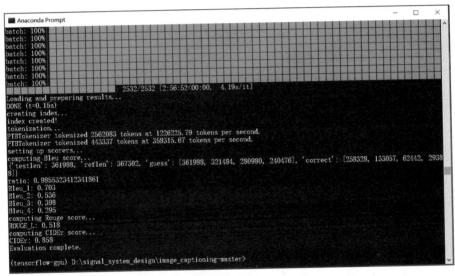

图4-5　Image-Caption模型验证结果

2. Poetry Generator 模型

由于评价古诗需要考虑的因素（如音韵、诗意、通畅性和主题切合度）过多，难以实现自动化，训练过程中损失值的变化如图4-6所示。

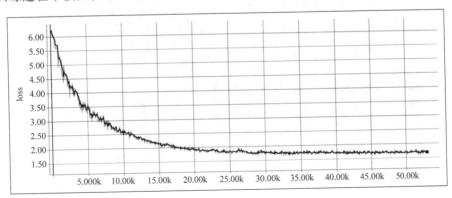

图4-6　损失值变化

4.4.2 模型效果

本部分包括 Image-caption 和 Poetry Generator 模型效果。

1. Image-Caption 模型效果

把测试图片放在 ./image_caption/test/images 路径下，调用指令便能返回标题结果。用画图工具绘制主题结果，效果如图 4-7 所示。

> python image_caption/train_imagecaption.py -- phase = test

图 4-7　Image-caption 模型效果

2. Poetry Generator 模型效果

调用如下指令，测试 Poetry Generator 模型效果。输入一句话，程序就会完成关键词的提取，并生成诗句，如图 4-8 所示。

> python image_caption/main_poem.py

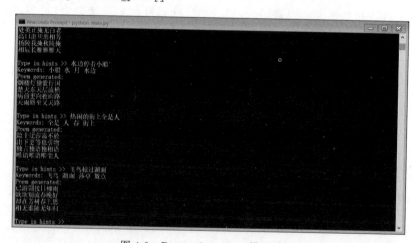

图 4-8　Poetry Generator 模型效果

4.4.3 整合应用

本部分包括应用方式、使用说明和测试结果。

1. 应用方式

切换目录到 Django 的根目录,输入以下指令:

> python manage.py runserver localhost:8000

待出现以下提示时,表示初始化完毕,可以进行访问:

```
$ System check identified no issues (0 silenced).
$ May 02, 2020 - 17:02:28
$ Django version 3.0, using settings 'image2poem.settings'
$ Starting development server at http://localhost:8000/
$ Quit the server with CTRL-BREAK.
```

单击"选择文件"按钮,选择一张图片后单击"上传"按钮,如图 4-9 所示。

图 4-9 Web 初始界面

测试结果(成功)显示界面如图 4-10 所示。

图 4-10 测试结果显示界面

2. 使用说明

（1）载入两个独立的神经网络模型，在启动应用之前确保系统内存大于 3GB，否则在使用过程中会存在内存泄漏并引发系统崩溃的风险。

（2）由于使用百度通用翻译 API 属于免费级别，每秒翻译请求上限数为 1，若超出将会被拒绝服务。因此，确保每次只有一个用户进行测试。

（3）Image-Caption 模块只接收 .jpg 与 .jpeg 格式的图片，若无法识别其他格式，则导致应用退出。

（4）图片路径不能包含非 ASCII 字符。

（5）应用初始化时间大约为 2min。

3. 测试结果

上传的图片和生成的古诗如表 4-2 所示。

表 4-2 测试样例

图 片	古 诗
	明旧掀堂一比舞 携到城人江人天 远说青云到向时 手见于中君见朝
	苍闲事带行风眼 乡色壁烟烟水西 海心孤道心行意 一在一鬓在鬓场
	仅钟曾房弓裙同 对天一腰复对中 何世何言何山箭 闲闲一厌坐无空

项目 5　歌曲人声分离

PROJECT 5

本项目通过 TensorFlow 构建 Bi-LSTM，针对唱歌软件的基本伴奏资源，使用 STFT（Short Time Fourier TransForm，短时傅里叶变换）进行处理，实现原曲获得音频质量较好的伴奏和纯人声音轨。

5.1　总体设计

本部分包括系统整体结构图和系统流程图。

5.1.1　系统整体结构图

系统整体结构如图 5-1 所示。

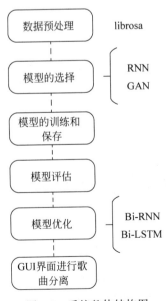

图 5-1　系统整体结构图

5.1.2 系统流程图

系统流程如图 5-2 所示。

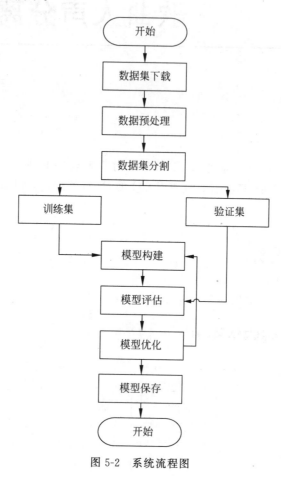

图 5-2 系统流程图

5.2 运行环境

本部分包括 Python 环境、TensorFlow 环境和 Jupyter Notebook 环境。

5.2.1 Python 环境

需要 Python 3.6 及以上配置,在 Windows 环境下推荐下载 Anaconda 完成 Python 所需的配置,下载地址为 https://www.anaconda.com/,默认下载 Python 3.7 版本。打开 Anaconda Prompt,安装开源的音频处理库 librosa,输入命令:

```
pip install librosa
```

安装完毕。

5.2.2 TensorFlow 环境

(1) 打开 Anaconda Prompt,输入清华仓库镜像,输入命令：

```
conda config -- add channels https://mirrors.tuna.tsinghua.edu.cn/anaconda/pkgs/free/
conda config - set show_channel_urls yes
```

(2) 创建 Python 3.6 环境,默认 3.7 版本和低版本 TensorFlow 存在不兼容问题,所以建立 Python 3.6 版本,输入命令：

```
conda create - n python36 python = 3.6
```

依据给出的相关提示,逐步安装。

(3) 在 Anaconda Prompt 中激活创建的虚拟环境,输入命令：

```
activate python36
```

(4) 安装 CPU 版本的 TensorFlow,输入命令：

```
conda install - upgrade -- ignore - installed tensorflow
```

安装完毕。

5.2.3 Jupyter Notebook 环境

(1) 首先打开 Anaconda Prompt,激活安装 TensoFlow 的虚拟环境,输入命令：

```
activate python36
```

(2) 安装 ipykernel,输入命令：

```
conda install ipykernel
```

(3) 将此环境写入 Jupyter Notebook 的 Kernel 中,输入命令：

```
python - m ipykernel install -- name python36 -- display - name "tensorflow(python36)"
```

(4) 打开 Jupyter Notebook,进入工作目录后,输入命令：

```
jupyter notebook
```

安装完毕。

5.3 模块实现

本项目包括 5 个模块：数据准备、数据预处理、模型构建、模型训练及保存、模型测试，下面分别给出各模块的功能介绍及相关代码，目录结构如图 5-3 所示。

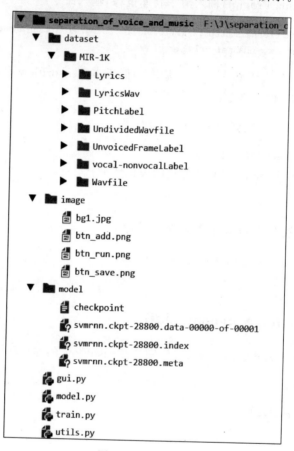

图 5-3　目录结构

(1) dataset：放置训练数据和验证数据。
(2) image：放置 GUI 制作所需要的图片素材。
(3) model：存储训练的模型。
(4) gui.py：模型的使用和 GUI 界面。
(5) model.py：构建网络模型。
(6) train.py：训练文件，包括数据的预处理、模型的训练及保存。
(7) utils.py：存放需要的工具函数。

5.3.1 数据准备

数据集适用于歌曲旋律提取、歌声分离、人声检测，如图 5-4 所示。

Wavfile 文件夹包含 1150 个歌曲文件，全部采用双声道.wav 格式。左声道为人声，右声道为伴奏。UndividedWavfile 文件夹中包含了 110 个歌曲文件，文件格式为单声道.wav。

Wavfile 文件夹中的所有数据作为训练集，UndividedWavfile 文件夹中的所有数据作为验证集。将 MIR-1K 下载解压至工作目录中的 dataset 文件夹中，完成数据集的准备工作。数据集下载地址为 http://mirlab.org/dataset/public/MIR-1K_for_MIREX.rar。

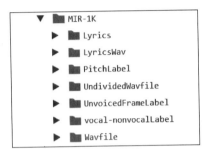

图 5-4　数据集结构

5.3.2 数据预处理

数据预处理主要完成如下功能：载入文件列表；将列表中的.wav 文件读取到内存中，并分解为单声道原曲、单声道伴奏、单声道纯人声；把处理好的音频进行短时傅里叶变化。具体包括载入数据集和验证集文件列表、读入文件并进行分离、.wav 文件进行 STFT 变换、处理频域文件、相关路径与参数的设定。

1. 载入数据集和验证集文件列表

将 dataset 文件夹中的训练集和验证集文件建立索引，保存到列表中，进行数据预处理。

```
# 导入操作系统模块
import os
# 导入文件夹中的数据
def load_file(dir):
    file_list = list()
    for filename in os.listdir(dir):
        file_list.append(os.path.join(dir, filename))
    # 将文件夹中的信息存入 file_list 列表中
    return file_list
# 设置数据集路径
dataset_train_dir = './dataset/MIR-1K/Wavfile'
dataset_validate_dir = './dataset/MIR-1K/UndividedWavfile'
train_file_list = load_file(dataset_train_dir)
valid_file_list = load_file(dataset_validate_dir)
```

2. 读入文件并进行分离

将列表中的文件读取到内存中，进行音频的分离操作。原数据集中的音频文件都是双声道文件。其中左声道为纯伴奏声轨，右声道为纯人声声轨，这里将左右声道和原文件分别保存。

```python
#导入音频处理库
import librosa
#数据集的采样率
mir1k_sr = 16000
#将 file_list 中的文件读入内存
def load_wavs(filenames, sr):
    wavs_mono = list()
    wavs_music = list()
    wavs_voice = list()
    #读取.wav文件(要求源文件是双声道的音频文件,一个声道是纯伴奏,另一个声道是纯人声)
    #将音频转换成单声道,存入 wavs_mono
    #将纯伴奏存入 wavs_music,
    #将纯人声存入 wavs_voice
    for filename in filenames:
        #librosa.load 函数:根据输入的采样率将文件读入
        wav, _ = librosa.load(filename, sr = sr, mono = False)
        assert(wav.ndim == 2) and (wav.shape[0] == 2), '要求 WAV 文件有两个声道!'
        #librosa.to_mono:将双声道转为单声道
        wav_mono = librosa.to_mono(wav) * 2
        wav_music = wav[0, :]
        wav_voice = wav[1, :]
        wavs_mono.append(wav_mono)
        wavs_music.append(wav_music)
        wavs_voice.append(wav_voice)
    #返回单声道原歌曲、纯伴奏、纯人声
    return wavs_mono, wavs_music, wavs_voice
    #导入训练数据集的.wav音频数据
#wavs_mono_train 存的是单声道音频,wavs_music_train 存的是纯伴奏,wavs_voice_train 存的是纯人声
wavs_mono_train, wavs_music_train, wavs_voice_train = load_wavs(filenames = train_file_list, sr = mir1k_sr)
    #导入验证集的.wav数据
wavs_mono_valid, wavs_music_valid, wavs_voice_valid = load_wavs(filenames = valid_file_list, sr = mir1k_sr)
```

3. wav 文件进行 STFT 变换

将读入的验证集和训练集音频文件从时域转化为频域。调用自定义的频域转换函数 wavs_to_specs()。但是训练集的数量较大，如果一次性全部转换，会导致内存占用过多。因此，每次随机抽取一个 batch_size 大小的文件进行频域转换。

```python
#导入 numpy 数学库
import numpy
#通过短时傅里叶变换将声音转到频域
#三组数据分别进行转换
```

```python
def wavs_to_specs(wavs_mono, wavs_music, wavs_voice, n_fft = 1024, hop_length = None):
    stfts_mono = list()
    stfts_music = list()
    stfts_voice = list()
    for wav_mono, wav_music, wav_voice in zip(wavs_mono, wavs_music, wavs_voice):
        #在librosa0.7.1及以上版本中,单声道音频文件必须为fortran-array格式,才能送入librosa.stft()进行处理
        #使用numpy.asfortranarray()函数,对单声道音频文件进行上述格式转换
        stft_mono = librosa.stft((numpy.asfortranarray(wav_mono)), n_fft = n_fft, hop_length = hop_length)
        stft_music = librosa.stft((numpy.asfortranarray(wav_music)), n_fft = n_fft, hop_length = hop_length)
        stft_voice = librosa.stft((numpy.asfortranarray(wav_voice)), n_fft = n_fft, hop_length = hop_length)
        stfts_mono.append(stft_mono)
        stfts_music.append(stft_music)
        stfts_voice.append(stft_voice)
    return stfts_mono, stfts_music, stfts_voice
#调用wavs_to_specs函数,转化验证集的数据
stfts_mono_valid, stfts_music_valid, stfts_voice_valid = wavs_to_specs(wavs_mono = wavs_mono_valid, wavs_music = wavs_music_valid, wavs_voice = wavs_voice_valid, n_fft = n_fft, hop_length = hop_length)
#定义batch_size大小
batch_size = 64
#定义n_fft大小,STFT的窗口大小
n_fft = 1024
#定义存储要转化训练集数据的数组
wavs_mono_train_cut = list()
wavs_music_train_cut = list()
wavs_voice_train_cut = list()
#从训练集中随机选取64个音频数据
for seed in range(batch_size):
    index = np.random.randint(0, len(wavs_mono_train))
    wavs_mono_train_cut.append(wavs_mono_train[index])
    wavs_music_train_cut.append(wavs_music_train[index])
    wavs_voice_train_cut.append(wavs_voice_train[index])
#短时傅里叶变换,将选取的音频数据转到频域
stfts_mono_train_cut, stfts_music_train_cut, stfts_voice_train_cut = wavs_to_specs(wavs_mono = wavs_mono_train_cut, wavs_music = wavs_music_train_cut, wavs_voice = wavs_voice_train_cut, n_fft = n_fft, hop_length = hop_length)
```

4. 处理频域文件

频域文件中的数据是复数,包含频率信息和相位信息,但是在训练时只需要考虑频率信息,所以将频率和相位信息分开,使用mini_batch的方法进行训练数据的输入。

```python
#获取频率
```

```python
def separate_magnitude_phase(data):
    return np.abs(data), numpy.angle(data)
# mini_batch 进行数据的输入
# stfts_mono:单声道 STFT 频域数据
# stfts_music:纯伴奏 STFT 频域数据
# stfts_music:纯人声 STFT 频域数据
# batch_size:batch 的大小
# sample_frames:获取多少帧数据
def get_next_batch(stfts_mono, stfts_music, stfts_voice, batch_size = 64, sample_frames = 8):
    stft_mono_batch = list()
    stft_music_batch = list()
    stft_voice_batch = list()
    #随机选择 batch_size 个数据
    collection_size = len(stfts_mono)
    collection_idx = numpy.random.choice(collection_size, batch_size, replace = True)
    for idx in collection_idx:
        stft_mono = stfts_mono[idx]
        stft_music = stfts_music[idx]
        stft_voice = stfts_voice[idx]
        #统计有多少帧
        num_frames = stft_mono.shape[1]
        assert num_frames >= sample_frames
        #随机获取 sample_frames 帧数据
        start = numpy.random.randint(num_frames - sample_frames + 1)
        end = start + sample_frames
        stft_mono_batch.append(stft_mono[:,start:end])
        stft_music_batch.append(stft_music[:,start:end])
        stft_voice_batch.append(stft_voice[:,start:end])
    #将数据转成 numpy.array,再对形状做一些变换
    # Shape: [batch_size, n_frequencies, n_frames]
    stft_mono_batch = numpy.array(stft_mono_batch)
    stft_music_batch = numpy.rray(stft_music_batch)
    stft_voice_batch = numpy.array(stft_voice_batch)
    #送入 RNN 的形状要求: [batch_size, n_frames, n_frequencies]
    data_mono_batch = stft_mono_batch.transpose((0, 2, 1))
    data_music_batch = stft_music_batch.transpose((0, 2, 1))
    data_voice_batch = stft_voice_batch.transpose((0, 2, 1))
    return data_mono_batch, data_music_batch, data_voice_batch
    #调用 get_next_batch()
    data_mono_batch, data_music_batch, data_voice_batch = get_next_batch(
stfts_mono = stfts_mono_train_cut, stfts_music = stfts_music_train_cut,
stfts_voice = stfts_voice_train_cut,batch_size = batch_size,
sample_frames = sample_frames)
#获取频率值
x_mixed_src, _ = separate_magnitude_phase(data = data_mono_batch)
y_music_src, _ = separate_magnitude_phase(data = data_music_batch)
y_voice_src, _ = separate_magnitude_phase(data = data_voice_batch)
```

5．设定相关路径与参数

```
# 可以通过命令设置的参数
# dataset_dir:数据集路径
# model_dir:模型保存的文件夹
# model_filename:模型保存的文件名
# dataset_sr:数据集音频文件的采样率
# learning_rate:学习率
# batch_size:小批量训练数据的长度
# sample_frames:每次训练获取多少帧数据
# iterations:训练迭代次数
# dropout_rate:丢弃率
def parse_arguments(argv):
    parser = argparse.ArgumentParser()
    parser.add_argument('-- dataset_train_dir', type = str, help = '数据集训练数据路径', default = './dataset/MIR - 1K/Wavfile')
    parser.add_argument('-- dataset_validate_dir', type = str, help = '数据集验证数据路径', default = './dataset/MIR - 1K/UndividedWavfile')
    parser.add_argument('-- model_dir', type = str, help = '模型保存的文件夹', default = 'model')
    parser.add_argument('--model_filename', type = str, help = '模型保存的文件名', default = 'svmrnn.ckpt')
    parser.add_argument('-- dataset_sr', type = int, help = '数据集音频文件的采样率', default = 16000)
    parser.add_argument('-- learning_rate', type = float, help = '学习率', default = 0.0001)
    parser.add_argument('-- batch_size', type = int, help = '小批量训练数据的长度', default = 64)
    parser.add_argument('-- sample_frames', type = int, help = '每次训练获取多少帧数据', default = 10)
    parser.add_argument('-- iterations', type = int, help = '训练迭代次数', default = 30000)
    parser.add_argument('-- dropout_rate', type = float, help = 'dropout率', default = 0.95)
    return parser.parse_args(argv)
```

5.3.3 模型构建

将数据加载进模型之后，需要定义结构、优化模型和模型实现。

1．定义结构

定义的架构为 1024 层循环神经网络，每层 RNN 的隐藏神经元个数从 1～1024 递增。最后是一个输入/输出的全连接层。在每层 RNN 之后，引入进行丢弃正则化，消除模型的过拟合问题。

模型初始化步骤：按顺序建立 1024 层 RNN，每层拥有从 1～1024 递增的 RNN 神经元个数。原始混合数据通过 RNN 以及丢弃正则化之后，由 ReLU 函数激活，并利用全连接层输出音频特征值为 513 的纯伴奏和纯人声的数据。

为了约束输出 y_music_src，y_voice_src 的大小，即使输出的纯伴奏数据 y_music_src

和纯人声数据 y_voice_src 与输入的混合数据 x_mixed_src 大小相同,输出需要进行以下变换:

$$y_m = \frac{d_m}{d_m + d_v + \alpha} \times x_m$$

$$y_v = \frac{d_v}{d_m + d_v + \alpha} \times x_m$$

$$y_m + y_v = \frac{d_m + d_v}{d_m + d_v + \alpha} \times x_m \approx x_m$$

其中,y_m 是 y_music_src,纯伴奏数据;y_v 是 y_voice_src,纯人声数据;d_m 是 y_dense_music_src,全连接层输出的纯伴奏数据;d_v 是 y_dense_voice_src,全连接层输出的人声数据;x_m 是 x_mixed_src,人声和伴奏的混合数据。

在分母上添加一个足够小的数 α,防止分母为 0。

```
    # 保存传入的参数
        self.num_features = num_features
        self.num_rnn_layer = len(num_hidden_units)
        self.num_hidden_units = num_hidden_units
    # 设置变量
    # 训练步数
        self.g_step = tf.Variable(0, dtype=tf.int32, name='g_step')
    # 设置占位符
    # 学习率
        self.learning_rate = tf.placeholder(tf.float32, shape=[], name='learning_rate')
    # 混合了伴奏和人声的数据
        self.x_mixed_src = tf.placeholder(tf.float32, shape=[None, None, num_features], name='x_mixed_src')
    # 伴奏数据
        self.y_music_src = tf.placeholder(tf.float32, shape=[None, None, num_features], name='y_music_src')
    # 人声数据
        self.y_voice_src = tf.placeholder(tf.float32, shape=[None, None, num_features], name='y_voice_src')
    # 保持丢弃,用于 RNN 网络
        self.dropout_rate = tf.placeholder(tf.float32)
    # 初始化神经网络
        self.y_pred_music_src, self.y_pred_voice_src = self.network_init(
    # 创建会话
        self.sess = tf.Session()
    # 构建神经网络
def network_init(self):
        rnn_layer = []
    # 根据 num_hidden_units 的长度来决定创建几层 RNN,每个 RNN 长度为 size
        for size in self.num_hidden_units:
    # 使用 LSTM 保证大数据集情况下的模型准确度
```

```
        # 加上丢弃,防止过拟合
            layer_cell = tf.nn.rnn_cell.LSTMCell(size)
            layer_cell = tf.contrib.rnn.DropoutWrapper(layer_cell, input_keep_prob = self.
dropout_rate)
            rnn_layer.append(layer_cell)
    # 创建多层 RNN
    # 为保证训练时考虑音频的前后时间关系,使用双向 RNN
        multi_rnn_cell = tf.nn.rnn_cell.MultiRNNCell(rnn_layer)
        outputs, state = tf.nn.bidirectional_dynamic_rnn(cell_fw = multi_rnn_cell, cell_bw =
multi_rnn_cell,
            inputs = self.x_mixed_src, dtype = tf.float32)
        out = tf.concat(outputs, 2)
    # 全连接层
    # 采用 ReLU 激活
        y_dense_music_src = tf.layers.dense(
            inputs = out,
            units = self.num_features,
            activation = tf.nn.relu,
            name = 'y_dense_music_src')
        y_dense_voice_src = tf.layers.dense(
            inputs = out,
            units = self.num_features,
            activation = tf.nn.relu,
            name = 'y_dense_voice_src')
        y_music_src = y_dense_music_src / (y_dense_music_src + y_dense_voice_src + np.finfo
(float).eps) * self.x_mixed_src
        y_voice_src = y_dense_voice_src / (y_dense_music_src + y_dense_voice_src + np.finfo
(float).eps) * self.x_mixed_src
        return y_music_src, y_voice_src
```

2. 优化模型

本项目使用的损失函数基于 reduce_mean(),计算输出的纯伴奏数据 y_music_src 及纯人声数据 y_voice_src 的方差。模型优化器选取 Adam 优化器,优化模型参数。

```
    # 损失函数
        def loss_init(self):
            with tf.variable_scope('loss') as scope:
                # 求方差(reduce_mean 方法)
                loss = tf.reduce_mean(
                    tf.square(self.y_music_src - self.y_pred_music_src)
                    + tf.square(self.y_voice_src - self.y_pred_voice_src), name = 'loss')
            return loss
    # 优化器
    # 采取常用的 Adam 优化器
        def optimizer_init(self):
            optimizer = tf.train.AdamOptimizer(learning_rate = self.learning_rate).minimize
```

```python
        (self.loss)
        return optimizer
```

3. 模型实现

构建好模型结构、损失函数及优化器之后,定义训练、测试及保存的相关函数,便于使用。

```python
#保存模型
    def save(self, directory, filename, global_step):
        #如果目录不存在,则创建
        if not os.path.exists(directory):
            os.makedirs(directory)
        self.saver.save(self.sess, os.path.join(directory, filename), global_step=global_step)
        return os.path.join(directory, filename)
    #加载模型,如果没有,则初始化所有变量
    def load(self, file_dir):
        #初始化变量
        self.sess.run(tf.global_variables_initializer())
        #如果没有模型,重新初始化
        kpt = tf.train.latest_checkpoint(file_dir)
        print("kpt:", kpt)
        startepo = 0
        if kpt != None:
            self.saver.restore(self.sess, kpt)
            ind = kpt.find("-")
            startepo = int(kpt[ind + 1:])
        return startepo
    #开始训练
    def train(self, x_mixed_src, y_music_src, y_voice_src, learning_rate, dropout_rate):
        #已经训练的步数
        #step = self.sess.run(self.g_step)
        _, train_loss = self.sess.run([self.optimizer, self.loss],
        feed_dict = {self.x_mixed_src: x_mixed_src, self.y_music_src: y_music_src, self.y_voice_src: y_voice_src,
        self.learning_rate:learning_rate, self.dropout_rate: dropout_rate})
        return train_loss
    #验证
    def validate(self, x_mixed_src, y_music_src, y_voice_src, dropout_rate):
        y_music_src_pred, y_voice_src_pred, validate_loss = self.sess.run([self.y_pred_music_src,
        self.y_pred_voice_src, self.loss],
        feed_dict = {self.x_mixed_src: x_mixed_src, self.y_music_src: y_music_src, self.y_voice_src: y_voice_src, self.dropout_rate: dropout_rate})
        return y_music_src_pred, y_voice_src_pred, validate_loss
    #测试
    def test(self, x_mixed_src, dropout_rate):
        y_music_src_pred, y_voice_src_pred = self.sess.run([self.y_pred_music_src, self.y_pred_
```

```
voice_src],
feed_dict = {self.x_mixed_src: x_mixed_src, self.dropout_rate: dropout_rate})
return y_music_src_pred, y_voice_src_pred
```

5.3.4 模型训练及保存

本项目使用训练集和测试集拟合,具体包括模型训练及模型保存。

1. 模型训练

模型训练具体操作如下:

```
#初始化模型
    model = SVMRNN(num_features = n_fft // 2 + 1, num_hidden_units = num_hidden_units)
    #加载模型
    #如果没有模型,则初始化所有变量
    startepo = model.load(file_dir = model_dir)
    print('startepo:' + str(startepo))
    #开始训练
    #index 是切割训练集位置的标识符
    index = 0
    for i in (range(iterations)):
    #从模型中断处开始训练
        if i < startepo:
            continue
        wavs_mono_train_cut = list()
        wavs_music_train_cut = list()
        wavs_voice_train_cut = list()
    #从训练集中随机选取 64 个音频数据
        for seed in range(batch_size):
            index = np.random.randint(0, len(wavs_mono_train))
            wavs_mono_train_cut.append(wavs_mono_train[index])
            wavs_music_train_cut.append(wavs_music_train[index])
            wavs_voice_train_cut.append(wavs_voice_train[index])
    #短时傅里叶变换,将选取的音频数据转到频域
        stfts_mono_train_cut, stfts_music_train_cut, stfts_voice_train_cut = wavs_to_specs(
        wavs_mono = wavs_mono_train_cut, wavs_music = wavs_music_train_cut, wavs_voice =
wavs_voice_train_cut,
        n_fft = n_fft, hop_length = hop_length)
    #获取下一批数据
        data_mono_batch, data_music_batch, data_voice_batch = get_next_batch( stfts_mono =
stfts_mono_train_cut, stfts_music = stfts_music_train_cut, stfts_voice = stfts_voice_train_cut,
        batch_size = batch_size, sample_frames = sample_frames)
    #获取频率值
        x_mixed_src, _ = separate_magnitude_phase(data = data_mono_batch)
        y_music_src, _ = separate_magnitude_phase(data = data_music_batch)
        y_voice_src, _ = separate_magnitude_phase(data = data_voice_batch)
```

```python
# 送入神经网络,开始训练
    train_loss = model.train(x_mixed_src = x_mixed_src, y_music_src = y_music_src, y_voice_src = y_voice_src,learning_rate = learning_rate, dropout_rate = dropout_rate)
# 每10步输出一次训练结果的损失值
    if i % 10 == 0:
        print('Step: %d Train Loss: %f' % (i, train_loss))
# 每200步输出一次测试结果
    if i % 200 == 0:
        print('==========================================')
        data_mono_batch, data_music_batch, data_voice_batch = get_next_batch(stfts_mono = stfts_mono_valid, stfts_music = stfts_music_valid,stfts_voice = stfts_voice_valid, batch_size = batch_size, sample_frames = sample_frames)
        x_mixed_src, _ = separate_magnitude_phase(data = data_mono_batch)
        y_music_src, _ = separate_magnitude_phase(data = data_music_batch)
        y_voice_src, _ = separate_magnitude_phase(data = data_voice_batch)
        y_music_src_pred, y_voice_src_pred, validate_loss = model.validate(x_mixed_src = x_mixed_src,
            y_music_src = y_music_src, y_voice_src = y_voice_src, dropout_rate = dropout_rate)
        print('Step: %d Validation Loss: %f' % (i, validate_loss))
        print('==========================================')
```

batch_size 是在一次前向/后向传播过程用到的训练样例数量,训练时随机选取 64 个数据并开始训练,总共训练 1110 个数据,迭代 30 000 步,如图 5-5 所示。

```
Step: 3600 Validation Loss: 1.106188
==========================================
Step: 3610 Train Loss: 0.808172
Step: 3620 Train Loss: 1.425123
Step: 3630 Train Loss: 0.829051
Step: 3640 Train Loss: 1.110434
Step: 3650 Train Loss: 1.124288
Step: 3660 Train Loss: 0.998556
Step: 3670 Train Loss: 0.996172
Step: 3680 Train Loss: 1.189133
Step: 3690 Train Loss: 1.257032
Step: 3700 Train Loss: 1.298231
Step: 3710 Train Loss: 1.067620
Step: 3720 Train Loss: 0.947821
Step: 3730 Train Loss: 0.974504
Step: 3740 Train Loss: 0.957819
Step: 3750 Train Loss: 0.742355
Step: 3760 Train Loss: 0.977313
Step: 3770 Train Loss: 0.933429
```

图 5-5　训练结果

2. 模型保存

模型保存有两种作用:一是为了在训练过程中出现意外而中断时,能够在上次保存的模型处开始;二是为了在应用中直接使用训练好的模型。

```python
# 每200步保存一次模型
if i % 200 == 0:
    model.save(directory = model_dir, filename = model_filename, global_step = i)
```

5.3.5 模型测试

采用 Python 自带的 Tkinter 库进行 GUI 设计,GUI 实现如下功能。

1. 批量选取歌曲和保存路径

定义 addfile() 和 choose_save_path() 两个函数,使用 Tkinter 中 filedialog 模块打开系统路径。

```
def addfile():
    #定义全局变量 music_path,用于添加音频文件
    global music_path
    paths = tk.filedialog.askopenfilenames(title = '选择要分离的歌曲')
    #保存选择的歌曲
    #遍历添加
    for path in paths:
        music_path.append(path)
    label_info['text'] = '\n'.join(music_path)
    #选择分离结束后保存的文件夹
def choose_save_path():
    global save_path
    save_path = tk.filedialog.askdirectory(title = '选择保存文件夹')
    save_info['text'] = save_path
```

2. 模型导入及调用

定义调用模型的函数 separate()。该函数完成如下功能:

(1) 将带转化的文件进行列表保存、数据分割、短时傅里叶变换,完成数据的预处理。

```
#加载音频文件
wavs_mono = list()
for filename in music_path:
    wav_mono, _ = librosa.load(filename, sr = dataset_sr, mono = True)
    wavs_mono.append(wav_mono)
#短时傅里叶变换的 fft 点数
#默认情况下,窗口长度 = fft 点数
n_fft = 1024
#冗余度
hop_length = n_fft // 4
#将音频数据转换到频域
stfts_mono = list()
for wav_mono in wavs_mono:
    stft_mono = librosa.stft(wav_mono, n_fft = n_fft, hop_length = hop_length)
    stfts_mono.append(stft_mono.transpose())
```

(2) 数据预处理后,调用训练好的模型,进行人声和伴奏的分离。

```
#初始化神经网络
```

```python
model = SVMRNN(num_features = n_fft // 2 + 1, num_hidden_units = num_hidden_units)
# 导入模型
model.load(file_dir = model_dir)
for wav_filename, wav_mono, stft_mono in zip(music_path, wavs_mono, stfts_mono):
    wav_filename_base = os.path.basename(wav_filename)
    # 单声道音频文件
    wav_mono_filename = wav_filename_base.split('.')[0] + '_mono.wav'
    # 分离后的纯伴奏音频文件
    wav_music_filename = wav_filename_base.split('.')[0] + '_music.wav'
    # 分离后的纯人声音频文件
    wav_voice_filename = wav_filename_base.split('.')[0] + '_voice.wav'
    # 要保存文件的相对路径
    wav_mono_filepath = os.path.join(save_path, wav_mono_filename)
    wav_music_hat_filepath = os.path.join(save_path, wav_music_filename)
    wav_voice_hat_filepath = os.path.join(save_path, wav_voice_filename)
        print('Processing %s...' % wav_filename_base)
        stft_mono_magnitude, stft_mono_phase = separate_magnitude_phase(data = stft_mono)
        stft_mono_magnitude = np.array([stft_mono_magnitude])
        y_music_pred, y_voice_pred = model.test(x_mixed_src = stft_mono_magnitude, dropout_rate = dropout_rate)
```

（3）将处理好的频率文件和原本的相位信息相加，进行傅里叶逆变换。

```python
# 根据振幅和相位，得到复数
# 信号 s(t) 乘上 e^(j * phases) 表示信号 s(t) 移动相位 phases
def combine_magnitude_phase(magnitudes, phases):
    return magnitudes * np.exp(1.j * phases)
```

（4）将时域文件写成 .wav 格式的歌曲保存。

```python
# 保存数据，使用 librosa.output.write_wav() 函数，将文件保存成 .wav 格式歌曲文件
librosa.output.write_wav(wav_mono_filepath, wav_mono, dataset_sr)
librosa.output.write_wav(wav_music_hat_filepath, y_music_hat, dataset_sr)
librosa.output.write_wav(wav_voice_hat_filepath, y_voice_hat, dataset_sr)
    # 检测在保存文件夹中是否生成了伴奏文件，若存在则自动打开该文件夹
    if os.path.exists(wav_music_hat_filepath):
        os.startfile(save_path)
        remind_window.destroy()
```

3. GUI 代码

GUI 相关代码如下：

```python
from tkinter import Tk, filedialog
import tkinter as tk
import librosa
import os
import numpy as np
from NewModel import SVMRNN
```

```python
from NewUtils import separate_magnitude_phase, combine_magnitude_phase
music_path = []
save_path = str()
wav_music_hat_filepath = str()
def addfile():
    # 定义全局变量 music_path,用于添加音频文件
    global music_path
    paths = tk.filedialog.askopenfilenames(title = '选择要分离的歌曲')
    # 保存选择的歌曲
    # 遍历添加
    for path in paths:
        music_path.append(path)
    label_info['text'] = '\n'.join(music_path)
    # 选择分离结束后保存的文件夹
def choose_save_path():
    global save_path
    save_path = tk.filedialog.askdirectory(title = '选择保存文件夹')
    save_info['text'] = save_path
    # 弹窗信息的定义
def pop_window():
    global wav_music_hat_filepath
    def separate():
        dataset_sr = 16000        # 采样率
        model_dir = './model'     # 模型保存文件夹
        dropout_rate = 0.95       # 丢弃率
        # 加载音频文件
        wavs_mono = list()
        for filename in music_path:
            wav_mono, _ = librosa.load(filename, sr = dataset_sr, mono = True)
            wavs_mono.append(wav_mono)
        # 短时傅里叶变换的 fft 点数
        # 默认情况下,窗口长度 = fft 点数
        n_fft = 1024
        # 冗余度
        hop_length = n_fft // 4
        # 用于创建 RNN 节点数
        num_hidden_units = [1024, 1024, 1024, 1024, 1024]
        # 将音频数据转换到频域
        stfts_mono = list()
        for wav_mono in wavs_mono:
            stft_mono = librosa.stft(wav_mono, n_fft = n_fft, hop_length = hop_length)
            stfts_mono.append(stft_mono.transpose())
        # 初始化神经网络
        model = SVMRNN(num_features = n_fft//2 + 1, num_hidden_units = num_hidden_units)
        # 导入模型
        model.load(file_dir = model_dir)
        for wav_filename, wav_mono, stft_mono in zip(music_path, wavs_mono, stfts_mono):
```

```python
            wav_filename_base = os.path.basename(wav_filename)
            # 单声道音频文件
            wav_mono_filename = wav_filename_base.split('.')[0] + '_mono.wav'
            # 分离后的纯伴奏音频文件
             wav_music_filename = wav_filename_base.split('.')[0] + '_music.wav'
            # 分离后的纯人声音频文件
            wav_voice_filename = wav_filename_base.split('.')[0] + '_voice.wav'
            # 要保存文件的相对路径
            wav_mono_filepath = os.path.join(save_path, wav_mono_filename)
            wav_music_hat_filepath = os.path.join(save_path, wav_music_filename)
            wav_voice_hat_filepath = os.path.join(save_path, wav_voice_filename)
            print('Processing %s ...' % wav_filename_base)
            stft_mono_magnitude, stft_mono_phase = separate_magnitude_phase(data=stft_mono)
            stft_mono_magnitude = np.array([stft_mono_magnitude])
            y_music_pred, y_voice_pred = model.test(x_mixed_src=stft_mono_magnitude,
dropout_rate=dropout_rate)
            # 根据振幅和相位,转为复数,用于下面的逆短时傅里叶变换
             y_music_stft_hat = combine_magnitude_phase(magnitudes=y_music_pred[0],
phases=stft_mono_phase)
             y_voice_stft_hat = combine_magnitude_phase(magnitudes=y_voice_pred[0],
phases=stft_mono_phase)
            y_music_stft_hat = y_music_stft_hat.transpose()
            y_voice_stft_hat = y_voice_stft_hat.transpose()
            # 通过逆短时傅里叶变换,将分离好的频域数据转换为音频,生成相对应的音频文件
            y_music_hat = librosa.istft(y_music_stft_hat, hop_length=hop_length)
            y_voice_hat = librosa.istft(y_voice_stft_hat, hop_length=hop_length)
            # 保存数据
librosa.output.write_wav(wav_mono_filepath, wav_mono, dataset_sr)
            librosa.output.write_wav(wav_music_hat_filepath, y_music_hat, dataset_sr)
            librosa.output.write_wav(wav_voice_hat_filepath, y_voice_hat, dataset_sr)
            if os.path.exists(wav_music_hat_filepath):
                os.startfile(save_path)
                remind_window.destroy()
    remind_window = tk.Toplevel()
    remind_window.title('提示')
    remind_window.minsize(width=400, height=200)
    tk.Label(remind_window, text='加载模型中,请勿关闭软件').place(x=70, y=60)
    tk.Button(remind_window, text='我知道了', font=('Fangsong', 14), command=separate).
place(x=150, y=100)
root = Tk()
# 窗口标题
root.title('歌曲人声分离')
# 大小不可调整
root.resizable(0, 0)
# 创建背景图片
canvas = tk.Canvas(root, width=800, height=900, bd=0, highlightthickness=0)
imgpath = 'image/bg1.jpg'
```

```python
img = Image.open(imgpath)
photo = ImageTk.PhotoImage(img)
# 设置背景图片在窗口显示的偏移量
canvas.create_image(750, 400, image = photo)
canvas.pack()
# 添加按钮
# 添加按钮的图片
btn_add = tk.PhotoImage(file = 'image/btn_add.png')
btn_addfile = tk.Button(root, command = addfile, image = btn_add)
# 将按钮摆放到窗口上
canvas.create_window(70,70, width = 84, height = 84, window = btn_addfile)
# 功能说明文字
canvas.create_text(70,130, text = '添加文件', fill = 'white', font = ('Fangsong', '15', 'bold'))
# 选择保存路径的按钮
pic_save = tk.PhotoImage(file = 'image/btn_save.png')
btn_save = tk.Button(root, command = choose_save_path, image = pic_save)
canvas.create_window(200, 70, width = 84, height = 84, window = btn_save)
canvas.create_text(200,130,text = '选择保存路径',fill = 'white', font = ('Fangsong', '15', 'bold'))
# 运行按钮
pic_run = tk.PhotoImage(file = 'image/btn_run.png')
btn_run = tk.Button(root, command = pop_window, image = pic_run)
canvas.create_window(330, 70, width = 84, height = 84, window = btn_run)
canvas.create_text(330, 130, text = '运行', fill = 'white', font = ('Fangsong', '15', 'bold'))
# 显示待分离歌曲
canvas.create_text(70, 180, text = '待分离的歌曲', fill = 'white', font = ('Fangsong', '14', 'bold'))
label_info = tk.Label(root, bg = 'white', anchor = 'nw', justify = 'left')
canvas.create_window(210, 300, width = 400, height = 200, window = label_info)
# 显示保存的路径
canvas.create_text(48, 430, text = '保存路径', fill = 'white', font = ('Fangsong', '14', 'bold'))
save_info = tk.Label(root, bg = 'white', anchor = 'nw', justify = 'left')
canvas.create_window(210, 470, width = 400, height = 50, window = save_info)
root.mainloop()
```

5.4 系统测试

本部分包括训练准确率、测试效果及模型应用。

5.4.1 训练准确率

训练迭代到靠后的步数,损失函数的值小于 0.5,这意味着这个预测模型训练比较成功。在整个迭代训练过程中,随着 epoch 的增加,模型损失函数的值在逐渐减小,并且在 17 000 步以后趋于稳定,如图 5-6 所示。

```
==========================================
Step: 17200 Validation Loss: 0.420042
==========================================
Step: 17210 Train Loss: 0.507203
Step: 17220 Train Loss: 0.502529
Step: 17230 Train Loss: 0.541619
Step: 17240 Train Loss: 0.580664
Step: 17250 Train Loss: 0.450910
Step: 17260 Train Loss: 0.628350
Step: 17270 Train Loss: 0.479427
Step: 17280 Train Loss: 0.497587
Step: 17290 Train Loss: 0.542923
Step: 17300 Train Loss: 0.432395
Step: 17310 Train Loss: 0.469025
Step: 17320 Train Loss: 0.444656
Step: 17330 Train Loss: 0.449266
```

图 5-6　训练准确率

5.4.2　测试效果

将数据代入模型进行测试，并将分离得到的纯伴奏和纯人声波形经过对比以及人耳辨别，得到验证：模型可以实现歌曲伴奏和人声分离。如图 5-7 和图 5-8 所示。

图 5-7　分离的纯人声波形

图 5-8　分离的纯伴奏波形

5.4.3　模型应用

使用说明包括以下 4 部分。

（1）打开 gui.py 文件，界面如图 5-9 所示。

（2）单击"添加文件"按钮，选择要获得伴奏的歌曲，在"待分离的歌曲"中显示添加歌曲的路径和名称，如图 5-10 所示。

图 5-9　主界面

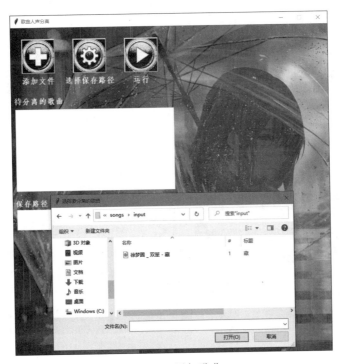

图 5-10　添加歌曲

（3）选择保存路径，如图 5-11 所示。

图 5-11　保存路径

（4）单击"运行"按钮后会弹出提示框，确认后程序开始运行，运行完毕后将自动打开保存文件夹，如图 5-12 所示。

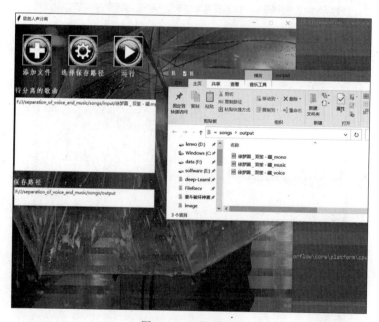

图 5-12　运行结果

项目 6 基于 Image Caption 的英语学习

PROJECT 6

本项目基于深度学习的 CNN 神经网络和 LSTM 循环神经网络,将计算机视觉和机器翻译技术相互结合,使用 BLEU(Bilingual Evaluation Understudy,双语评估替代)算法评估自然语言处理任务生成的文本。通过对图片描述,建立英语学习和具体生活场景的联系,实现一款辅助英语学习的应用。

6.1 总体设计

本部分包括系统整体结构图和系统流程图。

6.1.1 系统整体结构图

系统整体结构如图 6-1 所示。

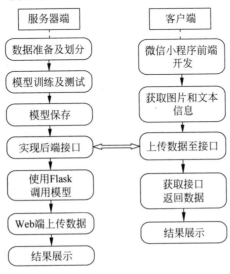

图 6-1 系统整体结构图

6.1.2 系统流程图

系统流程如图 6-2 所示。

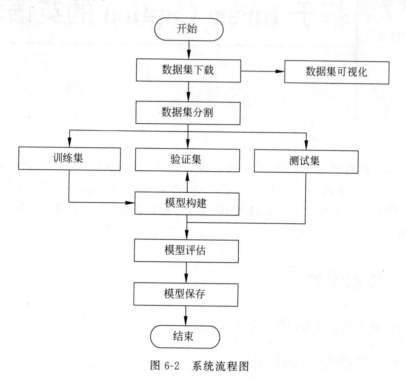

图 6-2 系统流程图

6.2 运行环境

本部分主要包括 Python 环境、TensorFlow 环境和微信开发者工具。

6.2.1 Python 环境

需要 Python 3.6 及以上配置,在 Windows 环境下推荐下载 Anaconda 完成 Python 所需的配置,下载地址为 https://www.anaconda.com/,也可以下载虚拟机在 Linux 环境下运行代码。在已经安装 Anaconda 的情况下,通过命令行添加新的 Python 环境,主要用到的库函数及其版本如下:

- 开发环境:Python 3.6;
- Python 库:TensorFlow 1.12.0、Flask 1.1.1、nltk 3.4.5、Numpy 1.18.1;
- Python IDE:PyCharm 2019.3。

6.2.2 TensorFlow 环境

打开 Anaconda Prompt，输入清华仓库镜像，输入命令：

```
conda config -- add channels https://mirrors.tuna.tsinghua.edu.cn/anaconda/pkgs/free/
conda config - set show_channel_urls yes
```

创建 Python 3.6 的环境，名称为 TensorFlow，此时 Python 版本和后面 TensorFlow 的版本有匹配问题，此步选择 Python 3.x，输入命令：

```
conda create -n tensorflow python=3.6
```

有需要确认的地方，都输入 y。

在 Anaconda Prompt 中激活 TensorFlow 环境，输入命令：

```
activate tensorflow
```

安装 CPU 版本的 TensorFlow，输入命令：

```
pip install - upgrade -- ignore - installed tensorflow
```

安装完毕。

6.2.3 微信开发者工具

微信开发者工具下载地址为 https://developers.weixin.qq.com/miniprogram/dev/devtools/devtools.html。

6.3 模块实现

本项目包括 5 个模块：准备数据、模型构建、模型训练及保存、模型调用、模型测试。下面分别给出各模块的功能介绍及相关代码。

6.3.1 准备数据

本部分包括数据下载和数据预处理。

1. 数据下载

获取图像下载地址为 http://mscoco.org/。COCO 数据集具有 5 种标签类型：目标检测、关键点检测、物体分割、多边形分割以及图像描述，这些标注数据以 json 格式存储。

若使用 Linux 系统，则运行 data 目录下的 download_and_preprocess_mscoco.sh 获取数据集和运行数据预处理程序。

Windows 系统下使用相关软件执行可以成功下载图像文件，但是解压时会出现脚本停止,采取手动进行解压,标注文件需要额外下载;也可以自行下载、解压或使用其他数据集。

Microsoft COCO 数据集下载地址为 http://cocodataset.org/#download。本项目使用 2014 版本的数据集,训练集下载地址为 http://images.cocodataset.org/zips/train2014.zip;验证集下载地址为 http://images.cocodataset.org/zips/val2014.zip;测试集下载地址为 http://images.cocodataset.org/zips/test2014.zip;标注下载地址为 http://images.cocodataset.org/annotations/annotations_trainval2014.zip。

部分数据集图片如图 6-3 所示。

图 6-3 部分数据集

2. 数据预处理

data 目录下 build_mscoco_data.py 文件的主要功能是将训练集和验证集合并之后按比例重新划分训练集、验证集、测试集,并产生词表文件(即 word_counts.txt)。预处理程序的参数设定如下:

```
tf.flags.DEFINE_string("train_image_dir","/tmp/train2014/","训练集目录(.jpg)")
tf.flags.DEFINE_string("val_image_dir", "/tmp/val2014","验证集目录(.jpg)")
tf.flags.DEFINE_string("train_captions_file", "/tmp/captions_train2014.json","训练集 JSON 标注文件")
tf.flags.DEFINE_string("val_captions_file", "/tmp/captions_val2014.json","验证集 JSON 标注文件")
tf.flags.DEFINE_string("output_dir", "/tmp/", "输出目录")
tf.flags.DEFINE_integer("train_shards", 256, "训练集 TFRecord 块文件数目")
tf.flags.DEFINE_integer("val_shards", 4, "验证集 TFRecord 块文件数目")
tf.flags.DEFINE_integer("test_shards", 8, "测试集 TFRecord 块文件数目")
tf.flags.DEFINE_string("start_word", "<S>","开始文本标签")
tf.flags.DEFINE_string("end_word", "</S>","结束文本标签")
tf.flags.DEFINE_string("unknown_word", "<UNK>","未知文本标签")
tf.flags.DEFINE_integer("min_word_count", 4, "出现在词汇表中的单词最小出现次数")
tf.flags.DEFINE_string("word_counts_output_file", "/tmp/word_counts.txt","词汇表文件输出目录")
tf.flags.DEFINE_integer("num_threads", 8, "线程数")
FLAGS = tf.flags.FLAGS
```

```
ImageMetadata = namedtuple("ImageMetadata",["image_id","filename","captions"])
```

预处理程序的 main() 函数,是将原数据集进行重新划分,使用测试集生成训练的标准 TFRecord 文件:

```
def _is_valid_num_shards(num_shards):
    #如果 num_shards 与 FLAGS.num_threads 兼容,则返回 True
    return num_shards < FLAGS.num_threads or not num_shards % FLAGS.num_threads
assert _is_valid_num_shards(FLAGS.train_shards), (
    "Please make the FLAGS.num_threads commensurate with FLAGS.train_shards")
assert _is_valid_num_shards(FLAGS.val_shards), (
    "Please make the FLAGS.num_threads commensurate with FLAGS.val_shards")
assert _is_valid_num_shards(FLAGS.test_shards), (
    "Please make the FLAGS.num_threads commensurate with FLAGS.test_shards")
if not tf.gfile.IsDirectory(FLAGS.output_dir):
    tf.gfile.MakeDirs(FLAGS.output_dir)
#从标注文件中加载图片三元组 metadata 类型的变量
mscoco_train_dataset = _load_and_process_metadata(FLAGS.train_captions_file,
                                                   FLAGS.train_image_dir)
mscoco_val_dataset = _load_and_process_metadata(FLAGS.val_captions_file,
                                                 FLAGS.val_image_dir)
#划分训练集、验证集和测试集
train_cutoff = int(0.85 * len(mscoco_val_dataset))
val_cutoff = int(0.90 * len(mscoco_val_dataset))
train_dataset = mscoco_train_dataset + mscoco_val_dataset[0:train_cutoff]
val_dataset = mscoco_val_dataset[train_cutoff:val_cutoff]
test_dataset = mscoco_val_dataset[val_cutoff:]
#从训练集标注中创建字典
train_captions = [c for image in train_dataset for c in image.captions]
vocab = _create_vocab(train_captions)
_process_dataset("train", train_dataset, vocab, FLAGS.train_shards)
_process_dataset("val", val_dataset, vocab, FLAGS.val_shards)
_process_dataset("test", test_dataset, vocab, FLAGS.test_shards)
```

6.3.2 模型构建

数据加载模型之后,需要定义结构、优化模型。

1. 定义结构

NIC 算法模型结构在训练时根据图像特征和单词向量定义损失函数并计算预测误差,而预测时根据图像内容由训练模型推算出最贴切的描述语句。因此,网络结构定义在训练和预测时是不同的,在程序内部实现中可以使用条件判断语句定义不同的网络结构。

在 show_and_tell_model.py 文件中的 build_model() 函数可以看到 LSTM 网络结构的实现程序。网络结构是利用 TensorFlow 当中最为基础的 LSTM,建立 LSTM cell 对象,以元组形式返回预测结果和状态值。通过分批数据包方式训练,所以使用 lstm_cell.zero_ state()函数创建输入图像特征时的零状态元组信息。

```python
# LSTM cell 输出为 new_c * sigmoid(o)
lstm_cell = tf.contrib.rnn.BasicLSTMCell(
    num_units = self.config.num_lstm_units, state_is_tuple = True)
if self.mode == "train":
    lstm_cell = tf.contrib.rnn.DropoutWrapper(
        lstm_cell,
        input_keep_prob = self.config.lstm_dropout_keep_prob,
        output_keep_prob = self.config.lstm_dropout_keep_prob)
with tf.variable_scope("lstm", initializer = self.initializer) as lstm_scope:
    # 使用 image embeddings 图片嵌入层初始化 LSTM 神经网络
    zero_state = lstm_cell.zero_state(
        batch_size = self.image_embeddings.get_shape()[0], dtype = tf.float32)
    _, initial_state = lstm_cell(self.image_embeddings, zero_state)
    # 允许 LSTM 变量被复用
    lstm_scope.reuse_variables()
    if self.mode == "inference":
        # 使用连接状态方便反馈和获取
        tf.concat(axis = 1, values = initial_state, name = "initial_state")
        # 创建用于反馈一批连接状态的占位符
        state_feed = tf.placeholder(dtype = tf.float32,
                                    shape = [None, sum(lstm_cell.state_size)],
                                    name = "state_feed")
        state_tuple = tf.split(value = state_feed, num_or_size_splits = 2, axis = 1)
        # 运行 LSTM 的一步
        lstm_outputs, state_tuple = lstm_cell(
            inputs = tf.squeeze(self.seq_embeddings, axis = [1]),
            state = state_tuple)
        # 连接结果
        tf.concat(axis = 1, values = state_tuple, name = "state")
    else:
        # 通过 LSTM 训练一批图片嵌入层
        sequence_length = tf.reduce_sum(self.input_mask, 1)
        lstm_outputs, _ = tf.nn.dynamic_rnn(cell = lstm_cell,
                                            inputs = self.seq_embeddings,
                                            sequence_length = sequence_length,
                                            initial_state = initial_state,
                                            dtype = tf.float32,
                                            scope = lstm_scope)
# 直接堆叠批次
lstm_outputs = tf.reshape(lstm_outputs, [-1, lstm_cell.output_size])
with tf.variable_scope("logits") as logits_scope:
    logits = tf.contrib.layers.fully_connected(
        inputs = lstm_outputs,
        num_outputs = self.config.vocab_size,
        activation_fn = None,
```

```
                weights_initializer = self.initializer,
                scope = logits_scope)
```

2. 优化模型

在训练时,将根据批量数据的所有单词序列预测结果计算损失函数值。因为不同图像描述语句长度不同,在具体的计算过程中使用 self.input_mask()作为有效单词的掩码矩阵 weights。函数 tf.contrib.losses.add_loss()将根据批量损失函数估计全局损失函数。

当模型中定义了多个损失函数节点时,使用该函数可以返回一个总的损失函数值。训练时加入 tf.scalar.summary()等统计函数有利于 TensorBoard 的可视化分析,便于验证模型的有效性。

```
if self.mode == "inference":
    tf.nn.softmax(logits, name = "softmax")
else:
    targets = tf.reshape(self.target_seqs, [ -1])
    weights = tf.to_float(tf.reshape(self.input_mask, [ -1]))
    #计算损失率
    losses = tf.nn.sparse_softmax_cross_entropy_with_logits(labels = targets, logits = logits)
    batch_loss = tf.div(tf.reduce_sum(tf.multiply(losses, weights)),
                        tf.reduce_sum(weights),
                        name = "batch_loss")
    tf.losses.add_loss(batch_loss)
    total_loss = tf.losses.get_total_loss()
    #添加到总结中
    tf.summary.scalar("losses/batch_loss", batch_loss)
    tf.summary.scalar("losses/total_loss", total_loss)
    for var in tf.trainable_variables():
        tf.summary.histogram("parameters/" + var.op.name, var)
    self.total_loss = total_loss
    self.target_cross_entropy_losses = losses  #在验证过程中使用
    self.target_cross_entropy_loss_weights = weights  #在验证过程中使用
```

6.3.3 模型训练及保存

在定义模型架构和编译之后,使用训练集和测试集拟合并保存模型。以下为 train.py 文件中的训练模型代码。参数设定部分:

```
tf.flags.DEFINE_string("input_file_pattern", "",
                       "TFRecord 训练文件目录")
tf.flags.DEFINE_string("inception_checkpoint_file", "",
                       "预训练的 inception_v3 模型文件目录")
tf.flags.DEFINE_string("train_dir", "",
                       "保存和导入模型 checkpoint 的目录")
tf.flags.DEFINE_boolean("train_inception", False,
```

```
                            "是否要训练 inception 子模型")
tf.flags.DEFINE_integer("number_of_steps", 1000000, "训练步数")
tf.flags.DEFINE_integer("log_every_n_steps", 1,
                        "输出损失和步数的频率")
```

运行 main() 函数,使用训练集进行训练,并且在指定目录当中保存模型,以便下次训练或者使用。

```
model_config = configuration.ModelConfig()
model_config.input_file_pattern = FLAGS.input_file_pattern
model_config.inception_checkpoint_file = FLAGS.inception_checkpoint_file
training_config = configuration.TrainingConfig()
#创建训练目录
train_dir = FLAGS.train_dir
if not tf.gfile.IsDirectory(train_dir):
    tf.logging.info("创建训练目录: %s", train_dir)
    tf.gfile.MakeDirs(train_dir)
#建立 TensorFlow 图形
g = tf.Graph()
with g.as_default():
    #建立模型
    model = show_and_tell_model.ShowAndTellModel(
        model_config, mode="train", train_inception=FLAGS.train_inception)
    model.build()
    #设置训练速率
    learning_rate_decay_fn = None
    if FLAGS.train_inception:
        learning_rate = tf.constant(training_config.train_inception_learning_rate)
    else:
        learning_rate = tf.constant(training_config.initial_learning_rate)     #学习率
        if training_config.learning_rate_decay_factor > 0:                     #学习率衰减
            num_batches_per_epoch = (training_config.num_examples_per_epoch /
                                     model_config.batch_size)
            decay_steps = int(num_batches_per_epoch *
                              training_config.num_epochs_per_decay)
            def _learning_rate_decay_fn(learning_rate, global_step):           #衰减函数
                return tf.train.exponential_decay(
                    learning_rate,
                    global_step,
                    decay_steps=decay_steps,
                    decay_rate=training_config.learning_rate_decay_factor,
                    staircase=True)
            learning_rate_decay_fn = _learning_rate_decay_fn
    #设置训练参数
    train_op = tf.contrib.layers.optimize_loss(
        loss=model.total_loss,
        global_step=model.global_step,
```

```
        learning_rate = learning_rate,
        optimizer = training_config.optimizer,
        clip_gradients = training_config.clip_gradients,
        learning_rate_decay_fn = learning_rate_decay_fn)
    #设置保存和存储模型 checkpoint
    saver = tf.train.Saver(max_to_keep = training_config.max_checkpoints_to_keep)
    #开始运行训练
tf.contrib.slim.learning.train(
    train_op,
    train_dir,
    log_every_n_steps = FLAGS.log_every_n_steps,
    graph = g,
    global_step = model.global_step,
    number_of_steps = FLAGS.number_of_steps,
    init_fn = model.init_fn,
    saver = saver)
```

当训练结果完成时，出现 4 个文件，用于模型后期的运行和应用，考虑到数据集的庞大和运算能力的限制，直接获取已经训练好的模型，运行程序。如图 6-4 所示，已经训练好的模型包含在代码压缩包中。

checkpoint	2018/8/6 16:12	文件	1 KB
model.ckpt-1000000.meta	2018/3/31 15:28	META 文件	146,293 KB
model.ckpt-1000000.data-00000-of-00001	2018/3/31 15:27	DATA-00000-OF...	145,511 KB
model.ckpt-1000000.index	2018/3/31 15:27	INDEX 文件	17 KB

图 6-4　训练好的模型文件

6.3.4　模型调用

本部分包括生成摘要、调用生成摘要与评分。

1. 生成摘要

摘要生成部分在 inference_util/caption_generator.py 中，相关代码如下：

```
class Caption(object):
    #一个完成的摘要生成部分
    def __init__(self, sentence, state, logprob, score, metadata = None):
        """
        初始化，参数：
            sentence：包含单词 ID 的一组列表
            state：在生成前一个单词后的模型状态
            logprob：摘要的对数概率
            score：摘要的评分
            metadata：可选与部分句子连接的元数据
        """
```

```python
        self.sentence = sentence
        self.state = state
        self.logprob = logprob
        self.score = score
        self.metadata = metadata
    def __cmp__(self, other):
        #使用分数对比摘要
        assert isinstance(other, Caption)
        if self.score == other.score:
            return 0
        elif self.score < other.score:
            return -1
        else:
            return 1
    #对Python3的兼容
    def __lt__(self, other):
        assert isinstance(other, Caption)
        return self.score < other.score
    #同样是对Python3的兼容
    def __eq__(self, other):
        assert isinstance(other, Caption)
        return self.score == other.score
class TopN(object):
    #保持更新集合中的前n个元素
    def __init__(self, n):
        self._n = n
        self._data = []
    def size(self):
        assert self._data is not None
        return len(self._data)
    def push(self, x):
        #添加新元素
        assert self._data is not None
        if len(self._data) < self._n:
            heapq.heappush(self._data, x)
        else:
            heapq.heappushpop(self._data, x)
    def extract(self, sort=False):
        #参数:sort 是否按照降序返回元素,返回集合中前n个元素
        assert self._data is not None
        data = self._data
        self._data = None
        if sort:
            data.sort(reverse=True)
        return data
    def reset(self):
        #重置,返回空状态
```

```python
    self._data = []
class CaptionGenerator(object):
    # img2txt 模型生成摘要
    def __init__(self,
                 model,
                 vocab,
                 beam_size = 3,
                 max_caption_length = 20,
                 length_normalization_factor = 0.0):
        """
        初始化生成器
        参数：
            model:压缩已经训练好的 image-to-text 模型,必须含有 feed_image()和 inference_step()方法
            vocab: 词汇表对象
            beam_size: 生成摘要的 beam search 一次处理大小
            max_caption_length: 最大的摘要长度
            length_normalization_factor: 如果不为 0,参数为 x 则代表摘要的分数为 logprob/length^x
        而不是 logprob,此参数会增加摘要长度对评分的影响
        """
        self.vocab = vocab
        self.model = model
        self.beam_size = beam_size
        self.max_caption_length = max_caption_length
        self.length_normalization_factor = length_normalization_factor
    def beam_search(self, sess, encoded_image):
        """
        运行一个图片的 beam search 生成器
        参数：
            sess: TensorFlow 的 Session 对象
            encoded_image: 解码后的图片字符串,返回降序的一组摘要
        """
        # 设置初始状态
        initial_state = self.model.feed_image(sess, encoded_image)
        initial_beam = Caption(
            sentence = [self.vocab.start_id],
            state = initial_state[0],
            logprob = 0.0,
            score = 0.0,
            metadata = [""])
        partial_captions = TopN(self.beam_size)
        partial_captions.push(initial_beam)
        complete_captions = TopN(self.beam_size)
        # 运行集束搜索
        for _ in range(self.max_caption_length - 1):
            partial_captions_list = partial_captions.extract()
            partial_captions.reset()
            input_feed = np.array([c.sentence[-1] for c in partial_captions_list])
```

```python
            state_feed = np.array([c.state for c in partial_captions_list])
            softmax, new_states, metadata = self.model.inference_step(sess,
                                                                        input_feed,
                                                                        state_feed)

            for i, partial_caption in enumerate(partial_captions_list):
                word_probabilities = softmax[i]
                state = new_states[i]
                #对于生成部分句子求出 beam_size 大小最可能的下一个单词
                words_and_probs = list(enumerate(word_probabilities))
                words_and_probs.sort(key=lambda x: -x[1])
                words_and_probs = words_and_probs[0:self.beam_size]
                #生成每个单词给出下一可能生成的摘要
                for w, p in words_and_probs:
                    if p < 1e-12:
                        continue  #避免 log(0)的情况出现
                    sentence = partial_caption.sentence + [w]
                    logprob = partial_caption.logprob + math.log(p)
                    score = logprob
                    if metadata:
                        metadata_list = partial_caption.metadata + [metadata[i]]
                    else:
                        metadata_list = None
                    if w == self.vocab.end_id:
                        if self.length_normalization_factor > 0:
                            score /= len(sentence)**self.length_normalization_factor
                        beam = Caption(sentence, state, logprob, score, metadata_list)
                        complete_captions.push(beam)
                    else:
                        beam = Caption(sentence, state, logprob, score, metadata_list)
                        partial_captions.push(beam)
            if partial_captions.size() == 0:
                #当 beam_size = 1,会耗尽所有的待选文本
                break

        if not complete_captions.size():
            complete_captions = partial_captions
        return complete_captions.extract(sort=True)
```

2. 调用生成摘要与评分

在 run.py 当中,使用训练好的 checkpoint 和词汇表 word_counts.txt,将图形转换成 TensorFlow 对应的处理形式,调用 inference_utils 目录下的 caption_generator.py 文件生成评分最高的3个摘要。BLEU 评分直接采用 NLTK 库中的 sentence_bleu()函数。

```python
checkpoint_path = 'data'  #模型存档目录
vocab_file = 'data/word_counts.txt'  #引用词汇表
tf.logging.set_verbosity(tf.logging.INFO)
def compute_bleu(reference, candidate):
```

```python
        score = sentence_bleu(reference, candidate)  # 调用NLTK库的BLEU函数
        return score
def runs(input_files = r"../../pic/dog.jpg"):
    # 创建引用图形
    s = []
    sa = []
    g = tf.Graph()
    with g.as_default():
        model = inference_wrapper.InferenceWrapper()
        restore_fn = model.build_graph_from_config(configuration.ModelConfig(),
                                                   checkpoint_path)
    g.finalize()
    # 建立词汇表
    vocab = vocabulary.Vocabulary(vocab_file)
    filenames = []
    for file_pattern in input_files.split(","):
        filenames.extend(tf.gfile.Glob(file_pattern))
    tf.logging.info("Running caption generation on %d files matching %s",
                    len(filenames), input_files)
    with tf.Session(graph = g) as sess:
        # 从存档点导入模型
        restore_fn(sess)
        # 准备摘要的生成,使用默认beam search参数,参考caption_generator.py文件
        generator = caption_generator.CaptionGenerator(model, vocab)
        for filename in filenames:
            with tf.gfile.GFile(filename, "rb") as f:
                image = f.read()
            captions = generator.beam_search(sess, image)
            # 输出生成的摘要
            print("Captions for image %s:" % os.path.basename(filename))
            for i, caption in enumerate(captions):
                # 忽略起始和结束标志
                sentence = [vocab.id_to_word(w) for w in caption.sentence[1:-1]]
                sa.append(sentence)
                sentence = " ".join(sentence)
                print("  %d) %s (p=%f)" % (i, sentence, math.exp(caption.logprob)))
                s.append(sentence)
    return (s, sa)
```

6.3.5 模型测试

本测试主要有后端接口实现、获取图片数据、结果展示。实现方式分别如下:一是使用Python的Flask框架搭建后端的请求接口,以便调用模型训练结果并向前端返回相应的结果;二是通过微信小程序调用摄像头和相册以获取数字图片;三是将数字图片转化为数据,通过接口输入到TensorFlow的模型中,并且获取输出。

1. 后端接口实现

后端生成两个接口，/uploadee 是接收微信小程序上传的图片，并返回模型对该图片给出相似度最高的 3 个描述句子数组；/getscore 是对应小程序给出用户输入的句子相对于机器给出句子的 BLEU 算法分数。在文件 web_performance.py 中可以看到，相关代码如下：

```python
#对接小程序前端的图片上传接口
@app.route('/uploadee', methods=['POST'])
def uploadee_file():
    f = request.files['Image']              #获取上传的图片
    filename = secure_filename(f.filename)
    p = os.path.join(app.config['UPLOAD_FOLDER'], filename)
    #将图片保存到本地 UPLOAD_FOLDER 目录
    f.save(p)
    (s,sa) = run.runs(p)                    #运行模型得到结果
    #sc = run.compute_bleu(sa, userS)
    info = {
        'image' : p,
        'sentence' : sa
    }
    print(info)
    return jsonify(info)
#对接小程序获取分数接口
@app.route('/getscore', methods=['GET', 'POST'])
def get_score():
    print('访问 getscore')
    if request.method == 'POST':
        post_data = request.get_json()
        print(post_data)
        ref = post_data.get('ref')                    #获取对照参考原文
        cand = post_data.get('cand').split()          #获取分析候选译文
        print(ref,cand)
        score = run.compute_bleu(ref, cand)           #计算 cand 相对于 ref 的 BLEU 分数
        return jsonify(score)
```

2. 获取图片数据

微信小程序当中，设计两个按钮分别实现相册权限和调用摄像头权限的获取，获取到的图片将显示在按钮下方。设计上传按钮，单击触发将图片传至后端接口。

(1) 在 enter.wxml 中实现页面布局，如下所示：

```
<!-- pages/enter/enter.wxml -->
<view class = "container">
    <view class = 'userbtn'>
        <button bindtap = 'bindViewTap'>选择图像</button>
```

```
        < button bindtap = "takePhoto">相机拍照</button >
        < button bindtap = "upload"> 开始上传 </button >
    </view >
    < view class = 'imagesize'>
        < image src = '{{tempFilePaths}}' mode = "widthFix"></image >
    </view >
</view >
```

按钮对应的触发函数分别为 bindViewTap、takePhoto、upload。image 设定 mode 为 widthFix,是在固定图片显示宽度后按原图比例显示。

(2) 在 enter.wxss 中实现对组件的样式参数设定,如下所示:

```
/* pages/enter/enter.wxss */
/* 图片展示样式 */
.imagesize{
  display:flex;
  width:100%;
  height: 400rpx;
  justify-content: center;
}
.imagesize image {
  width:80%;
}
/* 按键样式 */
.userbtn
{
    display:flex;
    width: 90%;
    margin-bottom: 30px;
    justify-content: center;
}
.userbtn button
{
    background-color: #70DB93;
    color: black;
    border-radius: 25rpx;
}
```

(3) 在 enter.js 中实现触发函数,获取权限及上传图片,相关代码如下:

```
//pages/enter/enter.js
import { $init, $digest } from '../../utils/common.util'
Page({
    //页面的初始数据
    data: {
        tempFilePaths: '',
        //image: [],
```

```javascript
    text:''
  },
  //生命周期函数--监听页面加载
  onLoad: function (options) {
    $ init(this)
  },
  //从相册中选择上传
  bindViewTap: function() {
    var that = this //!!!!!!!!!!"搭桥"
    //使用 API 从本地读取一张图片
    wx.chooseImage({
      count: 1,
      sizeType: ['original', 'compressed'],
      sourceType: ['album', 'camera'],
      success: function (res) {
        //var tempFilePaths = res.tempFilePaths
        //将读取的图片替换之前图片
        that.setData(
          {
            tempFilePaths: res.tempFilePaths,
          }
        )//通过 that 访问
        console.log(that.data.tempFilePaths)
      }
    })
  },
  //相机拍照上传
  takePhoto() {
    var that = this
    wx.chooseImage({
      count: 1, // 默认 9
      sizeType: ['original', 'compressed'],
      sourceType: ['camera'],
      success: function (res) {
        //var tempFilePaths = res.tempFilePaths
        that.setData(
          {
            tempFilePaths: res.tempFilePaths,
          })
        console.log("res.tempImagePath:" + that.data.tempFilePaths)
      }
    })
  },
  //上传照片
  upload: function () {
    var that = this
    wx.uploadFile({
```

```javascript
          url: 'http://127.0.0.1:5000/uploadee',
          //filePath: that.data.img_arr[0],
          filePath: that.data.tempFilePaths[0],
          name: 'Image',
          //formData: adds,
          success: function (res) {
                    var da = JSON.parse(res.data);
                    console.log(da);
          that.setData({
            text: da.sentence
          })
          console.log(that.data.tempFilePaths)
          if (res) {
                    var sentence = JSON.stringify(that.data.text);
          wx.navigateTo({
            url: '../show/show?tempFilePaths = ' + that.data.tempFilePaths[0] + '&text = ' + sentence,
          })
          }
        }
      })
    this.setData({
      formdata: ''
    })
  }
})
```

创建两个初始数据 tempFilePaths 和 text, 分别用于存储图片路径和模型生成的文本。在获取相册和摄像机权限时, 调用 wx.chooseImage() 函数, 详情参见 https://developers.weixin.qq.com/miniprogram/dev/api/media/image/wx.chooseImage.html。上传图片数据时, 调用官方 wx.uploadFile() 函数, 其中, url 后端接口地址为 https://developers.weixin.qq.com/miniprogram/dev/api/network/upload/wx.uploadFile.html。收到后端传回的数据, 页面跳转至测试结果展示页面(即显示页面), 同时传递参数 tempFilePaths 和 text 的值。

3. 测试结果展示

在页面中展示上传的照片, 并且让用户输入对图片的英文描述, 单击"分析"按钮后, 文本会上传至 getscore 接口并返回最后的分数, 显示模型生成的句子以及对比后的评分。

(1) 在 show.wxml 中实现了对页面布局的设计, 相关代码如下:

```
<!-- pages/show/show.wxml -->
< image src = "{{tempFilePaths }}" mode = "aspecFill" style = "width: 100 % ; height: 450rpx"/>
< view class = "text - input - area">
```

```html
<!-- 正文区域 -->
<view class="weui-cells weui-cells_after-title">
    <view class="weui-cell">
        <view class="weui-cell__bd">
            <!-- 多行输入框 -->
            <textarea value="{{content}}" class="weui-textarea" placeholder="请用英文输入对图片内容的描述." maxlength="300" placeholder-style="color:#b3b3b3;font-size:20px;" style="height: 6rem" bindinput="handleContentInput"/>
            <!-- 正文输入字数统计 -->
            <view class="weui-textarea-counter">{{contentCount}}/300 </view>
        </view>
    </view>
</view>
<!-- 评分及显示 -->
<view>
    <text class="textshow">{{text_show}}</text>
    <view>你的得分:<input class="scoreshow" value="{{score}}" style="height: 6rem"/></view>
</view>
<button bindtap="getscore">开始评分</button>
```

使用多行输入文本框 textarea 组件，并通过 bindinput 处理函数计算当前输入字数，采用 view 组件显示在文本框下。

(2) 在 show.wxss 中实现对组件样式参数的设计，相关代码如下：

```
/* pages/show/show.wxss */
/* 文本框样式 */
textarea {
    width: 700rpx;
    height: 500rpx;
    margin-left: 10rpx;
    margin-right: 10rpx;
    margin-top: 10rpx;}
.textarea-bg {
    background-color: #999;
    padding: 10rpx;
    font-size: 32rpx;}
.title-bg {
    font-size: 32rpx;
    margin-left: 10rpx;
    margin-right: 10rpx;
    margin-top: 10rpx;
    color: #43c729;}
```

(3) 在 show.js 中实现触发函数、文本数据的上传和参数传递，相关代码如下：

```
//pages/show/show.js
```

```js
import { $init, $digest } from '../../utils/common.util'
Page({
  //页面的初始数据
  data: {
    tempFilePaths: '',
    contentCount: 0, //正文字数
    content:'',
    text:[],
    score:'',
    text_show:'这里显示输出的语句'
  },
  //生命周期函数--监听页面加载
  onLoad: function (options) {
      console.log(options)
      this.setData({
          tempFilePaths: options.tempFilePaths,
          text: JSON.parse(options.text)
      }),
      $init(this)
  },
  //限制输入字数
  handleContentInput(e) {
    const value = e.detail.value
    this.data.content = value
    this.data.contentCount = value.length //计算已输入的正文字数
    $digest(this)
  },
  //获取分数
  getscore: function(){
    var that = this
      console.log('评价为' + that.data.content)
      wx.request({
          url: 'http://127.0.0.1:5000/getscore',
          data:{
              ref: that.data.text,
              cand: that.data.content
          },
          method:'POST',
        success: function (res) {
      console.log(res.data);
      that.setData({
        score: res.data
      })
                  if (res) {
                      for(var i = 0;i < that.data.text.length;i++){
                          that.data.text[i] = that.data.text[i].join(" ")
                      }
```

```
                    that.data.text = that.data.text.join('\n')
                 that.setData({
                   text_show:that.data.text
                 })
               }
            },
            fail: function(res){
              console.log(res.data);
              that.setData({
                score: 0,
                text_show:that.data.text
              })
            }
         })
      }
   })
```

在 OnLoad()函数中将上个页面传递的值赋给当前页面的数据。通过对象深层比较，将页面的数据进行批量、按需更新到视图层 WXML 中的功能。在 getscore()函数中调用了 API, wx.request(), 请求方式为 POST, 微信开发文档链接为 https://developers.weixin.qq.com/miniprogram/dev/api/network/request/wx.request.html。

4. 前端完整代码

程序文件目录如图 6-5 所示。

1) 小程序全局配置 app.json

用于决定页面文件的路径、窗口表现、设置网络超时等，相关代码如下：

```
{
  "pages": [ //展示的页面情况
    "pages/enter/enter",
    "pages/show/show"
  ],
  "window": { //小程序窗体显示的标题、顶栏等样式
    "backgroundTextStyle": "light",
    "navigationBarBackgroundColor": "#E9C2A6",
    "navigationBarTitleText": "ImageCaption",
    "navigationBarTextStyle": "black"
  },
  "sitemapLocation": "sitemap.json"
}
```

2) 小程序工具配置 project.config.json:

在工具上做的任何配置都会写入此文件中，相关代码如下：

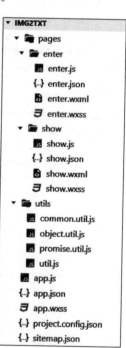

图 6-5　程序文件构成

```json
{
    "description": "项目配置文件",
    "packOptions": {
        "ignore": []
    },
    "setting": {
        "urlCheck": false,
        "es6": true,
        "postcss": true,
        "minified": true,
        "newFeature": true,
        "autoAudits": false,
        "checkInvalidKey": true
    },
    "compileType": "miniprogram",
    "libVersion": "2.7.1",
    "appid": "wx5892fca58ff03519",          //微信小程序 AppID
    "projectname": "IMG2TXT",                //微信小程序项目名称
    "debugOptions": {
        "hidedInDevtools": []
    },
    "isGameTourist": false,
    "simulatorType": "wechat",
    "simulatorPluginLibVersion": {},
    "condition": {
        "search": {
            "current": -1,
            "list": []
        },
        "conversation": {
            "current": -1,
            "list": []
        },
        "plugin": {
            "current": -1,
            "list": []
        },
        "game": {
            "currentL": -1,
            "list": []
        },
        "gamePlugin": {
          "current": -1,
          "list": []
        },
        "miniprogram": {                     //包含的页面情况
            "current": -1,
```

```
            "list": [
                {
                    "id": 0,
                    "name": "pages/show/show",
                    "pathName": "pages/show/show",
                    "query": "tempFilePaths = http://tmp/wx5892fca58ff03519.
o6zAJszRuevi8nlb8fGe….7I1gt4wbVQV4aa7661bffe46b27f3e2d6964e7af9a6e.jpg&text = aaaaa",
                    "scene": null
                },
                {
                    "id": -1,
                    "name": "pages/enter/enter",
                    "pathName": "pages/enter/enter",
                    "query": "tempFilePaths = http://tmp/wx5892fca58ff03519.
o6zAJszRuevi8nlb8fGe….7I1gt4wbVQV4aa7661bffe46b27f3e2d6964e7af9a6e.jpg&text = aaaaa",
                    "scene": null
                }
            ]
        }
}
```

6.4 系统测试

本部分包括训练准确率、测试效果及模型应用。

6.4.1 训练准确率

该模型数据集内容庞大(原始数据集的大小约为 20GB,经过预处理后 TFRecord 类型的数据将会达到 100GB),需要在 Linux 系统下进行训练才能兼顾效率。本文采用训练好的模型,且模型已经保存在源代码文件中,最终损失率大概维持在 2%,准确率在 98%,训练输出如图 6-6 所示。

```
INFO:tensorflow:global step 65: loss = 5.1898 (0.34 sec/step)
INFO:tensorflow:global step 66: loss = 5.1952 (0.27 sec/step)
INFO:tensorflow:global step 67: loss = 5.3461 (0.27 sec/step)
INFO:tensorflow:global step 68: loss = 5.9272 (0.47 sec/step)
INFO:tensorflow:global step 69: loss = 5.4241 (0.33 sec/step)
INFO:tensorflow:global step 70: loss = 5.1671 (0.21 sec/step)
INFO:tensorflow:global step 71: loss = 5.7036 (0.30 sec/step)
INFO:tensorflow:global step 72: loss = 5.4312 (0.36 sec/step)
INFO:tensorflow:global step 73: loss = 5.7431 (0.27 sec/step)
```

图 6-6 训练输出

6.4.2 测试效果

直接运行 run_inference.py(想要更改测试图片可以修改参数或以参数启动),可以看到输出相似度最高的三句词数数组,PyCharm 运行的输出结果如图 6-7 所示。

```
Captions for image dog.jpg:
['a', 'small', 'white', 'dog', 'sitting', 'on', 'a', 'table', '.']
  0) a small white dog sitting on a table . (p=0.000271)
['a', 'small', 'white', 'dog', 'sitting', 'on', 'a', 'chair', '.']
  1) a small white dog sitting on a chair . (p=0.000200)
['a', 'small', 'white', 'dog', 'sitting', 'on', 'a', 'table']
  2) a small white dog sitting on a table (p=0.000134)
```

图 6-7 运行输出结果

6.4.3 模型应用

本部分包括程序运行、应用使用说明和测试结果。

1. 程序运行

本项目使用网站前后端架构实现模型的应用。因此,前后端要分开运行,此外,测试均在本地进行,与服务器端运行过程相同。

后端使用 Python 的 Flask 框架实现各种接口,确保安装代码运行要求的库后,直接运行 web_performance.py 文件,如图 6-8 所示,表明后端接口已经运行在本地端口 5000。

```
* Debugger is active!
* Debugger PIN: 984-994-515
* Running on http://127.0.0.1:5000/ (Press CTRL+C to quit)
```

图 6-8 运行后端输出结果

前端通过微信开发者工具进行小程序的调试测试工作,使用测试 AppID 导入解压后的微信小程序代码,最后编译运行,如图 6-9 所示。

2. 应用使用说明

微信小程序界面如图 6-10 所示。

在首页通过一张图片和三个按钮进行选择,单击"选择图像"按钮后通过选取本地图片上传得到结果;而"相机拍照"则是调用手机摄像头获得图片展现在图片栏中;单击"开始上传"按钮,当选择的图片不为空时,图片数据会上传至后端接口,并跳转至结果展示页面,如图 6-11 所示。

进入到展示页面当中,用户输入自己的描述,即可得到对图片的英文描述和机器给出的英文描述对比 BLEU 得分。

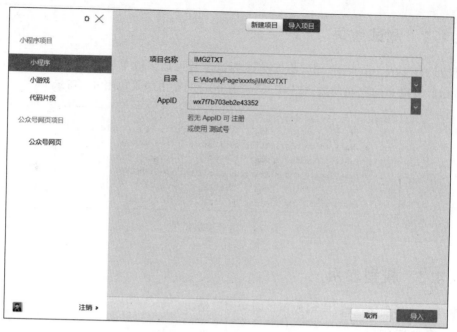

图 6-9　微信开发者工具导入

图 6-10　微信小程序界面

图 6-11　微信小程序提交图片后的结果展示页面

3. 测试结果

移动端测试结果如图 6-12 所示。

图 6-12　移动端测试结果

项目 7　智能聊天机器人

PROJECT 7

本项目基于微博公开的数据进行提取,通过注意力机制的 Seq2Seq 机器学习模型训练,使用 GloVe 构建词向量,实现智能聊天机器人。

7.1　总体设计

本部分包括系统整体结构图和系统流程图。

7.1.1　系统整体结构图

系统整体结构如图 7-1 所示。

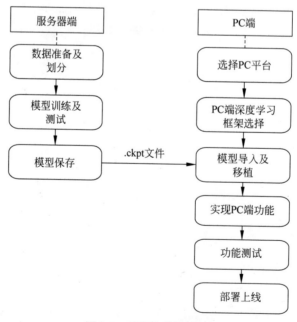

图 7-1　系统整体结构图

7.1.2 系统流程图

系统流程如图 7-2 所示。

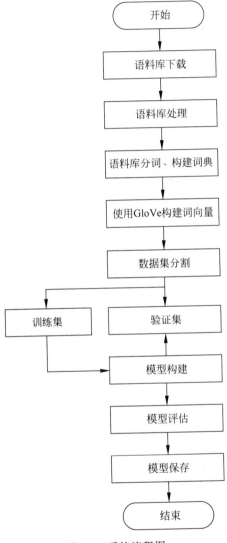

图 7-2 系统流程图

7.2 运行环境

本部分包括 Python 环境和 TensorFlow 环境。

7.2.1 Python 环境

Python 3.7 需要在 Windows 环境下载 Anaconda 完成 Python 所需的配置,下载地址为 https://www.anaconda.com/(注:必须选择 64 位,TensorFlow 不支持 32 位)。

7.2.2 TensorFlow 环境

要求配置 Nvidia 显卡且支持 CUDA,查看自己的显卡是否符合,如图 7-3 所示,下载地址为 https://developer.nvidia.com/cuda-gpus。

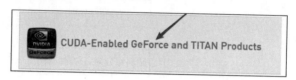

图 7-3 显卡要求图

1. CUDA 安装

CUDA 必须为 CUDA 10.1,下载链接为 https://developer.nvidia.com/cuda-toolkit-archive,下载后进行默认路径安装,安装的过程选择自定义。如果计算机本身有 Visual Studio Integration,将安装列表中的 Visual Studio Integration 取消勾选,单击 Driver comonents,Display Driver,前面显示 CUDA 本身包含的驱动版本是 411.31,如果目前安装的驱动比自带的版本号更高,则取消勾选,否则会安装失败。环境变量配置如下:

① C:\Program Files\NVIDIA GPU Computing Toolkit\CUDA\v10.1;
② C:\Program Files\NVIDIA GPU Computing Toolkit\CUDA\v10.1\lib\x64;
③ C:\Program Files\NVIDIA GPU Computing Toolkit\CUDA\v10.1\bin;
④ C:\Program Files\NVIDIA GPU Computing Toolkit\CUDA\v10.1\libnvvp。

测试 CUDA 是否安装成功,打开 cmd 输入命令:

```
nvcc -V
```

若输出 CUDA 10.1 则安装成功。

2. cuDNN 安装

cuDNN 下载的版本必须适配 CUDA 10.1。下载链接为 https://developer.nvidia.com/cudnn,下载后解压,将 cuDNN 压缩包里面的 bin、clude 和 lib 文件直接复制到 CUDA 的安装目录下,覆盖安装即可。

3. TensorFlow 2.1 安装

在开始菜单中打开 Anaconda Prompt,在命令行输入 conda create-n tf2.1 python=3.7 以

创建虚拟环境。

激活创建的虚拟环境，使用 conda activate tf 2.1，在命令行输入命令：

pip install tensorflow-gpu -i https://pypi.tuna.tsinghua.edu.cn/simple

这样会默认下载最新版本，等待安装完成即可。在虚拟环境下使用 pip，安装各种需要的包，至此，TensorFlow 2.1 安装完成。

7.3 模块实现

本项目包括3个模块：数据预处理、模型构建和模型测试。下面分别给出各模块的功能介绍及相关代码。

7.3.1 数据预处理

本部分包括语料库的获取与处理、中文分词、构建词典和创建数据集。

1. 语料库的获取与处理

语料库获取方式为百度网盘，链接如下 https://pan.baidu.com/s/13k5n-Wl18gOJlpnWCa4X_A，提取码为 h293。数据集使用微博语料库处理后的文件，包含 440 万条微博对话，加载语料库，通过 TensorFlow API 实现，相关代码如下：

```python
def load_data(self):
    if not os.path.exists(self.args.sr_word_id_path) or not os.path.exists(self.args.train_samples_path):
        # 读取文件
        print("开始读取数据")
        self.conversations = pd.read_csv(self.args.conv_path,
                                         nrows = 10000,
                                         names = ["first_conv", "second_conv"],
                                         sep = '\t', header = None)
        self.conversations.first_conv = self.conversations.first_conv.apply(lambda conv : self.word_tokenizer(conv))
        self.conversations.second_conv = self.conversations.second_conv.apply(lambda conv : self.word_tokenizer(conv))
        print("数据读取完毕")
```

2. 中文分词

本项目采用 jieba 进行分词，相关代码如下：

```python
def word_tokenizer(self, sentence):
    # jieba 分词
    rerule = '[【.+】|#.+#|(.+)|「.+」|[^\u4e00-\u9fa5],|.|!|?|……]'
    # 删除语料库中的英文字母
```

```python
        sentence = re.sub(rerule, "", sentence)
        words = pseg.cut(sentence, use_paddle=True)
        result = []
        for word, flag in words:
            result.append(word)
        return result
```

3. 构建词典

不论是 GloVe 模型构建词向量，还是带有注意力机制的 Seq2Seq 模型训练聊天机器人，输入数据集都必须是数字类型而非字符类型，故需要构建词典建立词与数之间的对应关系，相关代码如下：

```python
def build_word_dict(self):
    if not os.path.exists(self.args.sr_word_id_path):
        # 得到word2id 和 id2word 两个词典
        print("开始构建词典")
        words = pd.concat([self.conversations.first_conv,
                           self.conversations.second_conv],
                          ignore_index=True)
        words = words.values
        words = list(chain(*words))
        sr_words_count = pd.Series(words).value_counts()
        # 筛选出现次数大于1 的词作为词典
        sr_words_size = np.where(sr_words_count.values > self.args.vacab_filter)[0].size
        sr_words_index = sr_words_count.index[0:sr_words_size]
        self.sr_word2id = pd.Series(range(self.numToken, self.numToken + sr_words_size),
                                    index=sr_words_index)    # word 到 ID
        self.sr_id2word = pd.Series(sr_words_index, index=range(self.numToken, self.numToken + sr_words_size))    # ID 到 word
        self.sr_word2id[self.padToken] = 0
        self.sr_word2id[self.goToken] = 1
        self.sr_word2id[self.eosToken] = 2
        self.sr_word2id[self.unknownToken] = 3
        self.sr_id2word[0] = self.padToken
        self.sr_id2word[1] = self.goToken
        self.sr_id2word[2] = self.eosToken
        self.sr_id2word[3] = self.unknownToken
        print("词典构建完毕")
        with open(os.path.join(self.args.sr_word_id_path), 'wb') as handle:
            data = {
                'word2id': self.sr_word2id,
                'id2word': self.sr_id2word,
            }
            pickle.dump(data, handle, -1)
    else:
        print("从{}载入词典".format(self.args.sr_word_id_path))
```

```python
with open(self.args.sr_word_id_path, 'rb') as handle:
    data = pickle.load(handle)
    self.sr_word2id = data['word2id']
    self.sr_id2word = data['id2word']
```

4. 创建数据集

通过字典将语料库中的每句话映射成一个列表,并根据一定规则筛选出适合作为数据集的数据,相关代码如下:

```python
def replace_word_with_id(self, conv):
    conv = list(map(self.get_word_id, conv))    # 构建列表
    # temp = list(map(self.get_id_word, conv))
    return conv
def get_word_id(self, word):                    # 根据 word 获取 ID
    if word in self.sr_word2id:
        return self.sr_word2id[word]
    else:
        return self.sr_word2id[self.unknownToken]
def get_id_word(self, id):                      # 根据 ID 获取 word
    if id in self.sr_id2word:
        return self.sr_id2word[id]
    else:
        return self.unknownToken
def filter_conversations(self, first_conv, second_conv):
    # 筛选样本,将 encoder_input 或 decoder_input 大于 max_length 的会话过滤
    # 将 target 中包含有 UNK 的会话过滤
    valid = True
    valid &= len(first_conv) <= self.args.maxLength
    valid &= len(second_conv) <= self.args.maxLength
    valid &= second_conv.count(self.sr_word2id[self.unknownToken]) == 0
    return valid
def generate_conversations(self):
    if not os.path.exists(self.args.train_samples_path):
        # 将 word 替换为 ID
        # self.replace_word_with_id()
        print("开始生成训练样本")
        # 将 ID 与 line 作为字典,方便生成训练样本
        for line_id in tqdm(range(len(self.conversations.first_conv.values) - 1), ncols=10):
            first_conv = self.conversations.first_conv[line_id]
            second_conv = self.conversations.second_conv[line_id]
            first_conv = self.replace_word_with_id(first_conv)
            second_conv = self.replace_word_with_id(second_conv)
            valid = self.filter_conversations(first_conv, second_conv)
            if valid:
                temp = [first_conv, second_conv]
```

```python
                self.train_samples.append(temp)
        print("生成训练样本结束")
        with open(self.args.train_samples_path, 'wb') as handle:
            data = {
                'train_samples': self.train_samples  # 获取训练样本
            }
            pickle.dump(data, handle, -1)
    else:
        with open(self.args.train_samples_path, 'rb') as handle:
            data = pickle.load(handle)
            self.train_samples = data['train_samples']
        print("从{}导入训练样本".format(self.args.train_samples_path))
    def sen2enco(self, sentence):
        sentence = self.word_tokenizer(sentence)
        enco = [1] + self.replace_word_with_id(sentence) + [2]
        return enco
```

7.3.2 模型构建

本部分包括 GloVe 模型和 Seq2Seq 模型。

1. GloVe 模型

将数据集进行预处理后，定义 GloVe 模型结构、优化损失函数，并保存模型。

1）预处理

```python
#读取语料库,拼接问答
args = Args()
textData = TextData(args)
inp = np.array(textData.train_samples)[:,0]
targ = np.array(textData.train_samples)[:,1]
dataset = np.concatenate((inp,targ),axis = 0)
num_tokens = sum([len(st) for st in dataset])
#扁平化 list,统计分词频数
df = [x for tup in dataset for x in tup]
counter = Counter(df)
#二次采样随机丢弃
def discard(idx):
    return random.uniform(0, 1) < 1 - math.sqrt(
        1e-4 / counter[idx] * num_tokens)
subsampled_dataset = [[tk for tk in st if not discard(tk)] for st in dataset]
#随机窗口滑动产生中心词和背景词
def get_centers_and_contexts(dataset, max_window_size):
    centers, contexts, n_contexts = [], [], []
    for st in dataset:
        if len(st) < 2: #每个句子至少要有 2 个词才可能组成一对"中心词 - 背景词"
            continue
```

```python
            centers += st
            for center_i in range(len(st)):
                window_size = random.randint(1, max_window_size)
                indices = list(range(max(0, center_i - window_size),
                                     min(len(st), center_i + 1 + window_size)))
                contexts.append([st[idx] for idx in indices])
                indices.remove(center_i) #将中心词排除在背景词之外
                n_contexts.append([st[idx] for idx in indices])
        #contexts没有排除中心词是为了在共现矩阵计算距离,n_contexts用于训练
        return centers, contexts, n_contexts
    all_centers, all_contexts, contexts = get_centers_and_contexts(subsampled_dataset, 5)
#初始化共现矩阵
vocab_size = len(textData.sr_word2id)
cooccurrence = np.zeros([vocab_size, vocab_size], dtype="float32")
for i in range(0, len(all_centers)):
    #中心词第一次出现在句中的位置
    aim = all_contexts[i].index(all_centers[i])
    for j in range(0, len(all_contexts[i])):
        if j != aim:
            x = all_centers[i]
            y = all_contexts[i][j]
            #用背景词和中心词距离的倒数代替频数
            cooccurrence[x][y] += 1/abs(j-aim)
#统一contexts列表大小,不足时末尾补零
contexts = keras.preprocessing.sequence.pad_sequences(contexts, padding='post')
#将all_centers转化为张量,并调大小
centers = tf.convert_to_tensor(all_centers)
centers = tf.reshape(centers, [-1,1])
#根据共现矩阵,构建标签
labels = np.zeros([centers.shape[0], contexts.shape[1]], dtype="float32")
for i in range(0, centers.shape[0]):
    center = centers[i]
    context = contexts[i]
    for j in range(0, context.shape[0]):
        labels[i][j] = cooccurrence[int(context[j])][int(center)]
labels = tf.convert_to_tensor(labels)
#训练大小和词向量维度
BATCH_SIZE = 1024
units = 64
    dataset = tf.data.Dataset.from_tensor_slices((centers, contexts, labels))
dataset = dataset.batch(BATCH_SIZE, drop_remainder=True)
```

2) 定义GloVe模型结构

```python
class Glove(keras.Model):
    def __init__(self, vocab_size, embedding_dim):
        super(Glove, self).__init__()
```

```python
        # Embedding(嵌入层)将正整数(下标)转换为具有固定大小的向量,此处 vocab_size 为词大
        # 小, embedding_dim 为全连接嵌入的维度, ev 为该词作为中心词的向量表示
        self.ev = keras.layers.Embedding(vocab_size, embedding_dim)
        # eu 为该词作为背景词时的向量表示
        self.eu = keras.layers.Embedding(vocab_size, embedding_dim)
        # 中心词向量偏置
        self.b = keras.layers.Embedding(vocab_size, 1)
        # 背景词向量偏置
        self.c = keras.layers.Embedding(vocab_size, 1)
    def call(self, centers, contexts):
        # 将 self.ev(centers)进行转置,并且以[0,2,1]重新排列,为下面的矩阵相乘做准备
        v = tf.transpose(self.ev(centers),[0,2,1])
        u = self.eu(contexts)
        res = u@v
        bb = self.b(centers)
        bc = self.c(contexts)
        res = res + bb + bc
        return res
    # 返回最终的词向量权重
    def get_uaddv_weights(self):
        self.ev(tf.zeros(0,dtype="int32"))
        self.eu(tf.zeros(0,dtype="int32"))
        return [self.ev.weights[0] + self.eu.weights[0]]
```

3) 优化损失函数

Glove 损失函数如下:

$$J = \sum_{i,j=1}^{v} f(x_{ij})(w_i^T \tilde{w}_j + b_i + \tilde{b}_j - \log(X_{ij}))^2$$

在一个语料库中,存在很多单词,这些单词的权重比那些很少在一起出现的单词要大,所以这个函数应该是非递减函数;但也不希望权重过大,当到达一定程度之后应该不再增加;如果两个单词没有一起出现,那么应该不参与到损失函数的计算当中,也就是 $f(x)$ 要满足 $f(0)=0$。

```python
class Glove(keras.Model):
    def __init__(self, vocab_size, embedding_dim):
        super(Glove, self).__init__()
        """
        Embedding(嵌入层)将正整数(下标)转换为具有固定大小的向量,
        vocab_size 为词大小, embedding_dim 为全连接嵌入的维度
        ev 为该词作为中心词的向量表示
        """
        self.ev = keras.layers.Embedding(vocab_size, embedding_dim)
        # eu 为该词作为背景词时的向量表示
        self.eu = keras.layers.Embedding(vocab_size, embedding_dim)
        # 中心词向量偏置
```

```python
        self.b = keras.layers.Embedding(vocab_size, 1)
        #背景词向量偏置
        self.c = keras.layers.Embedding(vocab_size, 1)
    def call(self, centers, contexts):
        #将self.ev(centers)进行转置,并且以[0,2,1]重新排列输出维度,为矩阵相乘做准备
        v = tf.transpose(self.ev(centers),[0,2,1])
        u = self.eu(contexts)
        res = u@v
        bb = self.b(centers)
        bc = self.c(contexts)
        res = res + bb + bc
        return res
    #返回最终的词向量权重
    def get_uaddv_weights(self):
        self.ev(tf.zeros(0,dtype="int32"))
        self.eu(tf.zeros(0,dtype="int32"))
        return [self.ev.weights[0] + self.eu.weights[0]]
def loss_function(res, labels):
    labels = tf.reshape(lables,[res.shape[0],res.shape[1],1])
    #权重函数c取100
    h = tf.math.pow(labels/100,0.75)
    h = tf.clip_by_value(h,0,1)
    res -= tf.math.log1p(labels)
    res = tf.math.pow(res,2)
    res *= h
    loss_ = tf.reduce_mean(res)
    return loss_
```

Adam是常用的梯度下降方法,使用它来优化模型参数。

4)模型训练及保存

本部分实现GloVe模型的训练及保存。

```python
@tf.function
def train_step(centers, contexts, lables):
    loss = 0
    #GradientTape是eager模式下计算梯度
    with tf.GradientTape() as tape:
        res = glove(centers, contexts)
        loss += loss_function(res,lables)
    #批损失
    batch_loss = float(loss)
    variables = glove.trainable_variables
    gradients = tape.gradient(loss, variables)
    #更新gradients, variables的梯度,同时不在里面的变量梯度不变
    optimizer.apply_gradients(zip(gradients, variables))
    return batch_loss
#训练轮数
```

```python
EPOCHS = 20
for epoch in range(EPOCHS):
    start = time.time()
    total_loss = 0
    for (batch, (centers, contexts, labels)) in enumerate(dataset):
        batch_loss = train_step(centers, contexts, labels)
        total_loss += batch_loss
        if batch % 100 == 0:
            print('Epoch {} Batch {} Loss {:.8f}'.format(epoch + 1,
                                                         batch,
                                                         batch_loss.numpy()))
        print('Epoch {} Loss {:.8f}'.format(epoch + 1,
                                             total_loss / batch))
    print('Time taken for 1 epoch {} sec\n'.format(time.time() - start))
#模型权重保存
checkpoint = tf.train.Checkpoint(glove = glove)
checkpoint.save('./save_model/glove/glove.ckpt')
```

2. Seq2Seq 模型

本部分包括定义模型结构、优化损失函数、模型训练及保存。

1) 定义模型结构

相关代码如下。

```python
class Encoder(keras.Model):
    def __init__(self, vocab_size, embedding_dim, enc_units, batch_sz):
        super(Encoder, self).__init__()
        self.batch_sz = batch_sz
        self.enc_units = enc_units
        self.embedding = tf.keras.layers.Embedding(vocab_size, embedding_dim)
        self.gru = tf.keras.layers.GRU(self.enc_units,
                                        return_sequences = True,
                                        return_state = True,
                                        recurrent_initializer = 'glorot_uniform')
    def call(self, x, hidden):
        x = self.embedding(x)
        output, state = self.gru(x, initial_state = hidden)
        return output, state
    def initialize_hidden_state(self):
        return tf.zeros((self.batch_sz, self.enc_units))
class BahdanauAttention(keras.layers.Layer):
    def __init__(self, units):
        super(BahdanauAttention, self).__init__()
        self.W1 = tf.keras.layers.Dense(units)
        self.W2 = tf.keras.layers.Dense(units)
        self.V = tf.keras.layers.Dense(1)
    def call(self, query, values):
```

```python
        # 隐藏层的形状 == (批大小,隐藏层大小)
        # hidden_with_time_axis 的形状 == (批大小,1,隐藏层大小)
        # 这样做是为了执行加法以计算分数
        hidden_with_time_axis = tf.expand_dims(query, 1)
        # 分数的形状 == (批大小,最大长度,1)
        # 在最后一个轴上得到 1,因为把分数应用于 self.V
        # 在应用 self.V 之前,张量的形状是(批大小,最大长度,单位)
        score = self.V(tf.nn.tanh(
            self.W1(values) + self.W2(hidden_with_time_axis)))
        # 注意力权重(attention_weights)的形状 == (批大小,最大长度,1)
        attention_weights = tf.nn.softmax(score, axis=1)
        # 上下文向量(context_vector)求和之后的形状 == (批大小,隐藏层大小)
        context_vector = attention_weights * values
        context_vector = tf.reduce_sum(context_vector, axis=1)
        return context_vector, attention_weights
class Decoder(keras.Model):  # 解码
    def __init__(self, vocab_size, embedding_dim, dec_units, batch_sz):
        super(Decoder, self).__init__()
        self.batch_sz = batch_sz
        self.dec_units = dec_units
        self.embedding = tf.keras.layers.Embedding(vocab_size, embedding_dim)
        self.gru = tf.keras.layers.GRU(self.dec_units,
                                       return_sequences=True,
                                       return_state=True,
                                       recurrent_initializer='glorot_uniform')
        self.fc = tf.keras.layers.Dense(vocab_size)
        # 用于注意力
        self.attention = BahdanauAttention(self.dec_units)
    def call(self, x, hidden, enc_output):
        # 编码器输出(enc_output)的形状 = (批大小,最大长度,隐藏层大小)
        context_vector, attention_weights = self.attention(hidden, enc_output)
        # x 在通过嵌入层后的形状 == (批大小,1,嵌入维度)
        x = self.embedding(x)
        # x 在拼接(concatenation)后的形状 = (批大小,1,嵌入维度 + 隐藏层大小)
        x = tf.concat([tf.expand_dims(context_vector, 1), x], axis=-1)
        # 将合并后的向量传送到 GRU
        output, state = self.gru(x)
        # 输出的形状 = (批大小 * 1,隐藏层大小)
        output = tf.reshape(output, (-1, output.shape[2]))
        # 输出的形状 = (批大小,vocab)
        x = self.fc(output)
        return x, state, attention_weights
```

2) 优化损失函数

确定模型架构之后进行编译,这是多类别的分类问题,因此,使用交叉熵作为损失函数。由于所有的标签都带有相似的权重,经常使用精确度作为性能指标。Adam 是常用的梯度下降方法,使用它来优化模型参数。

3) 模型训练及保存

本部分实现 Seq2Seq 模型的训练及保存。

```python
@tf.function
def train_step(inp, targ, enc_hidden):                    #训练步长
    loss = 0
    with tf.GradientTape() as tape:
        enc_output, enc_hidden = encoder(inp, enc_hidden)
        dec_hidden = enc_hidden
        dec_input = tf.expand_dims([1] * BATCH_SIZE, 1)
        #将目标词作为下一个输入
        for t in range(1, targ.shape[1]):
            #将编码器输出传送至解码器
            predictions, dec_hidden, _ = decoder(dec_input, dec_hidden, enc_output)
            loss += loss_function(targ[:, t], predictions)
            dec_input = tf.expand_dims(targ[:, t], 1)
        batch_loss = (loss / int(targ.shape[1]))
        variables = encoder.trainable_variables + decoder.trainable_variables
        gradients = tape.gradient(loss, variables)
        optimizer.apply_gradients(zip(gradients, variables))
        return batch_loss
#载入 glove 词向量作为编码和解码嵌入层预训练
glove = Glove(vocab_size, embedding_dim)
checkpoint = tf.train.Checkpoint(glove = glove)
checkpoint.restore(tf.train.latest_checkpoint("./save_model/glove"))
embedding_weights = glove.get_uaddv_weights()
encoder.embedding(tf.zeros(0, dtype = "int32"))
decoder.embedding(tf.zeros(0, dtype = "int32"))
encoder.embedding.set_weights(embedding_weights)
decoder.embedding.set_weights(embedding_weights)
optimizer = tf.keras.optimizers.Adam()
loss_object = tf.keras.losses.SparseCategoricalCrossentropy(
    from_logits = True, reduction = 'none')
#保存模型
checkpoint = tf.train.Checkpoint(optimizer = optimizer,
                                  encoder = encoder,
                                  decoder = decoder)
EPOCHS = 100
for epoch in range(EPOCHS):                               #按照轮次
    start = time.time()
    enc_hidden = encoder.initialize_hidden_state()
    total_loss = 0
    for (batch, (inp, targ)) in enumerate(dataset):
        batch_loss = train_step(inp, targ, enc_hidden)
        total_loss += batch_loss                          #计算损失
```

```
            if batch % 100 == 0:
                print('Epoch {} Batch {} Loss {:.4f}'.format(epoch + 1,
                                                             batch,
                                                             batch_loss.numpy()))
        print('Epoch {} Loss {:.4f}'.format(epoch + 1,
                                            total_loss / batch))
        print('Time taken for 1 epoch {} sec\n'.format(time.time() - start))
        checkpoint.save('./save_model/seq2seq/seq2seq.ckpt')
```

7.3.3 模型测试

具体应用 Python 的 GUI 工具实现 wxPython,在图形界面的对话框输入聊天内容,经过序列化,调用 TensorFlow 模型,获取对话输出,最终显示在图形界面。主要包括 GUI 设置、模型导入及调用。

1. GUI 设置

```
#控件事件
self.tc2.Bind(wx.EVT_TEXT_ENTER, self.EVT_TEXT_ENTER)
```

其中,wx.StaticText()设置的标签内容、位置大小、排列居左;wx.TextCtrl()设置对话框位置、大小、显示仅可读、排列居左;self.tc2.Bind()绑定对话框输入回车时触发事件。

2. 模型导入及调用

将模型测试代码写入 test.py,使用 checkpoint 载入 Seq2Seq 模型,定义封装的评估函数。函数功能包括序列化字符串类型输入,调用模型,返回反序列化的字符串。在 GUI 文件中实现调用。

7.4 系统测试

本部分包括训练损失、测试效果及模型应用。

7.4.1 训练损失

本部分包括 Glove 词向量训练损失和 Seq2Seq 训练损失。

1. Glove 词向量训练损失

横轴表示步骤数,纵轴表示损失,词向量维度为 64,如图 7-4 所示,词向量维度为 256 时如图 7-5 所示。

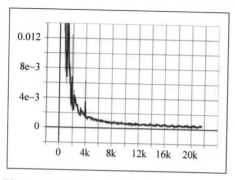 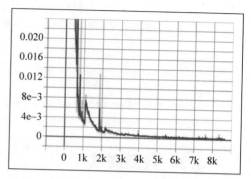

图 7-4　横轴步骤数,纵轴损失维度 64 示意图　　图 7-5　横轴步骤数,纵轴损失维度 256 示意图

2. Seq2Seq 训练损失

横轴表示训练轮数,纵轴表示损失,词向量维度为 64 时如图 7-6 所示,词向量维度为 256 时如图 7-7 所示。

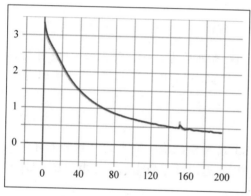 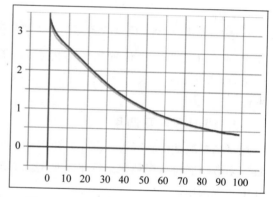

图 7-6　横轴轮数,纵轴损失维度 64 示意图　　图 7-7　横轴轮数,纵轴损失维度 256 示意图

7.4.2　测试效果

手动输入对话命令行测试,词向量维度 64,训练 100 轮次,如图 7-8 所示;词向量维度 64,训练 200 轮次,如图 7-9 所示;词向量维度 256,训练 100 轮次,如图 7-10 所示。

7.4.3　模型应用

程序运行 python window.py,初始界面如图 7-11 所示。

界面由 2 组标签和文本框组成,ChatBot 一栏对话框不可输入,对话输入一栏选中输入句子,按回车键显示结果,如图 7-12 所示。

```
input:你是谁
Building prefix dict from the default dictionary .
Dumping model to file cache /tmp/jieba.cache
Loading model cost 0.743 seconds.
Prefix dict has been built successfully.
2020-04-13 17:23:52.519391: I tensorflow/stream_ex
.so.7
2020-04-13 17:23:54.982916: I tensorflow/stream_ex
s.so.10
转发的,这老婆是谁?
input:你喜欢什么
哟,快来一系列的电影
input:我帅吗
我想要个什么?!我也很给力的
input:我很厉害
好恐怖。高手在一起
input:我会讲冷笑话
我卅。我卅,最后的手不要动。

我想知道为什么是为了掩盖改革。
input:我喜欢你
最近爱你,我喜欢你
input:你喜欢熊猫吗
喜欢喜欢,但我也想要。
```

图 7-8　词向量维度 64,训练 100 轮次示意图

```
input:你好聪明
Building prefix dict from the default dictionary ...
Loading model from cache /tmp/jieba.cache
Loading model cost 0.707 seconds.
Prefix dict has been built successfully.
2020-04-15 14:21:56.941233: I tensorflow/stream_execu
.so.7
2020-04-15 14:21:58.788193: I tensorflow/stream_execu
s.so.10
answer: 你好工作,楼下一个人!
input:你喜欢熊猫吗
answer: 好看是!有很多抵抗力
input:你的名字是
answer: 聊天包装中中中中中笑点
input:梅西是我兄弟
answer: 叫高警察和男篮的身体啊
input:名
answer: 街道未免也太残忍了。
input:明天会下雨吗
answer: 加油啊,大美人活那么你
input:我好累
answer: 搞得真好!求真相
input:我会讲冷笑话
answer: 怎么知道这个赛季好几天啊
input:今天天气很好
answer: 你们又是怎样的,参加着歌呢?
input:你喜欢唱歌吗
answer: 呵呵,我不喜欢这部电影么?
input:你喜欢看电影吗
answer: 你看恐怖片待我的脸,我是我的漫画都杯具了
input:我喜欢海贼王
answer: 喜欢第一件还是牛逼的生物
```

图 7-9　词向量维度 64,训练 200 轮次示意图

```
input:你是谁
answer: 应该是吧!不要打的,,,,,,,,,,,
input:你喜欢熊猫吗
answer: 这是一个小的是一样么?
input:今天天气真好
answer: 支持一下。为嘛就很有范儿
input:你喜欢吃冰棍吗
answer: 烤地瓜吃的啊,好可爱的吗?木有木有味道啊!
input:明天会下雨吗
answer: 老师还能见到你的地方吗
```

图 7-10　词向量维度 256,训练 100 轮次示意图

图 7-11　初始界面

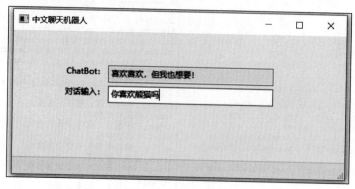

图 7-12 显示结果

项目 8　说唱歌词创作应用

PROJECT 8

本项目通过机器学习生成押韵且合理的中文说唱歌词,训练 NMT(Neural Machine Translation,神经机器翻译)的 Seq2Seq 模型,并对预测部分进行改进,实现贴合主题的说唱歌词。

8.1　总体设计

本部分包括系统整体结构图、系统流程图和前端流程图。

8.1.1　系统整体结构图

系统整体结构如图 8-1 所示。

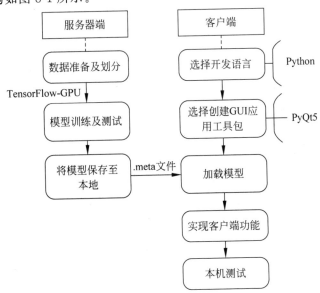

图 8-1　系统整体结构图

8.1.2 系统流程图和前端流程图

系统流程如图 8-2 所示,前端流程如图 8-3 所示。

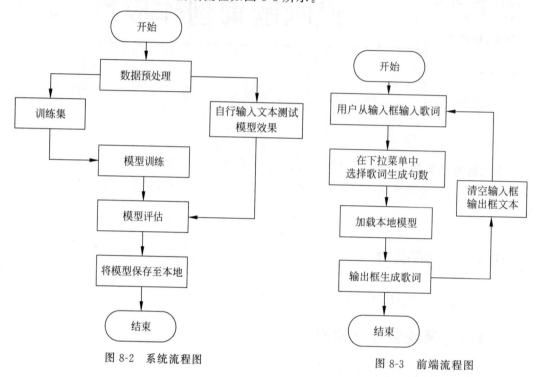

图 8-2　系统流程图　　　　　　　图 8-3　前端流程图

8.2 运行环境

本部分包括 Python 环境、TensorFlow 环境和其他环境。

8.2.1 Python 环境

需要 Python 3.6 及以上配置,在 Windows 环境下推荐下载 Anaconda 完成 Python 所需的配置,下载地址为 https://www.anaconda.com/。

8.2.2 TensorFlow 环境

模型训练基于 TensorFlow-GPU,版本为 1.11。在安装 TensorFlow-GPU 之前,需要确认 GPU 是否支持 CUDA,本项目所需为 CUDA 9.0 版本,下载地址为 https://developer.nvidia.com/cuda-toolkit-archive。

完成 cuDNN 的安装,对应 CUDA 9.0 的 cuDNN 版本为 7.0,下载地址为 https://

developer.nvidia.com/rdp/cuDNN-download，下载后复制、粘贴 cuDNN 路径下三个文件到 CUDA 的相应同名文件下：

① cuDNN\cuda\bin => CUDA\v10.0\bin；
② cuDNN\cuda\include => CUDA\v10.0\include；
③ cuDNN\lib\x64 => CUDA\v10.0\lib\x64。

完成 TensorFlow-GPU 的安装，输入 Win+R 后，输入 cmd 以打开命令提示符，输入以下命令：

```
pip install TensorFlow-GPU == 1.11
```

安装完毕。

8.2.3 其他环境

（1）PyQt5 安装：打开 Windows 命令提示符，输入命令：

```
pip install PyQt5
pip install PyQt5-tools
```

（2）jieba 分词库安装：打开 Windows 命令提示符，输入命令：

```
pip install jieba
```

安装完毕。

8.3 模块实现

本项目包括 4 个模块：数据预处理与加载、模型构建、模型训练及保存、模型测试。下面分别给出各模块的功能介绍及相关代码。

8.3.1 数据预处理与加载

数据下载链接为 https://drive.google.com/drive/folders/1QrO0JAti3A3vlZlUemouOW7jC3K5dFZr。数据集共包含 10 万条说唱歌词，使用 jieba 分词库进行分词，单句长度集中在 8～10 词。

1. 数据预处理

在创建模型并训练之前进行数据预处理，包括以下 4 步。

1）划分训练集、验证集、测试集

划分训练集、验证集、测试集的操作如下：

```
import pandas as pd
src_df = pd.read_csv("x.txt", header = None, names = ["src"])
```

```python
tgt_df = pd.read_csv("y.txt", header=None, names=["tgt"])
pair_df = pd.concat([src_df, tgt_df], axis=1)
pair_df = pair_df.sample(frac=1).reset_index(drop=True)
# 第 1 行至倒数第 200 行划为训练集
train_df = pair_df.iloc[:-200, :]
# 倒数第 200 行至倒数第 100 行划为验证集
dev_df = pair_df.iloc[-200:-100, :]
# 倒数第 100 行至最后一行划为测试集
test_df = pair_df.iloc[-100:, :]
```

2) 构建 x-y pair，使用上一句预测下一句，分割数据集

相关代码如下：

```python
# 训练集:构建源歌词集-目标歌词集
train_df["src"].to_csv("train.src", encoding='utf-8', header=False, index=False)
train_df["tgt"].to_csv("train.tgt", encoding='utf-8', header=False, index=False)
# 验证集:构建源歌词集-目标歌词集
dev_df["src"].to_csv("dev.src", encoding='utf-8', header=False, index=False)
dev_df["tgt"].to_csv("dev.tgt", encoding='utf-8', header=False, index=False)
# 测试集:构建源歌词集-目标歌词集
test_df["src"].to_csv("test.src", encoding='utf-8', header=False, index=False)
test_df["tgt"].to_csv("test.tgt", encoding='utf-8', header=False, index=False)
```

3) 歌词顺序颠倒

实现押韵效果需要知道生成歌词的结尾，即在哪个词押韵。本项目采取将输入模型的歌词倒过来，这样可确定对第一个词押韵，生成歌词后再将歌词反转，得到正常顺序的歌词。

```python
# 单句歌词顺序颠倒函数
def reverse_str(s):
    s_list = s.split()
    s_list.reverse()
    return " ".join(s_list)

def reverse_data(raw_path, reverse_path):
    # 读取文件
    df = pd.read_csv(raw_path, header=None, encoding="utf-8", names=["raw"])
    df["reverse"] = df.raw.map(lambda x: reverse_str(x))
    # 保存颠倒后的歌词
    df["reverse"].to_csv(reverse_path, encoding="utf-8", header=False, index=False)
# 执行函数
if __name__ == "__main__":
    reverse_data("v2/train.src", "v3/train.src")
    reverse_data("v2/train.tgt", "v3/train.tgt")
    reverse_data("v2/dev.src", "v3/dev.src")
    reverse_data("v2/dev.tgt", "v3/dev.tgt")
    reverse_data("v2/test.src", "v3/test.src")
    reverse_data("v2/test.tgt", "v3/test.tgt")
```

4)构建 word2idex 字典

处理自然语言问题时,使用深度学习方法搭建神经网络,需要将文本中的词或者字映射成数字 ID。本项目使用 TensorFlow 中的 Tokenizer 构建 word2idex 字典,获得词汇表。

```
#函数:构建字词表
def build_vocab(in_path, out_path, max_size = None, min_freq = 1, specials = Specials, tokenizer = None):
    #判断输出路径是否存在
    if not tf.gfile.Exists(out_path):
        print("Creating vocabulary {} from data {}".format(out_path, in_path))
        vocab = collections.Counter()
        #读取文件
        with tf.gfile.GFile(in_path, mode = 'r') as f:
            #逐行读取歌词并分词,存入容器
            for line in f:
                tokens = tokenizer(line) if tokenizer else naive_tokenizer(line)
                vocab.update(tokens)
        #将 collections 容器中的词排序,并逐个添加至列表 itos 中
        sorted_vocab = sorted(vocab.items(), key = lambda x: x[0])
        sorted_vocab.sort(key = lambda x: x[1], reverse = True)
        itos = list(specials)
        for word, freq in sorted_vocab:
            if freq < min_freq or len(itos) == max_size:
                break
            itos.append(word)
        #将词汇表写入输出文件
        with codecs.getwriter('utf-8')(tf.gfile.GFile(out_path, mode = 'wb')) as fw:
            for word in itos:
                fw.write(str(word) + '\n')
#构造词到索引的映射表
def create_vocab_tables(src_vocab_file, tgt_vocab_file, share_vocab = True):
    #构建词到索引的映射
    src_vocab_table = lookup_ops.index_table_from_file(
        src_vocab_file, default_value = UNK_ID)
    if share_vocab:
        tgt_vocab_table = src_vocab_table
    else:
        tgt_vocab_table = lookup_ops.index_table_from_file(
            tgt_vocab_file, default_value = UNK_ID)
    return src_vocab_table, tgt_vocab_table
```

2. 数据加载

数据预处理完成后载入模型,为训练做准备。首先,对数据的长度进行修正,限制源数据及目标数据的最大长度;其次,根据已经生成的 word2idex 字典,将字符转换为数字 ID,获得索引序列,再分别为源序列添加后缀< eos >,为目标序列添加前缀< sos >和后缀< eos >,

用于表示数据的开始和结束。

```python
# 加载训练数据到训练模型
if not output_buffer_size:
    output_buffer_size = batch_size * 1000
# 源数据和目标数据,每行使用 sos 和 eos 两个标记代表数据开始和结束,表示成 int32 类型整数
src_eos_id = tf.cast(src_vocab_table.lookup(tf.constant(eos)), tf.int32)
tgt_sos_id = tf.cast(tgt_vocab_table.lookup(tf.constant(sos)), tf.int32)
tgt_eos_id = tf.cast(tgt_vocab_table.lookup(tf.constant(eos)), tf.int32)
# 通过 zip 操作将源数据集和目标数据集合并在一起
src_tgt_dataset = tf.data.Dataset.zip((src_dataset, tgt_dataset))
if skip_count is not None:
    src_tgt_dataset = src_tgt_dataset.skip(skip_count)
# 随机打乱数据,切断相邻数据之间的联系
src_tgt_dataset = src_tgt_dataset.shuffle(
    output_buffer_size, random_seed, reshuffle_each_iteration)
src_tgt_dataset = src_tgt_dataset.map(
    lambda src, tgt: (
        tf.string_split([src]).values, tf.string_split([tgt]).values),
    num_parallel_calls=num_parallel_calls).prefetch(output_buffer_size)
# 过滤零长度序列
src_tgt_dataset = src_tgt_dataset.filter(
    lambda src, tgt: tf.logical_and(tf.size(src) > 0, tf.size(tgt) > 0))
# 限制源数据最大长度
if src_max_len:
    src_tgt_dataset = src_tgt_dataset.map(
        lambda src, tgt: (src[:src_max_len], tgt), num_parallel_calls=num_parallel_calls).prefetch(output_buffer_size)
# 限制目标数据的最大长度
if tgt_max_len:
    src_tgt_dataset = src_tgt_dataset.map(
        lambda src, tgt: (src, tgt[:tgt_max_len]), num_parallel_calls=num_parallel_calls).prefetch(output_buffer_size)
# 将字符串转换为数字 ID
src_tgt_dataset = src_tgt_dataset.map(
    lambda src, tgt: (tf.cast(src_vocab_table.lookup(src), tf.int32),
                      tf.cast(tgt_vocab_table.lookup(tgt), tf.int32)),
    num_parallel_calls=num_parallel_calls).prefetch(output_buffer_size)
# 为目标数据添加前缀<sos>和后缀<eos>
src_tgt_dataset = src_tgt_dataset.map(
    lambda src, tgt: (src,
                      tf.concat(([tgt_sos_id], tgt), 0),
                      tf.concat((tgt, [tgt_eos_id]), 0)),
    num_parallel_calls=num_parallel_calls).prefetch(output_buffer_size)
# 增加序列长度信息
src_tgt_dataset = src_tgt_dataset.map(
    lambda src, tgt_in, tgt_out: (
```

```
        src, tgt_in, tgt_out, tf.size(src), tf.size(tgt_in)),
        num_parallel_calls = num_parallel_calls).prefetch(output_buffer_size)
```

由于每行数据的长度不同,需要很大的运算量,因此,做数据对齐处理,对齐的同时,将数据集按照 batch_size 完成分批,最后加载数据进行训练,通过 make_initializable_iterator()迭代器实现,TensorFlow 提供 API。

```
#数据对齐
def batching_func(x):
    #调用 dataset 的 padded_batch 方法,对齐的同时,也对数据集进行分批
    return x.padded_batch(
        batch_size,
        #对齐数据的形状
        padded_shapes = (
            tf.TensorShape([None]),        #源数据
            tf.TensorShape([None]),        #目标输入
            tf.TensorShape([None]),        #目标输出
            tf.TensorShape([]),            #源长度
            tf.TensorShape([])),           #目标长度
                                           #对齐数据的值
        padding_values = (
            src_eos_id,                    #源标记
            tgt_eos_id,                    #目标输入标记
            tgt_eos_id,                    #目标输出标记
            0,
            0))
#将长度相近的数据放入相同存储单元 bucket 中,提高计算效率
#长度在[0,bucket_width) 范围内的归为 bucket0
#长度在[bucket_width, 2 * bucket_width) 范围内的归为 bucket1
if num_buckets > 1:
    def key_func(unused_1, unused_2, unused_3, src_len, tgt_len):
        #计算一个 bucket 的宽度 bucket_width
        if src_max_len:
            bucket_width = (src_max_len + num_buckets - 1) // num_buckets
        else:
            bucket_width = 10
        #按源歌词和目标歌词中的最大长度划分存储单元
        bucket_id = tf.maximum(src_len // bucket_width, tgt_len // bucket_width)
        return tf.to_int64(tf.minimum(num_buckets, bucket_id))
    def reduce_func(unused_key, windowed_data):
        return batching_func(windowed_data)      #分批数据集
    batched_dataset = src_tgt_dataset.apply(
        tf.contrib.data.group_by_window(
            key_func = key_func, reduce_func = reduce_func, window_size = batch_size))
else:
    batched_dataset = batching_func(src_tgt_dataset)
#通过迭代器分批获取数据
```

```
batched_iter = batched_dataset.make_initializable_iterator()
(src_ids, tgt_input_ids, tgt_output_ids, src_seq_len,
tgt_seq_len) = (batched_iter.get_next())
return BatchedInput(                          #返回分批输入数据
    initializer = batched_iter.initializer,
    source = src_ids,
    target_input = tgt_input_ids,
    target_output = tgt_output_ids,
    source_sequence_length = src_seq_len,
    target_sequence_length = tgt_seq_len)
```

8.3.2 模型构建

将数据加载进模型之后,需要定义模型结构,优化损失函数和设置押韵规则。

1. 定义模型结构

定义的架构为双向 RNN 结构,每个 RNN 单元选择长短期记忆和 LSTM 网络,整体构成双向 LSTM 网络。

在单向 RNN 中只考虑了上下文中的"上文",并未考虑后面的内容。可能会错过一些重要信息,使得预测的内容不够准确。双向 RNN 不仅从之前的时间步骤中学习,也从未来的时间步骤中学习,从而更好地理解上下文环境并消除歧义。双向 RNN 的结构和连接如图 8-4 所示,包括前向传播和后向传播。

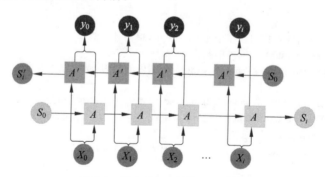

图 8-4 双向 RNN 结构和连接图

(1) 前向传播:从左向右移动,从初始时间步骤开始计算值,一直持续到达最终时间步骤为止;

(2) 后向传播:从右向左移动,从最后一个时间步骤开始计算值,一直持续到最终时间步骤。

用公式表示双向 RNN 过程如下:

从前向后:$h_i^1 = f(U^1 * X_i + W^1 * h_{i-1})$

从后向前:$h_i^2 = f(U^2 * X_i + W^2 * h_{i-1})$

输出： $y_i = \mathrm{softmax}(V * [h_i^1; h_i^2])$

LSTM是一种特殊的RNN网络,具有记忆单元,可以不断遗忘一些知识记忆,实现每一步的输出都考虑到之前所有的输入。双向RNN与LSTM模块相结合可以提高性能,提升歌词创作的准确性与合理性。在模型中引入丢弃进行正则化,用以消除过拟合问题。

```
#创建单层RNN单元:LSTM
def _build_single_cell(self, device_str = None):
    dropout = self.dropout if self.mode == tf.contrib.learn.ModeKeys.TRAIN else 0.0
    #RNN单元种类:LSTM
    utils.print_out(" LSTM, forget_bias = %g" % self.forget_bias, new_line = False)
    single_cell = tf.contrib.rnn.BasicLSTMCell(
        self.num_units,
        forget_bias = self.forget_bias)
    #设置丢弃,缓解过拟合
    if dropout > 0.0:
        single_cell = tf.contrib.rnn.DropoutWrapper(
            cell = single_cell, input_keep_prob = (1.0 - dropout))
        utils.print_out(" %s, dropout = %g " % (type(single_cell).__name__, dropout),
                    new_line = False)
    #确保这个单元在指定的设备上运行
    if device_str:
        single_cell = tf.contrib.rnn.DeviceWrapper(single_cell, device_str)
        utils.print_out(" %s, device = %s " %
                (type(single_cell).__name__, device_str), new_line = False)
    return single_cell
#创建多层RNN单元
def _build_rnn_cell(self, num_layers):
    cell_list = []
    for i in range(num_layers):
        utils.print_out(" cell %d" % i, new_line = False)
        #调用单层RNN单元创建函数
        single_cell = self._build_single_cell(self.get_device_str(i + self.base_gpu,
self.num_gpus))
        utils.print_out("")
        cell_list.append(single_cell)
    #单层
    if len(cell_list) == 1: # Single layer.
        return cell_list[0]
    #多层,调用API将多个BasicLSTMCell单元汇总为一个
    else: # Multi layers
        return tf.contrib.rnn.MultiRNNCell(cell_list)
#创建编码器单元
def _build_Encoder_cell(self, num_layers):
    return self._build_rnn_cell(num_layers)
#创建编码器Encoder
```

```python
def build_Encoder(self):
    num_layers = self.num_Encoder_layers
    with tf.variable_scope("Encoder") as scope:
        dtype = scope.dtype
        #编码器词嵌入矩阵输入
        self.Encoder_inputs_embedded = tf.nn.embedding_lookup(self.embedding_Encoder, self.Encoder_inputs)
        #创建一个双向层 LSTM
        num_bi_layers = int(num_layers / 2)
        utils.print_out(" num_bi_layers = %d" % num_bi_layers)
        #前向传播的 RNN
        fw_cell = self._build_Encoder_cell(num_bi_layers)
        #反向传播的 RNN
        bw_cell = self._build_Encoder_cell(num_bi_layers)
        #将两个单元作为参数传入双向动态 RNN 函数
        bi_outputs, bi_state = tf.nn.bidirectional_dynamic_rnn(
            fw_cell,
            bw_cell,
            self.Encoder_inputs_embedded,
            dtype = dtype,
            sequence_length = self.Encoder_inputs_length,
            time_major = False,
            swap_memory = True)
        self.Encoder_outputs, bi_Encoder_state = tf.concat(bi_outputs, -1), bi_state
        #保存前后向 RNN 最后的隐藏状态
        if num_bi_layers == 1:
            self.Encoder_state = bi_Encoder_state
        else:
            self.Encoder_state = []
            for layer_id in range(num_bi_layers):
                self.Encoder_state.append(bi_Encoder_state[0][layer_id]) # forward
                self.Encoder_state.append(bi_Encoder_state[1][layer_id]) # backward
            self.Encoder_state = tuple(self.Encoder_state)
#创建解码器单元
def _build_Decoder_cell(self, Encoder_state):
    cell = self._build_rnn_cell(self.num_Decoder_layers)
    #集束搜索
    if self.mode == tf.contrib.learn.ModeKeys.INFER and self.beam_width:
        Decoder_initial_state = tf.contrib.Seq2Seq.tile_batch(
            Encoder_state, multiplier = self.beam_width)
    else:
        Decoder_initial_state = Encoder_state
    return cell, Decoder_initial_state
```

2. 优化损失函数

确定模型架构之后进行编译,使用交叉熵作为损失函数。本项目使用梯度下降 Adam

算法，优化模型参数。在验证集和测试集上测试时，采用困惑度作为指标。

```python
# 损失函数
def _compute_loss(self, logits):
    Decoder_outputs = self.Decoder_outputs              # 解码输出
    max_time = tf.shape(Decoder_outputs)[1]
    crossent = tf.nn.sparse_softmax_cross_entropy_with_logits(
        labels = Decoder_outputs, logits = logits)      # 交叉熵
    target_weights = tf.sequence_mask(
        self.Decoder_inputs_length, max_time, dtype = logits.dtype)
    loss = tf.reduce_sum(
        crossent * target_weights) / tf.to_float(self.batch_size)
    return loss
# 优化器使用 Adam 算法进行训练模型的梯度更新
def build_optimizer(self):
    utils.print_out("# setting optimizer ...")
    params = tf.trainable_variables()
    opt = tf.train.AdamOptimizer(self.learning_rate)
    # 梯度计算及更新
    gradients = tf.gradients(self.loss, params)
    clipped_grads, _ = tf.clip_by_global_norm(gradients, self.max_gradient_norm)
    self.update = opt.apply_gradients(
        zip(clipped_grads, params), global_step = self.global_step)
# 测试模型在验证集和测试集时的困惑度
def run_internal_eval(model, global_step, sess, iterator, summary_writer, name):
    sess.run(iterator.initializer)
    total_loss = 0
    total_predict_count = 0
    start_time = time.time()
    while True:
        try:
            Encoder_inputs, Encoder_inputs_length, Decoder_inputs, Decoder_outputs, Decoder_inputs_length = \
                                                    # 编码输入，解码输出
                sess.run([
                    iterator.source,
                    iterator.source_sequence_length,
                    iterator.target_input,
                    iterator.target_output,
                    iterator.target_sequence_length])
            model.mode = "eval"
            loss, predict_count, batch_size = model.eval(sess, Encoder_inputs, Encoder_inputs_length,
                Decoder_inputs, Decoder_inputs_length, Decoder_outputs)
            total_loss += loss * batch_size     # 总的损失
            total_predict_count += predict_count
        except tf.errors.OutOfRangeError:
            break
```

```
perplexity = utils.safe_exp(total_loss / total_predict_count)    #困惑度
utils.print_time(" eval %s: perplexity %.2f" % (name, perplexity),
                 start_time)
utils.add_summary(summary_writer, global_step, "%s_ppl" % name, perplexity)
result_summary = "%s_ppl %.2f" % (name, perplexity)
#返回困惑度
return result_summary, perplexity
```

3. 设置押韵规则

说唱歌词的基本特点是句与句之间押韵，每句歌词最后一个字的单押效果。押韵方法是基于规则来实现，即获取输入歌词的最后一个词，分析韵脚，目标词只在相同韵脚的词中根据概率分布采样。

以输入歌词"你真美丽"为例，分词后，得到"你""真""美丽"，句尾词是"美丽"，模型分析出韵脚为 i，根据 i 的韵脚构建一个向量。在目标词中，将与"美丽"押韵的字词概率置为 1，非押韵的字词概率置为 0，得到押韵的概率分布。将该向量与原字词分布概率相乘，得到的分布 Distribution After Vector 便符合押韵的要求，同时保留了字词原来的分布概率。如图 8-5 所示，最终采样结果为"春泥"。

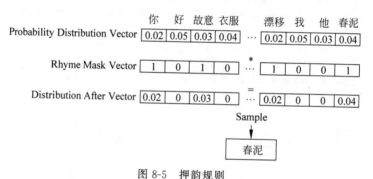

图 8-5　押韵规则

```
#押韵
def first_token_issue(self):
    #使用 sos 和 eos 两个标记代表目标数据的开始和结束，并表示成 int32 类型整数
    tgt_sos_id = tf.cast(self.target_vocab_table.lookup(tf.constant(self.sos)), tf.int32)
    tgt_eos_id = tf.cast(self.target_vocab_table.lookup(tf.constant(self.eos)), tf.int32)
    #设置最大解码长度
    maximum_iterations = self._get_infer_maximum_iterations(self.Encoder_inputs_length)
    #设置预测歌词和源歌词押韵
    #获取输入的最后一个词
    to_be_rhymed = self.Encoder_inputs[:, 0]
    to_be_rhymed = tf.expand_dims(to_be_rhymed, axis=1)
    #执行一个掩膜操作，通过切片获得和输入押韵的目标词索引范围
```

```
        rhyme_range = tf.gather_nd(self.table, to_be_rhymed)
        rhyme_range = tf.cast(rhyme_range, dtype = tf.int32)
        left = tf.slice(rhyme_range, [0, 0], [1, 1])
        left = tf.reshape(left, shape = [self.infer_batch_size])
        right = tf.slice(rhyme_range, [0, 1], [1, 1])
        right = tf.reshape(right, shape = [self.infer_batch_size])
        mask = magic_slice(left, right, self.tgt_vocab_size)
        mask = tf.cast(mask, dtype = tf.float32)
        #根据掩膜之后的概率分布进行采样
        first_inputs = tf.nn.embedding_lookup(
            self.embedding_Decoder,
            tf.fill([self.batch_size], tgt_sos_id))
        first_outputs, first_states = self.Decoder_cell(
            first_inputs, self.Decoder_initial_state)
        first_predictions = self.output_layer(first_outputs)
        first_logits = tf.multiply(first_predictions, mask)
        start_tokens = tf.argmax(first_logits, axis = 1)
        start_tokens = tf.cast(start_tokens, dtype = tf.int32)
        end_token = tgt_eos_id
        return maximum_iterations, start_tokens, end_token
```

8.3.3 模型训练及保存

在定义模型架构和编译之后,使用训练集训练模型,使模型根据输入歌词预测几句说唱歌词,并满足押韵要求。这里使用训练集和验证集拟合并保存模型。

1. 模型训练

模型训练的相关代码如下:

```
#模型训练
def train():
    #训练/验证/测试数据
    train_src_file = FLAGS.source_train_data
    train_tgt_file = FLAGS.target_train_data
    dev_src_file = FLAGS.source_dev_data
    dev_tgt_file = FLAGS.target_dev_data
    #词汇表
    src_vocab_file = FLAGS.src_vocab_file
    tgt_vocab_file = FLAGS.tgt_vocab_file
    #日志
    log_file = os.path.join(FLAGS.out_dir, "log_%d" % time.time())
    log_f = tf.gfile.GFile(log_file, mode = "a")
    utils.print_out("#log_file = %s" % log_file, log_f)
    config_proto = utils.get_config_proto(
        log_device_placement = FLAGS.log_device_placement,
        num_intra_threads = FLAGS.num_intra_threads,
```

```python
        num_inter_threads = FLAGS.num_inter_threads)
    #创建会话
    with tf.Session(config = config_proto) as train_sess:
        #构造词到索引的映射表
        src_vocab_table, tgt_vocab_table = vocab_utils.create_vocab_tables(
            src_vocab_file, tgt_vocab_file, share_vocab = FLAGS.share_vocab)
        #加载训练/验证数据
        train_iterator = load_data(train_src_file, train_tgt_file, src_vocab_table, tgt_vocab_table)
        dev_iterator = load_data(dev_src_file, dev_tgt_file, src_vocab_table, tgt_vocab_table)
        #训练模型
        model = hip_hop_model.Model(
            FLAGS,
            mode = tf.contrib.learn.ModeKeys.TRAIN,
            source_vocab_table = src_vocab_table,
            target_vocab_table = tgt_vocab_table,
            scope = None)
        loaded_train_model, global_step = model_helper.create_or_load_model(
            model, FLAGS.out_dir, train_sess, "train")
        #保存训练过程数据
        summary_writer = tf.summary.FileWriter(os.path.join(FLAGS.out_dir, "train_log"))
        #训练进程信息
        stats = init_stats()
        info = {"train_ppl": 0.0, "speed": 0.0, "avg_step_time": 0.0}
        #为了计量每轮训练时间,初始化一个时间
        start_train_time = time.time()
        utils.print_out("# Start step %d, %s" %
                        (global_step, time.ctime()), log_f)
        #开始训练
        utils.print_out("# Init train iterator.")
        train_sess.run(train_iterator.initializer)
        #训练循环:循环训练60轮,每轮训练1000次,每次随机抓取105条歌词
        epoch_idx = 0
        while epoch_idx < FLAGS.num_train_epochs:
            start_time = time.time()
            try:
                Encoder_inputs, Encoder_inputs_length, Decoder_inputs, Decoder_outputs, Decoder_inputs_length = \
                    train_sess.run([
                        train_iterator.source,
                        train_iterator.source_sequence_length,
                        train_iterator.target_input,
                        train_iterator.target_output,
                        train_iterator.target_sequence_length])
                loaded_train_model.mode = "train"
                step_result = loaded_train_model.train(train_sess, Encoder_inputs, Encoder
```

```
                _inputs_length,
                        Decoder_inputs, Decoder_inputs_length, Decoder_outputs)
                    FLAGS.epoch_step += 1
                except tf.errors.OutOfRangeError:
                    FLAGS.epoch_step = 0
                    epoch_idx += 1
                    utils.print_out(
                        "# Finished epoch % d, step % d." % (epoch_idx, global_step))
                    train_sess.run(train_iterator.initializer)
                    continue
                # 保存每一步的训练结果、日志等信息
                global_step, step_summary = update_stats(
                    stats, start_time, step_result)
                summary_writer.add_summary(step_summary, global_step)
                # 打印训练进程信息
                if global_step % FLAGS.steps_per_stats == 0:
                    process_stats(stats, info, FLAGS.steps_per_stats)
                    print_step_info(" ", global_step, info, log_f)
                    stats = init_stats()
        # 训练结束
        loaded_train_model.saver.save(
            train_sess,
            os.path.join(FLAGS.out_dir, "translate.ckpt"),
            global_step = global_step)
        # 打印模型在验证集上的表现
        result_summary, ppl = run_internal_eval(
                    loaded_train_model, global_step, train_sess, dev_iterator, summary_writer, "dev")
        print_step_info("# Final, ", global_step, info, log_f, result_summary)
        utils.print_time("# Done training!", start_train_time)
        summary_writer.close()
```

模型共训练 60Epoches,batch_size 取 105,即一次前向/后向传播过程随机抓取 105 条歌词进行训练,如图 8-6 所示。

2. 模型保存

为了能够随时读取模型,实现说唱歌词创作,用于前端程序读取利用 TensorFlow 中的 tf.train.Saver()模块进行模型的保存。

```
# 调用模块
self.saver = tf.train.Saver(
    tf.global_variables(), max_to_keep = flags.num_keep_ckpts)
# 保存训练模型
if global_step % FLAGS.steps_per_save == 0:
    loaded_train_model.saver.save(
        train_sess,
        os.path.join(FLAGS.out_dir, "translate.ckpt"),
        global_step = global_step)
```

```
step 67300 step-time 0.24s wps 6.68K ppl 1.78, Thu Apr 16 18:37:57 2020
step 67400 step-time 0.24s wps 6.53K ppl 1.72, Thu Apr 16 18:38:20 2020
# Finished epoch 18, step 67456.
step 67500 step-time 0.27s wps 5.73K ppl 1.71, Thu Apr 16 18:38:47 2020
step 67600 step-time 0.23s wps 6.70K ppl 1.61, Thu Apr 16 18:39:11 2020
step 67700 step-time 0.23s wps 6.79K ppl 1.61, Thu Apr 16 18:39:34 2020
step 67800 step-time 0.23s wps 6.78K ppl 1.64, Thu Apr 16 18:39:56 2020
step 67900 step-time 0.23s wps 6.75K ppl 1.73, Thu Apr 16 18:40:20 2020
step 68000 step-time 0.23s wps 6.68K ppl 1.67, Thu Apr 16 18:40:43 2020
step 68100 step-time 0.23s wps 6.73K ppl 1.70, Thu Apr 16 18:41:07 2020
step 68200 step-time 0.23s wps 6.74K ppl 1.76, Thu Apr 16 18:41:31 2020
step 68300 step-time 0.23s wps 6.78K ppl 1.72, Thu Apr 16 18:41:54 2020
# Finished epoch 19, step 68308.
step 68400 step-time 0.28s wps 5.51K ppl 1.63, Thu Apr 16 18:42:22 2020
step 68500 step-time 0.23s wps 6.91K ppl 1.60, Thu Apr 16 18:42:44 2020
step 68600 step-time 0.23s wps 6.84K ppl 1.60, Thu Apr 16 18:43:07 2020
step 68700 step-time 0.23s wps 6.84K ppl 1.71, Thu Apr 16 18:43:30 2020
step 68800 step-time 0.23s wps 6.79K ppl 1.66, Thu Apr 16 18:43:53 2020
step 68900 step-time 0.23s wps 6.80K ppl 1.67, Thu Apr 16 18:44:16 2020
step 69000 step-time 0.23s wps 6.84K ppl 1.75, Thu Apr 16 18:44:39 2020
step 69100 step-time 0.23s wps 6.81K ppl 1.69, Thu Apr 16 18:45:03 2020
# Finished epoch 20, step 69160.
step 69200 step-time 0.27s wps 5.82K ppl 1.70, Thu Apr 16 18:45:29 2020
```

图 8-6　训练结果

模型保存后,可以被重用,也可以移植到其他环境中使用。

8.3.4　模型测试

模型测试主要体现在前后端之间的数据交互,在歌词创作器输入一句说唱歌词,后端接收和将歌词输入到基于 TensorFlow 的双向 RNN 模型中,进行歌词创作,反馈到前端。具体包括前后端数据交互、模型预测和 GUI 设计。

1. 前后端数据交互

本部分主要包括 GUI 主函数部分加载模型、定义文本获取函数、按键与文本获取槽函数关联。

(1) 在 GUI 主函数部分加载模型操作如下。

```
if __name__ == "__main__":
    from PyQt5 import QtCore
    sess = tf.Session(config=utils.get_config_proto())
    loaded_model = inference.load_model(sess)
```

(2) 定义文本获取函数操作如下。

```
def getText(self):
text = self.lineEdit.text()
value = self.qj
```

```
output = inference.inference_n(loaded_model, sess, [text], int(value))
self.textEdit.setPlainText(output)  #将文本显示在输出框
```

(3) 将按键与文本获取槽函数关联操作如下。

```
self.pushButton.clicked.connect(self.getText)
```

2. 模型预测

在模型预测部分,进行了算法的创新,舍弃原算法中的贪婪搜索,采用集束搜索的方法实现歌词创作。

集束搜索在每个时间步都选择概率最大的前 k 个序列,得到一个候选输出序列的集合,再根据相应的函数选择最优解。集束搜索是一种启发式搜索算法,减少了搜索所占用的空间和时间,以较少的代价在相对受限的搜索空间中找出最优解。

1) 集束搜索算法

```
#解码器单元
def _build_Decoder_cell(self, Encoder_state):
    cell = self._build_rnn_cell(self.num_Decoder_layers)
    #预测阶段使用集束搜索
    if self.mode == tf.contrib.learn.ModeKeys.INFER and self.beam_width:
        Decoder_initial_state = tf.contrib.Seq2Seq.tile_batch(
            Encoder_state, multiplier = self.beam_width)
    else:
        Decoder_initial_state = Encoder_state
    return cell, Decoder_initial_state
#预测阶段模型及算法
def helper_and_dynamic_decoding(self, maximum_iterations, start_tokens, end_token, Decoder_scope):
    #集束搜索以获得更好的准确性
    inference_Decoder = tf.contrib.Seq2Seq.BeamSearchDecoder(
        cell = self.Decoder_cell,
        embedding = self.embedding_Decoder,
        start_tokens = start_tokens,
        end_token = end_token,
        initial_state = self.Decoder_initial_state,
        beam_width = self.beam_width,
        output_layer = self.output_layer)
    #动态解码,接收 inference_Decoder 类,依据编码进行解码,实现序列的生成
    outputs, final_context_state, _ = tf.contrib.Seq2Seq.dynamic_decode(
        Decoder = inference_Decoder,
        maximum_iterations = maximum_iterations,
        output_time_major = False,
        swap_memory = True,
        scope = Decoder_scope)
    return outputs
```

2）歌词预测

```python
# 删除行中的汉字间距
def del_chs_space(line):
    """delete space between Chinese characters in line."""
    pattern = u"((?<=[\u4e00-\u9fa5])\s+(?=[\u4e00-\u9fa5])|^\s+|\s+$)"
    res = re.sub(pattern, '', line)
    return res
# 调用 jieba 库分词
def tokenizer(line):
    return " ".join(jieba.cut(line))
def naive_tokenizer(line):
    return line.strip().split()
# 将输入歌词顺序颠倒
def reverse_str(line):
    res = line.split(" ")
    res.reverse()
    return " ".join(res)
# 将字符串转换为数字 ID
def convert_to_infer_data(line, src_vocab_table):
    src = [line]
    src = tf.convert_to_tensor(src)
    src = tf.string_split(src).values
    # 使用 word2idex 字典实现字符到 ID 的映射
    src = tf.cast(src_vocab_table.lookup(src), tf.int32)
    src = tf.expand_dims(src, 0)
    src_length = tf.size(src)
    src_length = tf.expand_dims(src_length, 0)
    return src, src_length
# 加载预测模型
def load_model(session, name="infer"):
    start_time = time.time()
    # 加载已保存的训练模型
    ckpt = tf.train.latest_checkpoint(FLAGS.out_dir)
    # 词汇表
    src_vocab_table, tgt_vocab_table = vocab_utils.create_vocab_tables(
        FLAGS.src_vocab_file, FLAGS.tgt_vocab_file, FLAGS.share_vocab)
    reverse_tgt_vocab_table = tf.contrib.lookup.index_to_string_table_from_file(
        FLAGS.tgt_vocab_file, default_value=vocab_utils.UNK)
    model = hip_hop_model.Model(
        FLAGS,
        mode=tf.contrib.learn.ModeKeys.INFER,
        source_vocab_table=src_vocab_table,
        target_vocab_table=tgt_vocab_table,
        reverse_target_vocab_table=reverse_tgt_vocab_table,
        scope=None)
    model.saver.restore(session, ckpt)
```

```python
    #初始化所有表
    session.run(tf.tables_initializer())
    utils.print_out(
        " loaded %s model parameters from %s, time %.2fs" %
        (name, ckpt, time.time() - start_time))
    return model
#解码器输出转换成文本
def get_translation(NMT_outputs, tgt_eos):
    if tgt_eos: tgt_eos = tgt_eos.encode("utf-8")
    output = NMT_outputs[0, :].tolist()
    #识别结束符
    if tgt_eos and tgt_eos in output:
        output = output[:output.index(tgt_eos)]
    translation = utils.format_text(output)
    return translation
#歌词预测函数,n表示预测次数
def _inference(model, session, line, n):
    num_translations_per_input = max(min(FLAGS.num_translations_per_input, FLAGS.beam_width), 1)
    #获取输入的歌词
    start_token = line.split()[0]
    #输入歌词顺序颠倒
    line = reverse_str(line)
    results = []
    #将文字转换为ID,用于模型预测
    source, source_sequence_length = convert_to_infer_data(line, model.source_vocab_table)
    #调用预测模型开始创作歌词
    for i in range(n - 1):
        Encoder_inputs, Encoder_inputs_length = session.run([source, source_sequence_length])      #输入
        NMT_outputs = model.infer(session, Encoder_inputs, Encoder_inputs_length)   #输出
        for beam_id in range(num_translations_per_input):
            translation = get_translation(NMT_outputs[beam_id], tgt_eos=FLAGS.eos)  #翻译
            translation = translation.decode("utf-8")   #翻译解码
            new_line = translation
            res = reverse_str(new_line)
            results.append(res)
            source, source_sequence_length = convert_to_infer_data(new_line, model.source_vocab_table)
    return results
```

3. GUI 设计

GUI 设计的相关代码如下。

```python
from PyQt5 import QtCore, QtGui, QtWidgets
from PyQt5.QtWidgets import *
```

```python
from PyQt5.QtGui import QIcon,QFont,QPalette
from PyQt5.QtCore import Qt
import sys
import tensorflow as tf
import misc_utils as utils
import inference
#定义主窗口类
class Ui_MainWindow(object):
    def setupUi(self, MainWindow):                          #设置窗口
        MainWindow.setObjectName("MainWindow")
        MainWindow.resize(400, 300)
        self.centralwidget = QtWidgets.QWidget(MainWindow)
        self.centralwidget.setObjectName("centralwidget")
        self.frame = QtWidgets.QFrame(self.centralwidget)
        self.frame.setGeometry(QtCore.QRect(120, 20, 241, 91))
        self.frame.setFrameShape(QtWidgets.QFrame.StyledPanel)
        self.frame.setFrameShadow(QtWidgets.QFrame.Raised)
        self.frame.setObjectName("frame")
        self.horizontalLayout = QtWidgets.QHBoxLayout(self.frame)
        self.horizontalLayout.setObjectName("horizontalLayout")
        self.lineEdit = QtWidgets.QLineEdit(self.frame)
        self.lineEdit.setObjectName("edit1")
        self.horizontalLayout.addWidget(self.lineEdit)
        self.pushButton = QtWidgets.QPushButton(self.frame)
        self.pushButton.setObjectName("btn1")
        self.horizontalLayout.addWidget(self.pushButton)
        MainWindow.setCentralWidget(self.centralwidget)
        self.menubar = QtWidgets.QMenuBar(MainWindow)
        self.menubar.setGeometry(QtCore.QRect(0, 0, 364, 18))
        self.menubar.setObjectName("menubar")
        MainWindow.setMenuBar(self.menubar)
        self.statusbar = QtWidgets.QStatusBar(MainWindow)
        self.statusbar.setObjectName("statusbar")
        MainWindow.setStatusBar(self.statusbar)
        self.retranslateUi(MainWindow)
        QtCore.QMetaObject.connectSlotsByName(MainWindow)
        #清零按钮
        btn = QPushButton(MainWindow)
        btn.setText("一键清空")
        btn.move(298,80)
        btn.resize(56,17)
        btn.clicked.connect(self.clear)
        #输入按键关联函数
        self.pushButton.clicked.connect(self.getText)
        #为按钮添加提示信息
        QToolTip.setFont(QFont('SansSerif',10))
        self.pushButton.setToolTip('单击完成 rap 词输入')
```

```python
        #设置标签
        label1 = QLabel(MainWindow)
        label1.setText("< font color = black >输入 rap 词:</font >")
        label1.resize(50,20)
        label1.move(75,57)
        label1.setAutoFillBackground(True)
        patette = QPalette()
        label1.setPalette(patette)
        label1.setAlignment(Qt.AlignCenter)
        box = QHBoxLayout()
        box.addWidget(label1)
        label2 = QLabel(MainWindow)
        label2.setText("< font color = black >选择生成 rap 词的行数:</font >")
        label2.resize(100, 20)
        label2.move(77, 80)
        #显示输出内容框
        self.textEdit = QTextEdit(MainWindow)
        self.textEdit.move(120,100)
        self.textEdit.resize(200,100)
        #设置下拉选项框
        self.cb = QComboBox(MainWindow)
        self.cb.move(170,81)
        self.cb.resize(50,15)
        self.cb.addItems(['1','2','3','4','5','6','7','8'])
        #设置选项框信号传递
        self.cb.currentIndexChanged[str].connect(self.value)
        self.qj = 1
        self.flag = 0
    def value(self,i):                           #编号
        self.qj = i
    def getText(self):                           #获取文本
        text = self.lineEdit.text()
        value = self.qj
        output = inference.inference_n(loaded_model, sess, [text], int(value))
        self.textEdit.setPlainText(output)
    def retranslateUi(self, MainWindow):         #界面更新
        _translate = QtCore.QCoreApplication.translate
        #设置窗口标题
        MainWindow.setWindowTitle(_translate("MainWindow", "rap 词生成器"))
        self.pushButton.setText(_translate("MainWindow", "输入"))
        self.status = MainWindow.statusBar()
        #下方消息栏显示 5s
        self.status.showMessage('欢迎使用本程序',5000)
        #按钮单击事件,退出程序
    def clear(self):
        self.lineEdit.clear()
```

```python
            self.textEdit.clear()
if __name__ == "__main__":                          # 主程序
    from PyQt5 import QtCore
    sess = tf.Session(config=utils.get_config_proto())
    loaded_model = inference.load_model(sess)
    QtCore.QCoreApplication.setAttribute(QtCore.Qt.
AA_EnableHighDpiScaling)
    #创建一个QApplication,也就是要开发的App软件
    app = QtWidgets.QApplication(sys.argv)
    app.setWindowIcon(QIcon('./rap.png'))
    #创建一个QMainWindow,用来装载需要的各种组件、控件
    MainWindow = QtWidgets.QMainWindow()
    #ui 是 Ui_MainWindow()类的实例化对象
    ui = Ui_MainWindow()
    #执行类中的setupUi方法,方法的参数是第二步中创建的QMainWindow
    ui.setupUi(MainWindow)
    #执行QMainWindow的show()方法,显示这个QMainWindow
    MainWindow.show()
    #使用exit()或者单击关闭按钮退出QApplication
    sys.exit(app.exec_())
```

8.4 系统测试

本部分包括模型困惑度和模型应用。

8.4.1 模型困惑度

对于语言模型,一般使用困惑度指标来衡量模型的优劣。困惑度越接近1,表明在预测中真实的下一个字符被成功预测的概率越大,语言模型就越好。根据训练日志,随着训练次数的增加,困惑度在不断下降,最终稳定在1.25左右,如图8-7所示,表明模型训练效果较好。

8.4.2 模型应用

本部分包括应用使用说明和应用示例。

1. 应用使用说明

运行程序后自动加载本地已训练好的模型,用户可在输入框中输入一句预想好的歌词,并用空格对一句歌词完成划分。在下拉框中选择 AI 生成说唱歌词的行数,单击"输入"按键可在下方的输出框获得歌词结果。单击"一键清空"按钮可以将输入与输出框中的文字清空。

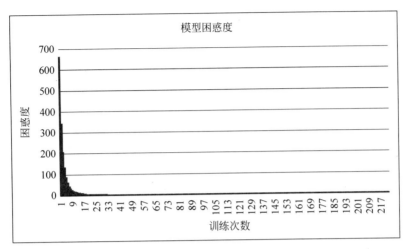

图 8-7　模型困惑度

2. 应用示例

输入一句歌词"你曾感受冰冷的风"后,如图 8-8 所示。

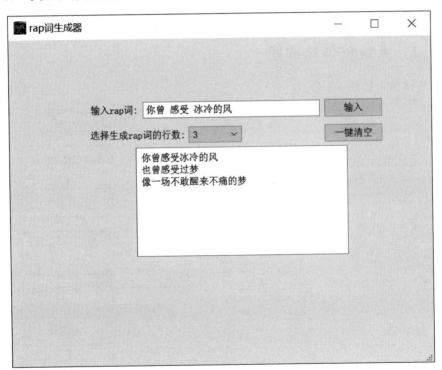

图 8-8　测试效果

项目 9 基于 LSTM 的语音/文本/情感识别系统

PROJECT 9

本项目使用 Google 公司的 Word2Vec 模型将单词转化为向量,并通过 LSTM 以及百度 API 的调用,完成机器的训练与学习,实现从语音到文本以及情感分析的综合功能。

9.1 总体设计

本部分包括系统整体结构图、系统流程图和网页端配置流程图。

9.1.1 系统整体结构图

系统整体结构如图 9-1 所示。

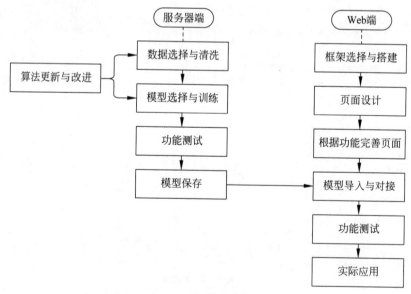

图 9-1 系统整体结构图

9.1.2 系统流程图

系统流程如图9-2所示。

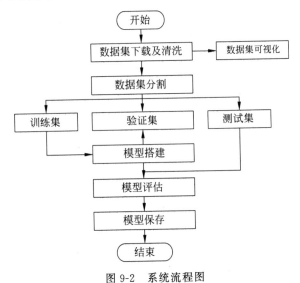

图 9-2 系统流程图

9.1.3 网页端配置流程图

网页端配置流程如图9-3所示。

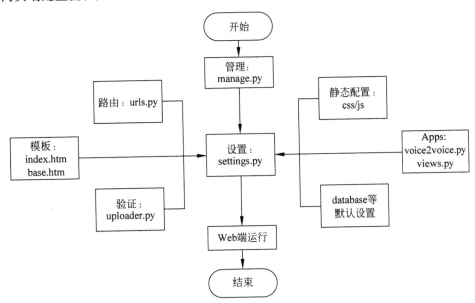

图 9-3 网页端配置流程图

9.2 运行环境

本部分包括 Python 环境、TensorFlow 环境和网页端环境（Django）。

9.2.1 Python 环境

本项目需要 Python 3.5 及以上配置完成 TensorFlow 的环境。有三种主要方法：一是通过 Anaconda 直接完成配置；二是在本地建立镜像源通过控制台完成配置；三是通过 Linux 虚拟机完成配置。

9.2.2 TensorFlow 环境

使用 TensorFlow 1.1.0 版本，输入 conda install tensorflow=1.1.0 命令进行安装。

9.2.3 网页端环境框架——Django

主要有两种途径：一是先安装虚拟环境后配置；二是在 Anaconda 下直接配置。

1. 虚拟环境配置

安装虚拟环境并启动，进入安装环境目录，在 cmd 命令行下输入：

```
pip install virtualenv
```

创建命令：

```
python -m venv xxx
```

运行 script 目录下的 activate.bat 文件启动虚拟环境，再安装 Django，输入命令：

```
pip install Django
```

在哪个目录下运行这句命令，就会安装在哪个目录下。

2. Anaconda 配置

创建虚拟环境 TensorFlow，输入命令：

```
conda create --name "tensorflow" python=3.5
```

创建 Python 版本 3.5 的虚拟环境，激活虚拟环境，输入命令：

```
conda activate tensorflow
```

激活对应的虚拟环境。

9.3 模块实现(服务器端)

本部分包括 4 个模块:数据处理、调用 API、模型构建、模型训练及保存。下面分别给出各模块的功能介绍及相关代码。

9.3.1 数据处理

情感分析数据集下载地址为 http://ai.stanford.edu/~amaas/data/sentiment/。本部分包括音频处理、文字处理和数据处理。

1. 音频处理

数据主要是音频文件,受百度 API 的限制,需要将输入文件转换成 .wav 格式,此处使用 ffmpeg 进行格式转换。

ffmpeg 可以记录、转换数字音频、视频,并将其转化为流的开源计算机程序。相关代码如下:

```
ffmpeg - i [input] [output.wav]
```

2. 文字处理

句子输入神经网络需要先将单词转换为词向量,使用 Google 公司的 word2vec 模型。相关代码如下:

```python
import numpy as np
wordsList = np.load('wordsList.npy')
print('Loaded the word list!')
wordsList = wordsList.tolist()
wordsList = [word.decode('UTF-8') for word in wordsList]
#将所有单词转化为词向量
wordVectors = np.load('wordVectors.npy')
```

3. 数据处理

训练使用 imdb 数据集,其中有 25 000 条正向和 25 000 条负向的评价,需要对这些句子去除标点符号、空格等。

```python
import re
strip_special_chars = re.compile("[^A-Za-z0-9 ]+")
def cleanSentences(string):
    string = string.lower().replace("<br />", " ")
    return re.sub(strip_special_chars, "", string.lower())
```

9.3.2 调用 API

本部分包括申请网络密钥和调用 API。

1. 申请网络密钥

在百度 AI 平台中申请短语音识别业务和翻译业务,选择英语语音识别业务和英文转中文的翻译业务,申请成功后获得相应的密钥。

2. 调用 API

百度语音识别使用 http://vop.baidu.com/server_api 和密钥进行调用。

(1) 语音转文字相关代码如下:

```
#需要识别的文件
AUDIO_FILE = dir    #只支持pcm/wav/amr 格式,极速版额外支持m4a 格式
#文件格式
FORMAT = AUDIO_FILE[-3:]    #文件后缀只支持pcm/wav/amr
CUID = '123456PYTHON'
#采样率
RATE = 16000    #固定值
#普通版
DEV_PID = 1737    #根据文档填写PID,选择语言及识别模型
ASR_URL = 'http://vop.baidu.com/server_api'    #API 调用地址
SCOPE = 'audio_voice_assistant_get'    #有语音识别能力
```

(2) 英文转中文相关代码如下:

```
httpClient = None
myurl = '/api/trans/vip/translate'            #API 调用地址
fromLang = 'auto'                              #原文语种
toLang = 'zh'                                  #译文语种
salt = random.randint(32768, 65536)
q = text
sign = appid + q + str(salt) + secretKey
sign = hashlib.md5(sign.encode()).hexdigest()    #校验编码
myurl = myurl + '?appid=' + appid + '&q=' + urllib.parse.quote(q) + '&from=' + fromLang
        + '&to=' + toLang + '&salt=' + str(salt) + '&sign=' + sign
```

9.3.3 模型构建

本部分包括训练数据导入、训练数据向量化和模型构建。

1. 训练数据导入

训练使用 imdb 数据集,包括 25 000 条正向和 25 000 条负向的评价。

```
from os import listdir
```

```python
from os.path import isfile, join
#导入积极情感数据集
positiveFiles = ['positiveReviews/' + f for f in listdir('positiveReviews/') if isfile(join('positiveReviews/', f))]
#导入消极情感数据集
negativeFiles = ['negativeReviews/' + f for f in listdir('negativeReviews/') if isfile(join('negativeReviews/', f))]
numWords = []
for pf in positiveFiles:
    with open(pf, "r", encoding='utf-8') as f:
        line = f.readline()
        counter = len(line.split())
        numWords.append(counter)
print('Positive files finished')          #正向数据
for nf in negativeFiles:
    with open(nf, "r", encoding='utf-8') as f:
        line = f.readline()
        counter = len(line.split())
        numWords.append(counter)
print('Negative files finished')          #反向数据
#查看数据集大小
numFiles = len(numWords)
print('The total number of files is', numFiles)
print('The total number of words in the files is', sum(numWords))
print('The average number of words in the files is', sum(numWords)/len(numWords))
```

2. 训练数据向量化

训练数据向量化相关代码如下:

```python
ids = np.zeros((numFiles, maxSeqLength), dtype='int32')
fileCounter = 0
#将积极情绪数据集全部向量化
for pf in positiveFiles:
   with open(pf, "r") as f:
        indexCounter = 0
        line = f.readline()
        cleanedLine = cleanSentences(line)
        split = cleanedLine.split()
        for word in split:
            try:
                ids[fileCounter][indexCounter] = wordsList.index(word)
            except ValueError:
                ids[fileCounter][indexCounter] = 399999   #未知词向量
            indexCounter = indexCounter + 1
            if indexCounter >= maxSeqLength:
```

```python
            break
        fileCounter = fileCounter + 1
#将消极情绪数据集全部向量化
for nf in negativeFiles:
    with open(nf, "r") as f:
        indexCounter = 0
        line = f.readline()
        cleanedLine = cleanSentences(line)
        split = cleanedLine.split()
        for word in split:
            try:
                ids[fileCounter][indexCounter] = wordsList.index(word)
            except ValueError:
                ids[fileCounter][indexCounter] = 399999 #未知词向量
            indexCounter = indexCounter + 1
            if indexCounter >= maxSeqLength:
                break
        fileCounter = fileCounter + 1
#保存处理后的矩阵
np.save('idsMatrix', ids)
```

3. 模型构建

将带有 LSTM 单元的 RNN 网络进行训练，RNN 模型在处理上下文有关联的情况时效果良好。

```python
#设置超参数
batchSize = 24
lstmUnits = 64
numClasses = 2
iterations = 100000
import tensorflow as tf
tf.reset_default_graph()
#设置节点
labels = tf.placeholder(tf.float32, [batchSize, numClasses])
input_data = tf.placeholder(tf.int32, [batchSize, maxSeqLength])
data = tf.Variable(tf.zeros([batchSize,maxSeqLength, numDimensions]),dtype=tf.float32)
data = tf.nn.embedding_lookup(wordVectors,input_data)
#引入 LSTM 单元
lstmCell = tf.contrib.rnn.BasicLSTMCell(lstmUnits)
lstmCell = tf.contrib.rnn.DropoutWrapper(cell=lstmCell, output_keep_prob=0.75)
value, _ = tf.nn.dynamic_rnn(lstmCell, data, dtype=tf.float32)
#初始化
weight = tf.Variable(tf.truncated_normal([lstmUnits, numClasses]))
bias = tf.Variable(tf.constant(0.1, shape=[numClasses]))
value = tf.transpose(value, [1, 0, 2])
last = tf.gather(value, int(value.get_shape()[0]) - 1)
```

```
prediction = (tf.matmul(last, weight) + bias)
correctPred = tf.equal(tf.argmax(prediction,1), tf.argmax(labels,1))
accuracy = tf.reduce_mean(tf.cast(correctPred, tf.float32))
#测试准确率
loss = tf.reduce_mean(tf.nn.softmax_cross_entropy_with_logits(logits = prediction, labels = labels))
optimizer = tf.train.AdamOptimizer().minimize(loss)
import datetime
tf.summary.scalar('Loss', loss)
tf.summary.scalar('Accuracy', accuracy)
merged = tf.summary.merge_all()
logdir = "tensorboard/" + datetime.datetime.now().strftime("%Y%m%d-%H%M%S") + "/"
writer = tf.summary.FileWriter(logdir, sess.graph)
```

9.3.4 模型训练及保存

本部分包括模型训练及模型保存。

1. 模型训练

模型训练相关代码如下：

```
sess = tf.InteractiveSession()        #交互会话
saver = tf.train.Saver()
sess.run(tf.global_variables_initializer())
for i in range(iterations):
    #下一批次
    nextBatch, nextBatchLabels = getTrainBatch();
    sess.run(optimizer,{input_data:nextBatch,labels: nextBatchLabels})
    #将总结写入Tensorboard
    if (i % 50 == 0):
        summary = sess.run(merged, {input_data: nextBatch, labels: nextBatchLabels})
        writer.add_summary(summary, i)
    if (i % 10000 == 0 and i != 0):
        save_path = saver.save(sess, "models/pretrained_lstm.ckpt", global_step = i)
        print("saved to %s" % save_path)
writer.close()
```

2. 模型保存

模型保存相关代码如下：

```
sess = tf.InteractiveSession()
saver = tf.train.Saver()
saver.restore(sess, tf.train.latest_checkpoint('models'))
```

9.4 网页实现(前端)

本部分包括 Django 的管理脚本、Django 的核心脚本、网页端模板的组成、Django 的接口验证脚本、Django 中 URL 模板的连接器、Django 中 URL 配置。

9.4.1 Django 的管理脚本

不论是在何种环境下运行 Django 模式下的 Web 服务器,执行的命令一定为 python manage.py runserver,即通过运行 manage.py 实现其他所有配置与功能。相关代码如下:

```python
import os
import sys
def main():
    os.environ.setdefault('DJANGO_SETTINGS_MODULE', 'aitrans.settings')
    #通过运行 manage.py 完成 settings.py 中所有配置的执行
    try:                              #异常捕获
        from django.core.management import execute_from_command_line
    except ImportError as exc:
        raise ImportError(
            "Couldn't import Django. Are you sure it's installed and "
            "available on your PYTHONPATH environment variable? Did you "
            "forget to activate a virtual environment?"
        ) from exc
    execute_from_command_line(sys.argv)
if __name__ == '__main__':            #主函数
    main()
```

9.4.2 Django 的核心脚本

基于 Django 的 Web 端框架依托于 settings.py 实现所有的配置,接口(嵌套)最终都要以某种方式出现在 settings.py 中,只有这样才能被 manage.py 调用并执行,默认的有 database(数据库)、views(视图)、urls(路由)等,除默认配置外,需要添加的都在 settings.py 中写明。相关代码如下:

```python
import os, sys
BASE_DIR = '.'
#快速开发设置
#参见 https://docs.djangoproject.com/en/3.0/howto/deployment/checklist/
#安全秘钥
SECRET_KEY = 'yzhdg*0532@fz4$4i%x^($pqf+ec9vt!ot4kpyudjv2jryj^6#'
DEBUG = True
ALLOWED_HOSTS = ['*']
```

```python
# 应用定义,核心代码放在 apps(voice2voice)
sys.path.insert(0, os.path.join(BASE_DIR, 'apps'))
INSTALLED_AppS = [
    'django.contrib.admin',
    'django.contrib.auth',
    'django.contrib.contenttypes',
    'django.contrib.sessions',
    'django.contrib.messages',
    'django.contrib.staticfiles',
    'voice2voice'
]
# 系统默认
MIDDLEWARE = [
    'django.middleware.security.SecurityMiddleware',
    'django.contrib.sessions.middleware.SessionMiddleware',
    'django.middleware.common.CommonMiddleware',
    # 'django.middleware.csrf.CsrfViewMiddleware',
    'django.contrib.auth.middleware.AuthenticationMiddleware',
    'django.contrib.messages.middleware.MessageMiddleware',
    'django.middleware.clickjacking.XFrameOptionsMiddleware',
]
# 指定 urls.py(路由)的路径,调用 v2vservice 中的类模块
ROOT_URLCONF = 'aitrans.urls'
# 模板部分,除特殊说明外,本块代码也为系统默认
TEMPLATES = [
    {
        'BACKEND': 'django.template.backends.django.DjangoTemplates',
        'DIRS': [os.path.join(BASE_DIR, 'templates')],
# 指向模板存放的路径——templates 文件夹,方便调用 index.htm 以及更高配置的 base.htm
        'App_DIRS': True,
        'OPTIONS': {
            'context_processors': [
                'django.template.context_processors.debug',
                'django.template.context_processors.request',
                'django.contrib.auth.context_processors.auth',
                'django.contrib.messages.context_processors.messages',
                'django.template.context_processors.media',
            ],
            'builtins':['django.templatetags.static'], # import static tag
        },
    },
]
WSGI_AppLICATION = 'aitrans.wsgi.application'
# Database,系统默认分配的数据库部分
DATABASES = {
    'default': {
        'ENGINE': 'django.db.backends.sqlite3',
```

```python
        'NAME': os.path.join(BASE_DIR, 'db.sqlite3'),
    }
}
# Password validation,系统默认分配的登录系统
AUTH_PASSWORD_VALIDATORS = [
    {
        'NAME': 'django.contrib.auth.password_validation.UserAttributeSimilarityValidator',
    },
    {
        'NAME': 'django.contrib.auth.password_validation.MinimumLengthValidator',
    },
    {
        'NAME': 'django.contrib.auth.password_validation.CommonPasswordValidator',
    },
    {
        'NAME': 'django.contrib.auth.password_validation.NumericPasswordValidator',
    },
]
# 国际化,系统默认
LANGUAGE_CODE = 'zh-hans'           # 配置简体中文,默认语音
TIME_ZONE = 'Asia/Shanghai'         # 时区
USE_I18N = True                     # 多语言
USE_L10N = True                     # 多语言,两种标准
USE_TZ = True                       # 时区使用
# Static files (CSS, JavaScript, Images),网页静态的高级配置
# 位于 aitrans 外面的 static 目录,存放 index.html 调用的.css 文件和.js 文件
# 从客户端调用的路径,可以指向不同位置(一对多的存放路径)
STATIC_URL = '/static/'
# 项目的特定路径,具体指向需要调用文件的目录
STATICFILES_DIRS = [
    os.path.join('/Users/bondsam/Downloads/workspace','static/'),
    os.path.join(BASE_DIR, 'js'),
]
# 上传图片和文件路径
MEDIA_URL = '/media/' # uploader.py 中匹配,上传完成后显示的网址(云端)
MEDIA_ROOT = os.path.join(BASE_DIR,'media')
# uploader.py 中匹配,所以把文件保存到 BASE_DIR 和 media 的拼接路径目录下(本地)
```

9.4.3 网页端模板的组成

index.htm 和 base.htm 是网页端模板的基本组成。index.htm 承载着出现给所有用户显示界面的基本框架(模板),即开发者预先设计好所有的基本显示,使用 HTML 语言进行编写。base.htm 调用项目外部 static 目录中存放 Web 端更高级的 css、js 等配置,完成 index.htm 中基本显示的组合与嵌套。相关代码如下:

index.htm:

```html
{% extends "home/base.htm" %} <!-- 引用 base.htm -->
{% block title %}{{ model.verbose_name }}{% endblock %} <!-- 如果需要,可填入实际名称替换 base.htm 中相同位置的名称 -->
{% block content %} <!-- 块级 -->
<div class="invoice p-3 mb-3 no-border invoice-rev">
    <!-- 主标题层设置 --><!-- i 为图标,可选择 -->
    <div class="row">
        <div class="col-12">
            <h4>
                <i class="fa fa-assistive-listening-systems" style="color:cornflowerblue;"></i> Voice2Voice App
            </h4>
        </div>
        <!-- /.col -->
    </div>
    <!-- 副标题层设置 --><!-- i 为图标,可选择 -->
    <div class="row">
        <div class="col-12" style="margin-top:10px;">
            <h6><i class="fa fa-angle-down bg-gray"></i> 说明:</h6>
            <!-- ul/ol 为列表 -->
            <ul>
                <ol class="list-unstyled" style="margin-top: 10px;">
                    <li>1.上传英文语音音频文件,文件类型,大小,</li>
                    <li>2.语音转文本</li>
                    <li>3.文本情感分析</li>
                    <li>4.文本在线翻译为中文</li>
                    <li>5.中文文本输出音频</li>
                </ol>
            </ul>
        </div>
        <!-- /.col -->
    </div>
    <!-- 功能按键层设置 -->
    <div class="card">
        <div class="card-header d-flex p-0">
            <h3 class="card-title p-3">Tabs</h3>
            <ul class="nav nav-pills ml-auto p-2">
                <li class="nav-item"><a class="nav-link active" href="#tab_1" data-toggle="tab">上传文件 >> </a></li>
                <li class="nav-item"><a class="nav-link" href="#tab_2" data-toggle="tab">解析文本 >> </a></li>
                <li class="nav-item"><a class="nav-link" href="#tab_3" data-toggle="tab">翻译与输出 >> </a></li>
            </ul>
        </div><!-- 功能实现框设置 -->
        <div class="card-body">
```

```html
                    <div class="tab-content">
                        <div class="tab-pane active" id="tab_1">
                            <div class="col-12">
                                <div class="input-group rounded-0">
                                    <!-- 上传文件:输入框显示了上传之后的路径名 -->
                                    <input type="text" name="audiopath" class="form-control" value="" id="audiopath" />
                                    placeholder="上传音频文件小于1MB,文件格式为WAV"
                                    <div class="input-group-append">
                                        <label class="btn btn-default btn-flat" for="uploadfield_btn">
                                            <i class="fa fa-upload"></i>
                                            <!-- 上传控制输入框 -->
                                            <input hidden type="file" name="file" id="uploadfield_btn">
                                        </label>
                                    </div>
                                </div>
                            </div>
                            {% csrf_token %} <!-- 使用自动生成的token防止被csrf攻击 -->
                        </div>
                        <!-- /.tab-pane -->
                        <div class="tab-pane" id="tab_2">
                            <div class="row" style="padding-bottom: 10px;">
                                <div class="col-12 form-group">
                                    <!-- 解析后文本框 -->
                                    <textarea name="parse_text" class="form-control form-control-rev"></textarea>
                                </div>
                            </div>
                            <div class="row" style="padding-bottom: 10px;">
                                <div class="col-12">
                                    <!-- 情感分析输入框 -->
                                    <input class="form-control form-control-rev" name="sentiment" value="" readonly />
                                </div>
                                <div class="col-12" style="margin-top: 10px;">
                                    <button type="button" class="btn btn-sm btn-primary btn-flat" id="parse_text_btn">解析</button>
                                    <button type="button" class="btn btn-sm btn-primary btn-flat pull-right" id="sentiment_btn">情感分析</button>
                                </div>
                            </div>
                        </div>
                        <div class="tab-pane" id="tab_3">
                            <div class="row" style="padding-bottom: 10px;">
```

```html
                    <div class="col-12">
                      <h6>
                        <i class="fa fa-angle-down bg-gray"></i> 中文文本：
                      </h6>
                    </div>
                    <div class="col-12">
                      <!-- 翻译后文本框 -->
                      <textarea class="form-control form-control-rev" name="trans_text"></textarea>
                    </div>
                    <div class="col-12" style="margin-top:10px;">
                      <!-- h5 音频播放器 -->
                      <audio preload="auto" controls="controls" src="" id="audio-src">
                      </audio>
                    </div>
                    <div class="col-12" style="margin-top: 10px;">
                      <button type="button" class="btn btn-sm btn-primary btn-flat pull-right" id="read_text_btn">语音</button>
                      <button type="button" class="btn btn-sm btn-primary btn-flat" id="trans_btn">翻译</button>
                    </div>
                  </div>
                </div>
                <div class="tab-pane" id="tab_4s">
                </div>
                <!-- /.tab-pane -->
              </div>
              <!-- /.tab-content -->
            </div><!-- /.card-body -->
          </div>
        </div>
{% endblock %}
{% block attach_script %}
<script src="{% static 'app.js' %}"></script>
<script src="{% static 'validate.js' %}"></script>
<script>
    $(function(){
        //jquery 框架通过 ID 名实现功能
        //上传文件按钮单击后可能出现的事件
        $("#uploadfield_btn").change(function(){
            var file = $(this)[0].files[0];
            if(!Validator.check_audioupload(file)){
                return false;
            }
            //查到 csrftoken 并放在表头, 便于识别
            var csrftoken = $('input[name="csrfmiddlewaretoken"]').val();
```

```javascript
            GLOBALS.audioupload(file,csrftoken,GLOBALS.audiouploadcallback);
        });
        //解析按钮单击后可能出现的事件
        $("#parse_text_btn").click(function(){
            var audiopath = $.trim($("#audiopath").val());
            if(audiopath != ''){
                //查到csrftoken并放在表头,便于识别
                var csrftoken = $('input[name="csrfmiddlewaretoken"]').val();
                //加载文本
$Ajax({method:"POST",url:'/v2v/parse/',formdata:{"uploadpath":audiopath},csrftoken:csrftoken,callback:GLOBALS.parsecallback})
            }else{
                alert("请先指定上传文件路径");
            }
        });
        //情感分析按钮单击后可能出现的事件
        $("#sentiment_btn").click(function(){
          var parse_text = $.trim($("textarea[name='parse_text']").val());
          if( parse_text == ''){
            alert("解析文本内容不能为空!");
            return false;
          }
            //查到csrftoken并放在表头,便于识别
            var csrftoken = $('input[name="csrfmiddlewaretoken"]').val();
            //加载文本
$Ajax({method:"POST",url:'/v2v/senti/',formdata:{"parse_text":parse_text},csrftoken:csrftoken,callback:GLOBALS.senticallback})
        });
        //翻译按钮单击后可能出现的事件
        $("#trans_btn").click(function(){
          var parse_text = $.trim($("textarea[name='parse_text']").val());
          if( parse_text == ''){
            alert("解析文本内容不能为空!");
            return false;
          }
            //查到csrftoken并放在表头,便于识别
            var csrftoken = $('input[name="csrfmiddlewaretoken"]').val();
            //加载文本
$Ajax({method:"POST",url:'/v2v/trans/',formdata:{"parse_text":parse_text},csrftoken:csrftoken,callback:GLOBALS.transcallback})
        });
        //单击语音按钮后可能出现的事件
        $("#read_text_btn").click(function(){
            var text = $.trim($("textarea[name='trans_text']").val());
            if(text == ''){
                alert("翻译文本内容不能为空")
                return false;
```

```
                    }
                    //查到csrftoken并放在表头,便于识别
                    var csrftoken = $('input[name="csrfmiddlewaretoken"]').val();
                    //加载文本
$Ajax({method:"POST",url:'/v2v/read/',formdata:{"trans_text":text},csrftoken:csrftoken,
callback:GLOBALS.readcallback});
                });
            });
        </script>
{% endblock %} <!-- 块结束 -->
```

base.htm:

```
        <!DOCTYPE html> <!-- 识别为html -->
        <html>
        <head>
            <meta charset="utf-8">
            <meta http-equiv="X-UA-Compatible" content="IE=edge">
            <title>{% block title %}Home 模板{% endblock %}</title>
            <!-- Tell the browser to be responsive to screen width -->
            <meta content="width=device-width, initial-scale=1, maximum-scale=1, user-scalable=no" name="viewport">
            <!-- Font Awesome -->
            <link rel="stylesheet" href="{% static 'plugins/font-awesome/css/font-awesome.min.css' %}">
            <!-- Ionicons -->
            <link rel="stylesheet" href="{% static 'plugins/Ionicons/css/ionicons.min.css' %}">
            <!-- Bootstrap 4 -->
            <link rel="stylesheet" href="{% static 'plugins/tempusdominus-bootstrap-4/css/tempusdominus-bootstrap-4.min.css' %}">
            <!-- AdminLTE Skins. Choose a skin from the css/skins folder instead of downloading all of them to reduce the load. -->
            <link rel="stylesheet" href="{% static 'plugins/daterangepicker/daterangepicker.css' %}">
            <!-- bootstrap-table -->
            <link rel="stylesheet" href="{% static 'plugins/bootstrap-table/dist/bootstrap-table.min.css' %}">
            <link rel="stylesheet" href="{% static 'plugins/jquery-ui/jquery-ui.min.css' %}">
            <!-- Theme style -->
            <link rel="stylesheet" href="{% static 'dist/css/adminlte.min.css' %}">
            <link rel="stylesheet" href="{% static 'dist/css/style.css' %}">
            <link rel="shortcut icon" href="{% static "favicon.ico" %}" />
            {% block attach_css %}{% endblock %}
            <link href="https://fonts.googleapis.com/css?family=Source+Sans+Pro:300,400,400i,700" rel="stylesheet">
            <![endif]-->
        </head>
        <!-- ADD THE CLASS sidebar-collapse TO HIDE THE SIDEBAR PRIOR TO LOADING THE SITE -->
        <body class="hold-transition sidebar-mini layout-fixed sidebar-collapse">
```

```html
<!-- 包装器 开始 -->
<div class="wrapper">
    {% include "home/navbar.htm" %}
    <!-- 左边栏 开始. sidebar-dark-primary -->
    <aside class="main-sidebar elevation-4 sidebar-light-cyan">
        <!-- website logo -->
        <a href="/" class="brand-link navbar-cyan">
            <img src="{% static 'dist/img/AdminLTELogo.png' %}" alt="NLPStudio Logo" class="brand-image img-circle elevation-3"
                 style="opacity: .8">
            <span class="brand-text font-weight-light">ChatBot</span>
        </a>
        <div class="sidebar">
            <!-- user info -->
            <a href="#">
                <div class="user-panel mt-3 pb-3 mb-3 d-flex">
                    <div class="pull-left image">
                        <img src="{% static 'img/avatar04.png' %}" class="img-circle" alt="User Image">
                    </div>
                    <div class="pull-left info">
                        <p>测试用户</p>
                        <i class="fa fa-circle text-success"></i> 在线 <!-- 离线... -->
                    </div>
                </div>
            </a>
            <!-- 侧边栏菜单: : style can be found in sidebar.less -->
            {% include "home/sidebarMenu.htm" %}
        </div>
    </aside>
    <!-- 头部内容 (Page header) -->
    <div class="content-wrapper">
        {% block content-header %}
        <section class="content-header">
            <div class="container-fluid">
                <div class="row">
                    <div class="col-6">
                        {% block module-title %}{% endblock %}
                    </div>
                    <!-- 面包屑导航 -->
                    <div class="col-6">
                        <ol class="breadcrumb float-sm-right">
                            {% block breadcrumb %}{% endblock %}
                        </ol>
                    </div>
                </div>
            </div>
```

```html
        </section>
        <!-- 主内容 -->
    {% endblock %}
    {% block content-body %}
    <section class="content">
        <div class="container-fluid">
            <div class="row" style="flex-wrap: nowrap">{# 设置不折行 #}
                <div class="col-9 nowrap border-top">
                    <div class="row" style="flex-wrap: nowrap">{# 设置不折行 #}
                        <div class="col nowrap" id="ui-layout-center">
                            {% block content %}{% endblock %}
                        </div>
                        {# 显示和隐藏右边栏布局 #}
                        <div class="col-3 nowrap" style="display:none;" id="ui-layout-east">
                            {% block rightbar %}{% endblock %}
                        </div>
                    </div>
                </div>
            </div>
        </div>
    </section>
    {% endblock %}
    </div>
    {% block content-footer %}
    <footer class="main-footer">
        <strong>Copyright &copy; 2019-2022 <a href="#">Voice2Voice Studio</a>.</strong>
        All rights reserved.
        <div class="float-right d-none d-sm-inline-block">
            <b>Version</b> 1.0.1
        </div>
    </footer>
    <!-- 侧边栏背景. 此DIV必须立即放置在控制器侧边栏之后 -->
    <aside class="control-sidebar control-sidebar-dark">
        <!-- Control sidebar content goes here -->
    </aside>
</div>
{% endblock %}
<!-- ./包装器 结束 -->
</body>
</html>
    <script src="{% static 'plugins/jquery/jquery.min.js' %}"></script>
    <script src="{% static 'plugins/jquery-ui/jquery-ui.min.js' %}"></script>
    <script src="{% static 'plugins/bootstrap/js/bootstrap.bundle.min.js' %}"></script>
    <script src="{% static 'plugins/moment/moment.min.js' %}"></script>
    <script src="{% static 'plugins/daterangepicker/daterangepicker.js' %}"></script>
```

```
                <script src="{% static 'plugins/tempusdominus-bootstrap-4/js/tempusdominus-
bootstrap-4.min.js' %}"></script>
                <script src="{% static 'plugins/bootstrap-table/dist/bootstrap-table.min.js'
 %}"></script>
                <script src="{% static 'plugins/bootstrap-table/dist/locale/bootstrap-table-
zh-CN.min.js' %}"></script>
                <script src="{% static 'dist/js/adminlte.min.js' %}"></script>
    {% block attach_script %}
{% endblock %}                        <!--块结束-->
```

9.4.4 Django 的接口验证脚本

uploader.py 文件是接口验证脚本,正常情况下,要执行某一特定功能时,应事先判断是否满足执行此功能的最基本要求,即剔除那些不满足条件的对象并节省资源。将音频文件上传到服务端,需要接口验证脚本避免错误的输入。相关代码如下:

```
import os,sys
from django.conf import settings
class AudioUploader(object):
    def __init__(self):
        self.allowtypes = ["audio/wav"]       #类型
        self.allowsize = 1024*1024            #大小
        self.filename = ""                    #名称
        self.savepath = os.path.join(settings.MEDIA_ROOT,"uploads")
#和 settings.py 建立联系
        self.urlpath = os.path.join(settings.MEDIA_URL,"uploads/");
#和 settings.py 建立联系
#检测文件类型
    def check_filetype(self,filetype):
        return filetype in self.allowtypes
#检测文件大小
    def check_filesize(self,filesize):
        if(filesize > self.allowsize):
            return False
        else:
            return True
#保存文件
    def savefile(self,file):
        savepath = os.path.join(self.savepath, file.name);
        with open(savepath,"wb") as fp:
            for chunk in file.chunks():
                fp.write(chunk)
```

9.4.5 Django 中 URL 模板的连接器

views.py 是 Django 中 URL 模板的连接器,在 Django 中,视图的作用相当于 URL 和模板的连接器,在浏览器中输入 URL 后,Django 通过视图找到相应的模板,然后返回浏览器并最终显示到服务器端。

执行过程:运行 manage.py,找到 settings.py,通过 ROOT_URLCONF 找到项目中 url.py 的配置和相应的视图函数,即 views.py 中与后端算法对接的功能函数,根据 views.py 的配置输出到服务器端。

相关代码请扫描二维码获取。

代码

9.4.6 Django 中 URL 配置

urls.py 为 Django 中 URL 的配置。URL,即统一资源定位符,提供一个地址,让编写好的网页在网页端运行并接收访问。相关代码如下:

```
from django.contrib import admin
from django.urls import path, re_path
from django.views.static import serve
from django.views.generic import TemplateView, RedirectView
from .settings import MEDIA_ROOT
from voice2voice.views import *        #导入项目视图模块中系统输出的部分
urlpatterns = [
    path('admin/', admin.site.urls),
    re_path(r'^ $ ', TemplateView.as_view(template_name = "index.htm"), name = "index"),
#直接调用本系统在线模板,即 index.htm,在 templates 里
    re_path('^media/(?P<path>.*)', serve, {"document_root":MEDIA_ROOT}),
    re_path(r'^v2v/upload/ $ ', UploadSrcView.as_view(),
    name = "upload_list"),          #调用 views.py 中前后端以及与 URL 对接的功能函数
re_path(r'^v2v/parse/ $ ', ParseTextView.as_view(), name = "parsetext_list"),
re_path(r'^v2v/senti/ $ ', SentimentView.as_view(), name = "sentiment_list"),
    re_path(r'^v2v/trans/ $ ', TranslateView.as_view(), name = "trans_list"),
    re_path(r'^v2v/read/ $ ', ReadView.as_view(), name = "read_list"),
]
```

9.5 系统测试

本部分包括训练准确率和效果展示。

9.5.1 训练准确率

相关代码如下:

```
iterations = 10
for i in range(iterations):
    nextBatch, nextBatchLabels = getTestBatch();
```

```
    print("Accuracy for this batch:", (sess.run(accuracy, {input_data: nextBatch, labels:
nextBatchLabels})) * 100)
```

模型训练准确率的结果如下：

```
Accuracy for this batch: 87.5
Accuracy for this batch: 87.5
Accuracy for this batch: 83.3333313465
Accuracy for this batch: 83.3333313465
Accuracy for this batch: 75.0
Accuracy for this batch: 83.3333313465
Accuracy for this batch: 75.0
Accuracy for this batch: 95.8333313465
Accuracy for this batch: 79.1666686535
Accuracy for this batch: 83.3333313465
```

9.5.2 效果展示

上传文件界面如图 9-4 所示，语音识别及情感分析界面如图 9-5 所示，翻译及语音输出界面如图 9-6 所示。

图 9-4 上传文件界面

图 9-5　语音识别及情感分析界面

图 9-6　翻译及语音输出界面

项目 10 基于人脸检测的表情包自动生成器

PROJECT 10

本项目基于 OpenCV 计算机视觉库下的 Haar 级联分类器算法做人脸检测，通过 TensorFlow 的三层卷积神经网络做人脸朝向预测，实现能够适应多种环境下自动生成每个人专属表情包的娱乐性应用。

10.1 总体设计

本部分包括系统整体结构图、系统流程图和文件结构。

10.1.1 系统整体结构图

系统整体结构如图 10-1 所示。

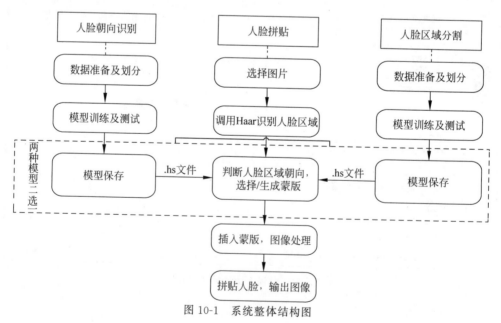

图 10-1 系统整体结构图

10.1.2 系统流程图

系统流程如图 10-2 所示。

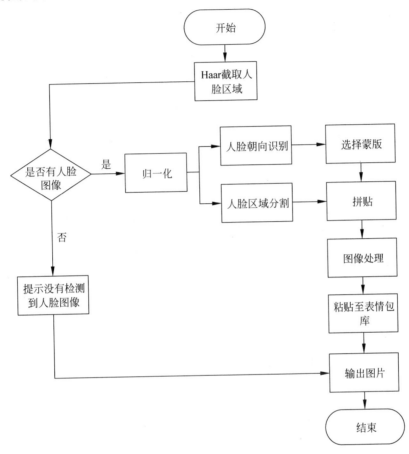

图 10-2　系统流程

10.1.3 文件结构

本项目所在文件夹结构如下。

```
AutoEmoticons
├──data          （存放 opencv 人脸图像分类器，训练好的人脸朝向模型）
├──materials     （存放表情背景和错误提示照片）
│  ├──background （存放背景照片，可进行用户定制）
│  └──mask       （存放蒙版）
├──save          （存放用户保存的照片）
```

```
└──(temp)            （临时文件夹，关闭程序、打开新照片将删除或更新）
    ├──(emoticons)   （临时文件夹，存放临时生成表情文件）
    └──(faces)       （临时文件夹，存放截取的人脸图像文件）
```

10.2 运行环境

本部分包括 Python 环境、TensorFlow 环境、OpenCV 环境和 Pillow 环境。

10.2.1 Python 环境

Python 3.7 及以上配置，在 Windows 环境下推荐下载 Anaconda 完成 Python 所需的配置，下载地址为 https://www.anaconda.com/。

10.2.2 TensorFlow 环境

环境为：TensorFlow 1.14.0。打开 Anaconda Prompt，输入清华仓库镜像，输入命令：

```
conda config -- add channels https://mirrors.tuna.tsinghua.edu.cn/anaconda/pkgs/free/
conda config - set show_channel_urls yes
```

创建 Python 3.7 的环境，名称为 tf1.14.0，此时 Python 版本和 TensorFlow 的版本有匹配问题，此步选择 Python 3.x，输入命令：

```
conda create - n tf1.14.0 python = 3.7
```

有需要确认的地方，都输入 y。
在 Anaconda Prompt 中激活 TensorFlow 环境，输入命令：

```
activate tf1.14.0
```

安装 CPU 版本的 TensorFlow，输入命令：

```
pip install - upgrade -- ignore - installed tensorflow == 1.14.0
```

安装完毕。

10.2.3 OpenCV 环境

需要安装 OpenCV 3.4.2 环境，使用 Anaconda 进行安装，输入命令：

```
conda install opencv - python
```

10.2.4 Pillow 环境

需要安装 Pillow 6.1.0 环境，使用 Anaconda 进行安装，输入命令：

```
conda install Pillow
```

10.3 模块实现

本项目包括 4 个模块：图形用户界面、人脸检测与标注、人脸朝向识别、人脸处理与表情包合成。下面分别给出各模块的功能介绍及相关代码。

10.3.1 图形用户界面

该模块能够实现系统在图片模式和摄像头模式下进行相互切换，支持在图形用户界面上输出实时视频。

1. 定义主框架类

定义一个主框架用于加载基础图形界面，并初始化变量与路径，控制界面主循环。

```
class Application():
    # 主窗口框架,控制界面主循环
    global rootPath, testMode
    def __init__(self):
        self.root = Tk()
        self.root.rootPath = rootPath
        print(self.root.rootPath)
        self.root.title("Auto Emoticons")
        self.root.geometry('780x370 + 600 + 300')
        self.root.maxsize(780, 370)
        self.root.minsize(780, 370)
        # 变量
        self.root.testMode = testMode                    # 测试模式
        self.root.picAddr = StringVar()                  # picAddr.get()为导入图像地址
        self.root.page = 0              # 当前展示的表情位于列表中的位置
        self.root.selectModel = self.reloadModel()       # 加载图像
        self.root.continueVideo = IntVar()               # 持续开启摄像头循环
        self.root.continueVideo.set(1)
        # 路径
        self.root.tempPath = os.path.join(
            self.root.rootPath, 'temp')                  # 临时文件保存路径
        self.root.emoticonPath = os.path.join(
            self.root.tempPath, 'emoticons')             # 临时生成表情保存路径
        self.root.bgPath = os.path.join(
            self.root.rootPath, 'materials', 'background')   # 表情背景保存路径
        self.root.maskPath = os.path.join(
            self.root.rootPath, 'materials', 'mask')     # 蒙版保存路径
        # 表情背景图片数目(-1是因为从0开始编号)
        self.root.bgsize = len(os.listdir(self.root.bgPath)) - 1
        picMode(self.root)
```

```python
    def reloadModel(self, model_name = './data/CovNet_0.945.h5'):
        #导入模型(提前加载模型,可免去每次导入图片都重新加载,大幅加快处理速度)
        model = load_model(model_name)
        if self.root.testMode:
            model.summary()
        return model
    def run(self):
        self.root.mainloop()
```

2. 定义照片模式类

继承主框架类：针对图片模式的功能特点定义照片模式类，要求能够实现打开图片传入识别算法、检测图片中的人脸、识别与处理人脸、生成表情包、表情和图片展示的翻页功能、保存表情、切换至摄像头模式。

相关代码请扫描二维码获取。

代码

3. 定义摄像头模式类

继承主框架类，根据摄像头模式的功能特点，定义摄像头模式类，实现实时读取数据、检测人脸、制作表情包、摄像头循环、保存表情以及切换至照片模式。

相关代码请扫描二维码获取。

代码

10.3.2 人脸检测与标注

搭建图形界面框架，实现各功能部件、照片或视频中的人脸检测与标注。人脸检测模块基于OpenCV库的Haar级联正脸分类算法，通过加载已训练好的人脸检测器haarcascade_frontalface_alt.xml，直接调用级联算法，便可检测传入的算法图像中是否有该人脸图像，若有则用方框标记出来，便于系统进行下一步的人脸图像处理。该模块可应用于照片和摄像头模式，具体输入参数与检测到人脸图像后的处理方式因模式的要求不同而略有差异。

```python
    def detect(self, image_path, picType):
        #OpenCV 检测是否有人脸图像
        data = './data/haarcascade_frontalface_alt.xml'   #人脸图像检测器
        face_cascade = cv2.CascadeClassifier(data)    #获取训练好的人脸图像参数数据
        image = cv2.imread(image_path)    #读取图片
        gray = cv2.cvtColor(image, cv2.COLOR_BGR2GRAY)    #灰度处理
        #探测图片中的人脸图像
        faces = face_cascade.detectMultiScale(
            gray, scaleFactor = 1.15, minNeighbors = 5, minSize = (5, 5), flags = cv2.CASCADE_SCALE_IMAGE)
        picNum = 0
        os.mkdir(os.path.join(self.master.rootPath, 'temp', 'faces'))
        for (x, y, w, h) in faces:  #faces 中可能包含多张人脸图像
            a, b, c, d = int(x), int(y), int(w), int(
                h)  #将 int32 转化为 int 类型,便于用列表裁剪图片
```

```
            face = image[b:b + d, a:a + c]    #按识别框裁剪人脸图片
            save = os.path.join(self.master.rootPath, 'temp',
                            'faces', 'tempPicFace' + str(picNum) + '.png')
            cv2.imwrite(save, face, None)    #保存裁剪后的人脸图像
            picNum += 1
        for (x, y, w, h) in faces:
            #保存带识别框的图像(两次for循环是为了先保存不带识别框的人脸图像)
            cv2.rectangle(image, pt1 = (x, y), pt2 = (x + w, y + h),
                            color = (255, 150, 0), thickness = 5)    #加入识别框
        cv2.imwrite(image_path, image)
        return picNum
```

人脸图像检测与标注结果如图 10-3 所示。

10.3.3 人脸朝向识别

检测与裁剪人脸之后,需要人脸朝向识别模块做朝向预测。

1. 制作人脸朝向数据集

本项目爬取了近 300 张公众人物照片作为数据集,通过 Haar 级联分类器检测与裁剪人脸,再经人工筛查分类对数据做正向、朝左及朝右三类的划分制作了人脸朝向数据集,如图 10-4~图 10-6 所示。

图 10-3 人脸检测与标注

图 10-4 正向人脸数据

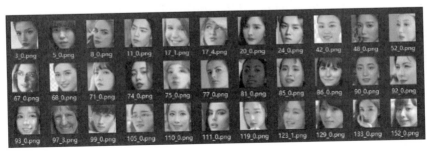

图 10-5 朝左人脸数据

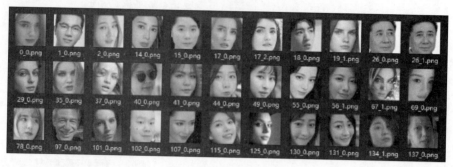

图 10-6　朝右人脸数据

2. 数据预处理

在模型训练之前,先划分训练集和验证集,将图像处理为统一标准后输入神经网络。相关代码如下:

```python
from tensorflow.keras.preprocessing.image import ImageDataGenerator
#路径
base_dir = './google_faces_train_val'
train_dir = os.path.join(base_dir, 'train')
val_dir = os.path.join(base_dir, 'val')
train_1_dir = os.path.join(train_dir, '1')
train_2_dir = os.path.join(train_dir, '2')
train_3_dir = os.path.join(train_dir, '3')
val_1_dir = os.path.join(val_dir, '1')
val_2_dir = os.path.join(val_dir, '2')
val_3_dir = os.path.join(val_dir, '3')
train_1_fnames = os.listdir(train_1_dir)
train_2_fnames = os.listdir(train_2_dir)
train_3_fnames = os.listdir(train_3_dir)
#统一图像大小
#图片缩放1./255
train_datagen = ImageDataGenerator(rescale = 1./255)
val_datagen = ImageDataGenerator(rescale = 1./255)
#分批读入训练数据与验证数据
#使用train_datagen生成器按20个批次训练图像
train_generator = train_datagen.flow_from_directory(
    train_dir,      #这是训练图像的源目录
    target_size = (150, 150),   #所有图像将调整为150 * 150
    batch_size = 20,
    #如果使用binary_crossentropy损失,需要二进制标签
    #class_mode = 'binary',
)
#使用val_datagen生成器按20个批次验证图像
validation_generator = val_datagen.flow_from_directory(
    val_dir,
```

```
    target_size = (150, 150),
    batch_size = 20,
    #class_mode = 'binary',
)
```

3. 网络结构定义

为实现人脸朝向分类，且要求网络结构简单，判别高效，通过对两个目标做权衡，在对比不同网络后，本项目选择了性价比更高的三层卷积神经网络，相关代码如下：

```
def cnn(input_shape = (150, 150, 3), classes = 3):
    #输入 150 * 150 像素的三通道 RGB 图像
    img_input = layers.Input(shape = input_shape)
    #第一层卷积
    x = layers.Conv2D(16, 3, activation = 'relu')(img_input)
    x = layers.MaxPooling2D(2)(x)
    #第二层卷积
    x = layers.Conv2D(32, 3, activation = 'relu')(x)
    x = layers.MaxPooling2D(2)(x)
    #第三层卷积
    x = layers.Convolution2D(64, 3, activation = 'relu')(x)
    x = layers.MaxPooling2D(2)(x)
    #展开图像成一维张量
    x = layers.Flatten()(x)
    #全连接层
    x = layers.Dense(512, activation = 'relu')(x)
    #0.5 是图 10 - 9 的随机损失率
    x = layers.Dropout(0.5)(x)
    #单节点输出层
    output = layers.Dense(classes, activation = 'softmax')(x)
    #定义与编译模型
    model = Model(img_input, output)
    return model
```

卷积神经网络结构如图 10-7 所示。

4. 模型训练

预处理数据集并定义好网络结构以后，将数据集输入神经网络中进行训练获得模型。模型训练输出结果如图 10-8 所示，相关代码如下：

```
checkpoint = ModelCheckpoint(       #检查点
    filepath = './tmp/direction_cnn_{epoch:03d}_{val_acc:.5f}.h5', monitor = 'val_acc', mode
= 'auto', save_best_only = 'True')       #文件路径
callbacks = [checkpoint]
model = cnn()
model.compile(loss = 'binary_crossentropy',
              optimizer = RMSprop(lr = 0.0001),
```

```
                        metrics = ['acc'])
        history = model.fit_generator(        #参数
            train_generator,
            steps_per_epoch = 20,
            epochs = 100,
            callbacks = callbacks,
            validation_data = validation_generator,
            validation_steps = 10)
```

Layer (type)	Output Shape	Param #
input_9 (InputLayer)	[(None, 150, 150, 3)]	0
conv2d_24 (Conv2D)	(None, 148, 148, 16)	448
max_pooling2d_24 (MaxPooling	(None, 74, 74, 16)	0
conv2d_25 (Conv2D)	(None, 72, 72, 32)	4640
max_pooling2d_25 (MaxPooling	(None, 36, 36, 32)	0
conv2d_26 (Conv2D)	(None, 34, 34, 64)	18496
max_pooling2d_26 (MaxPooling	(None, 17, 17, 64)	0
flatten_8 (Flatten)	(None, 18496)	0
dense_16 (Dense)	(None, 512)	9470464
dropout_8 (Dropout)	(None, 512)	0
dense_17 (Dense)	(None, 3)	1539

图 10-7　卷积神经网络结构

```
20/20 [==============================] - 9s 470ms/step - loss: 0.6001 - acc: 0.6733 - val_loss: 0.5726 - val_acc: 0.7300
Epoch 2/100
20/20 [==============================] - 7s 373ms/step - loss: 0.4939 - acc: 0.7583 - val_loss: 0.4622 - val_acc: 0.7616
Epoch 3/100
20/20 [==============================] - 8s 402ms/step - loss: 0.3680 - acc: 0.8575 - val_loss: 0.3399 - val_acc: 0.8481
Epoch 4/100
20/20 [==============================] - 8s 378ms/step - loss: 0.2819 - acc: 0.9058 - val_loss: 0.3057 - val_acc: 0.8650
Epoch 5/100
20/20 [==============================] - 8s 422ms/step - loss: 0.2309 - acc: 0.9275 - val_loss: 0.2549 - val_acc: 0.9051
```

图 10-8　模型训练输出结果

5. 模型评估与保存

通过评估模型得知,在测试集上的精度可达到 94.5%,Acc-Loss 图像如图 10-9 所示。

```
model.evaluate(validation_generator, verbose = 2)
acc = history.history['acc']
val_acc = history.history['val_acc']
# 为每一轮设置损失
loss = history.history['loss']
val_loss = history.history['val_loss']
# 获得训练轮次
epochs = range(len(acc))
```

```python
#每轮输出一次正确率图像
plt.plot(epochs, acc, label = 'acc')
plt.plot(epochs, val_acc, label = 'val_acc')
plt.legend()
plt.title('Training and validation accuracy')
plt.figure()
#每轮输出一次损失图像
plt.plot(epochs, loss, label = 'loss')
plt.plot(epochs, val_loss, label = 'val_loss')
plt.legend()
plt.title('Training and validation loss')
plt.figure()
#输出准确率-损失图像
plt.plot(epochs,acc,label = 'acc')
plt.plot(epochs,loss,label = 'loss')
plt.legend()
plt.title('Acc-Loss')
```

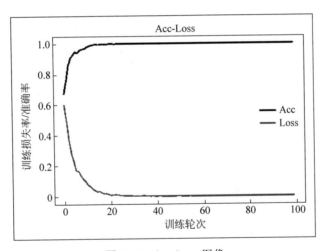

图 10-9　Acc-Loss 图像

10.3.4　人脸处理与表情包合成

得到裁剪人脸图像和朝向分类器后对其做不同的处理。

1. 人脸朝向检测

加载训练好的人脸朝向分类器,传入人脸图像做朝向判断,并回传判断结果:

```python
def recognizeDirection(pic_path, model):
    #识别人脸朝向
    classes = {1: 'straght', 2: 'left', 3: 'right'}
    image_path = pic_path
```

```python
        img = Image.open(image_path)
        img = img.resize((150, 150))
        img = np.expand_dims(img, axis = 0)
        result = model.predict(img, batch_size = 1)
        #print(result)
        result = result.tolist()
        mostLike = result[0].index(max(result[0])) + 1
        mostLikeInterpret = classes[mostLike]
        #print(mostLike)
    return mostLike, mostLikeInterpret
```

2. 表情包合成

基于人脸朝向识别步骤,选择合适的人脸蒙版遮盖脸外轮廓,再结合不同的光源环境选择不同的人脸五官处理方案以达到最佳效果:

```python
def draw(img, blur = 25, alpha = 1.0):
    #人脸图像明暗均衡处理
    #img1 = img.convert('L')
    img1 = img
    img2 = img1.copy()
    img2 = ImageOps.invert(img2)
    for i in range(blur):
        img2 = img2.filter(ImageFilter.BLUR)
    #img2 = img2.filter(MyGaussianBlur(radius = 1))       #高斯模糊
    width, height = img1.size
    for x in range(width):
        for y in range(height):
            a = img1.getpixel((x, y))
            b = img2.getpixel((x, y))
            img1.putpixel((x, y), min(int(a * 255/(256 - b * alpha)), 255))
    return img1
def emoticoning(self, filepath):
    #生成表情
    #global testMode
    os.mkdir(self.master.emoticonPath)
    face_path = os.path.join(self.master.rootPath,
                             'temp', 'faces')         #截取的人脸图片路径
    i = 0
    for image_name in os.listdir(face_path):
        image_path = os.path.join(face_path, image_name)
        #传入识别算法
        maskNum, maskType = recognizeDirection(
            image_path, self.master.selectModel)
        scheme = 4       #人脸图像五官提取方案选择
        if self.master.testMode:
            print(maskType)
```

```python
            maskPath = os.path.join(
                self.master.rootPath, 'materials', 'mask', 'maskTest{}.png'.format(maskNum))
        else:
            maskPath = os.path.join(
                self.master.rootPath, 'materials', 'mask', 'mask{}.png'.format(maskNum))
    image = Image.open(image_path).convert('L')          # 灰度图读取
    # 人脸图像五官提取方案
    if scheme == 1:
        image = image.filter(MyGaussianBlur(radius = 0.5))     # 高斯模糊
        image = image.filter(ImageFilter.CONTOUR)              # 轮廓检测
        image = image.filter(MyGaussianBlur(radius = 1))       # 高斯模糊
        image = ImageEnhance.Contrast(image).enhance(10.0)     # 图像增强
    elif scheme == 2:
        image = image.filter(MyGaussianBlur(radius = 0.7))     # 高斯模糊
        image = image.filter(ImageFilter.CONTOUR)              # 轮廓检测
        image = image.filter(MyGaussianBlur(radius = 0.5))     # 高斯模糊
        image = ImageEnhance.Contrast(image).enhance(15.0)     # 图像增强
        image = image.filter(MyGaussianBlur(radius = 0.5))     # 高斯模糊
    elif scheme == 3:
        image = image.filter(MyGaussianBlur(radius = 0.7))     # 高斯模糊
        image = image.filter(ImageFilter.EDGE_ENHANCE)         # 轮廓检测
        image = image.filter(MyGaussianBlur(radius = 0.5))     # 高斯模糊
        image = ImageEnhance.Contrast(image).enhance(15.0)     # 图像增强
        image = image.filter(MyGaussianBlur(radius = 0.5))     # 高斯模糊
    elif scheme == 4:
        image = draw(image)                                    # 明暗均衡处理
    image = image.resize((90, 90)).crop(
        (15, 20, 75, 80))     # 二次调整人脸大小及位置(可依据人脸方向进行偏向性调整)
    if maskPath:     # 添加人脸边缘蒙版
        image = image.convert('RGBA')
        mask = Image.open(maskPath).convert('RGBA')
        image.paste(mask, (0, 0, 60, 60), mask)
    # bg = random.randint(0, self.master.bgsize)      # 随机选择表情背景
    bg = 25
    back_ground = Image.open(os.path.join(
        self.master.bgPath, str(bg) + '.jpg'))
    back_ground.paste(image, (80, 60, 140, 120))
    save = os.path.join(self.master.emoticonPath,
                        'emoticon{}.png'.format(i))     # 裁剪后图片路径
    back_ground.save(save)
    i += 1
```

不同光源下的人脸五官特征处理方案对比如图 10-10 所示,在不同光源环境下,处理方案对五官轮廓的强度不同,各具优势。较暗的光线环境下,方案 1 和方案 2 无法呈现清晰的五官轮廓;而在较亮的环境下,方案 3 也丢失过多五官信息;并且在输入图像尺寸不同时,

由于人脸清晰度不同,标注并裁剪下来的图像质量不一,在处理质量低的小图像时,方案2更占优势。综合看,方案4对图像做了明暗均衡处理,虽然在耗时上比前三个方案更长,但是其对应用场景的普适性更强。

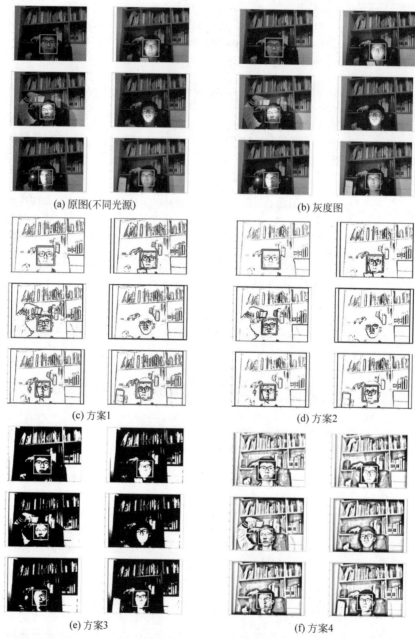

图 10-10　不同光源下人脸五官特征处理方案对比

10.4 系统测试

将图像输入系统中,开启测试模式,识别人脸位置并判断人脸朝向,选择合适的蒙版,系统效果演示如图10-11所示。

(a) 原图

(b) 放大图

(c) 选择蒙版

图 10-11 系统效果演示

10.4.1 确定运行环境符合要求

在运行本系统时,应满足一定的环境才能顺利运行:
- Windows 10;
- Python 3.7.x;
- OpenCV 3.4.2;
- Pillow 6.1.0;
- Numpy 1.16.4;
- Tensorflow 1.14.0;
- Keras 2.3.1;
- Tkinter 8.6.8。

若需使用摄像头模式,则需在带有摄像头的设备上运行本程序。

10.4.2 应用使用说明

打开 App,初始界面如图10-12所示,默认打开为照片模式。

界面菜单栏按钮依次为打开照片、摄像头模式、上一张、下一张、保存,并且配备键盘快捷键。

单击打开照片进行表情包制作,打开的照片在左下方导入照片处展示,生成的表情在右侧展示,如图10-13所示。

若一张照片中包含多个人脸,单击上一张和下一张可以切换展示不同人脸生成的表情,如图10-14所示,识别出的人脸依图像清晰度与人脸密集程度展现出质量差异。

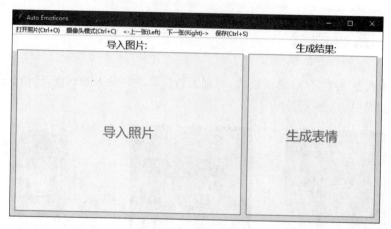

图 10-12　初始界面（照片模式）

图 10-13　打开照片

图 10-14　多人脸的切换展示

保存表情提示信息如图 10-15 所示。

图 10-15　保存表情提示信息

单击摄像头模式可以切换至使用摄像头生成实时表情，如图 10-16 所示。

图 10-16　摄像头模式

项目 11 AI 作曲

PROJECT 11

本项目基于 TensorFlow 开发环境使用 LSTM 模型,通过搜集 MIDI 文件,进行特征筛选和提取,训练生成合适的机器学习模型,实现 Magenta 原理,从而进行人工智能作曲。

11.1 总体设计

本部分包括系统整体结构图和系统流程图。

11.1.1 系统整体结构图

系统整体结构如图 11-1 所示。

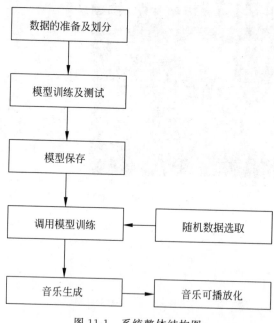

图 11-1 系统整体结构图

11.1.2 系统流程图

系统流程如图 11-2 所示。

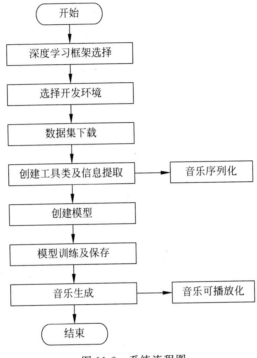

图 11-2 系统流程图

11.2 运行环境

本部分包括 Python 环境、虚拟机环境、TensorFlow 环境、Python 类库及项目软件。

11.2.1 Python 环境

需要 Python 2.7 及以上配置,推荐下载虚拟机在 Ubuntu 16.04 环境下运行代码。

11.2.2 虚拟机环境

安装 VirtualBox,下载地址为 https://www.virtualbox.org,选择 OS X hosts 下的 5.2.36 版本。

安装 Ubuntu 系统,下载地址为 https://ubuntu.com/download,版本号为 16.04 的 Ubuntu 系统为长期支持版(LTS)。下载 Ubuntu 后,需要下载 Ubuntu GNOME 桌面,网

站地址为 http://ubuntugnome.org。创建虚拟机,打开 VirtualBox,如图 11-3 所示。

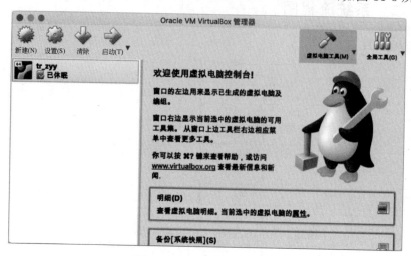

图 11-3　VirtualBox 主界面

单击"新建"按钮,出现如图 11-4 所示的对话框。

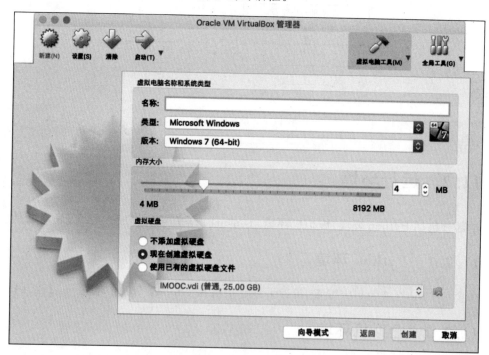

图 11-4　虚拟机创建界面

"名称"可自行定义;"类型"选择 Linux;"版本"选择 Ubuntu(64-bit);"内存大小"可自行设置,建议设置为 2048MB 及以上;"虚拟硬盘"选项选择默认选项,即"现在创建虚拟硬盘",之后单击"创建"按钮,在文件位置和大小对话框中将虚拟硬盘更改为 20GB,虚拟机映像文件创建完成。对该映像文件单击右键进行设置,单击"存储"按钮,如图 11-5 所示。

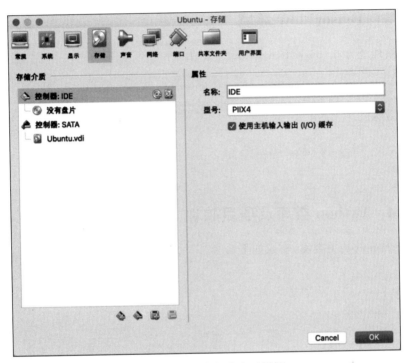

图 11-5　添加虚拟光盘对话框

依次选择没有盘片→分配光驱→选择一个虚拟光盘文件,添加下载好的 Ubuntu GnomeISO 镜像文件,单击 OK 按钮后,选择 install Ubuntu GNOME、Continue→Install Now→Continue→Continue,在 Keyboard layout 对话框中选择 Chinese,单击 Continue 按钮,等待安装完成后单击 Restart Now 按钮即可。

（1）进行 Ubuntu 的基本配置；

（2）打开 Terminal,安装谷歌输入法,输入命令：

```
sudo apt install fcitx fcitx-googlepinyin im-config
```

（3）安装 VIM,输入命令：

```
sudo apt install vim
```

（4）创建与主机共享的文件夹,输入命令：

```
mkdir share_folder
sudo apt install virtualbox-guest-utils
```

（5）创建主机文件夹 AIMM_Shared，建立主机与虚拟机的共享路径，输入命令：

```
sudo mount -t vboxsf AIMM_Shared home/share_folder
```

11.2.3 TensorFlow 环境

参考网站地址为 https://tensorflow.google.cn/install，打开 Terminal，输入命令：

```
sudo apt-get install python-pip python-dev python-virtualenv
virtualenv --system-site-packages tensorflow
source ~/tensorflow/bin/activate
easy_install -U pip
pip install --upgrade tensorflow
deactivate
```

11.2.4 Python 类库及项目软件

安装 Python 的相关类库，输入如下命令：

```
pip install numpy
pip install pandas
pip install matplotlib
sudo pip install keras
sudo pip install music21
sudo pip install h5py
sudo apt install ffmpeg
sudo apt install timidity
```

11.3 模块实现

本项目包括 5 个模块：数据预处理、信息提取、模型构建、模型训练及保存、音乐生成。下面分别给出各模块的功能介绍及相关代码。

11.3.1 数据预处理

数据来自于互联网下载的 70 首音乐文件（格式为 MIDI），如图 11-6 所示。百度网盘链接为 https://pan.baidu.com/s/1dQcdfXlSvDc0YIYLZ6UhSw，提取码为 e7sj。

11.3.2 信息提取

数据准备完成后，需要进行文件格式转换及音乐信息提取。

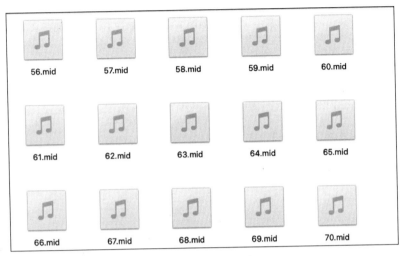

图 11-6　训练数据集

1. 文件格式转换

使用 Timidity 软件，实现将 MIDI 文件格式转换为 MP3 等其他流媒体格式的操作。

```
import os
import subprocess
import pickle  # 读取文件
import glob  # 读取文件,glob: 匹配所有符合条件的文件,并以 list 的形式返回
from music21 import converter, instrument, note, chord, stream
# 将神经网络生成的 MIDI 文件转成 MP3 文件
def convert_midi_to_mp3():
    input_file = 'output.mid'
    output_file = 'output.mp3'
    # 判断路径是否存在
    if not os.path.exists(input_file):
        raise Exception("MIDI 文件 {} 不在此目录下,请确保此文件被正确生成".format(input_file))
    print('将 {} 转换为 MP3'.format(input_file))
    # 用 timidity 把文件提取出来再用 ffmpeg 转成 MP3
    command = 'timidity {} -Ow -o - | ffmpeg -i - -acodec libmp3lame -ab 64k {}'.format(input_file, output_file)
    return_code = subprocess.call(command, shell=True)
    if return_code != 0:
        print('转换时出错,请查看出错信息')
    else:
        print('转换完毕. 生成的文件是 {}'.format(output_file))
# 从 music_midi 目录中的所有 MIDI 文件里提取 note(音符)和 chord(和弦)
# 确保包含所有 MIDI 文件的 music_midi 文件夹在所有 Python 文件的同级目录下
def get_notes():
```

```python
        if not os.path.exists("music_midi"):
            raise Exception("包含所有 MIDI 文件的 music_midi 文件夹不在此目录下,请添加")
    notes = []
    # glob:匹配所有符合条件的文件,并以 list 的形式返回
    for midi_file in glob.glob("music_midi/*.mid"):
        stream = converter.parse(midi_file)
        # converter 是 Music21 的一个类,parse 方法用于解析文件
        parts = instrument.partitionByInstrument(stream)
        # 如果有乐器部分,取第一个
        if parts:
            notes_to_parse = parts.parts[0].recurse()
        else:
            notes_to_parse = stream.flat.notes
        for element in notes_to_parse:
            # 如果是 Note 类型,那么取它的音调
            if isinstance(element, note.Note):
                # isinstance()函数来判断一个对象是否是已知的类型
                # 格式,例如: E6
                notes.append(str(element.pitch))
            # 如果是 Chord 类型,那么取它各个音调在映射表里对应的数字序号
            elif isinstance(element, chord.Chord):
                # 转换后格式,例如: 4.15.7
                notes.append('.'.join(str(n) for n in element.normalOrder))
    # 如果 data 目录不存在,创建此目录
    if not os.path.exists("data"):
        os.mkdir("data")
    # 将数据写入 data 目录下的 notes 文件
    with open('data/notes', 'wb') as filepath:
        pickle.dump(notes, filepath) # pickle.dump 用于管理文件
    return notes
# 用神经网络"预测"的音乐数据来生成 MIDI 文件,再转成 MP3 文件
def create_music(prediction):
    offset = 0    # 偏移,使添加的音符不会重叠,而是通过偏移让音符有先后顺序
    output_notes = []
    # 生成 Note(音符)或 Chord(和弦)对象
    for data in prediction:
        # 是 Chord 格式,例如: 4.15.7
        # 判断 data 是否含有".";isdigit()用于检测字符串是否只由数字组成
        if ('.' in data) or data.isdigit():
            notes_in_chord = data.split('.')
            notes = []
            for current_note in notes_in_chord:
                new_note = note.Note(int(current_note))
                new_note.storedInstrument = instrument.Piano()
                # 乐器用钢琴 (piano)
                notes.append(new_note)
            new_chord = chord.Chord(notes)
```

```python
            new_chord.offset = offset
            output_notes.append(new_chord)
        # 是 Note
        else:
            new_note = note.Note(data)
            new_note.offset = offset
            new_note.storedInstrument = instrument.Piano()
            output_notes.append(new_note)
        # 每次迭代都将偏移增加,这样才不会交叠覆盖
        offset += 0.5
# 创建音乐流(Stream)
midi_stream = stream.Stream(output_notes)
# 写入 MIDI 文件
midi_stream.write('midi', fp = 'output.mid')
# 将生成的 MIDI 文件转换成 MP3
convert_midi_to_mp3()
```

2. 音乐信息提取

需要将 MIDI 文件中的音符数据全部提取,包括 note 和 chord 的处理;note 是指音符,而 chord 是指和弦。所使用的软件是 Music21,它可以对 MIDI 文件进行数据提取或者写入。

```python
import os
from music21 import converter, instrument
def print_notes():
    if not os.path.exists("1.mid"):
        raise Exception("MIDI 文件 1.mid 不在此目录下,请添加")
    # 读取 MIDI 文件,输出 Stream 流类型
    stream = converter.parse("1.mid") # 解析 1.mid 的内容
    # 获得所有乐器部分
    parts = instrument.partitionByInstrument(stream)
    if parts:  # 如果有乐器部分,取第一个乐器部分,先采取一个音轨
        notes = parts.parts[0].recurse() # 递归获取
    else:
        notes = stream.flat.notes
    # 打印出每一个元素
    for element in notes:
        print(str(element))
if __name__ == "__main__":
    print_notes()
```

11.3.3 模型构建

数据加载后,需要进行定义模型结构、优化损失函数。

1. 定义模型结构

如图 11-7 所示为图形化的神经网络搭建模型,共 9 层,只使用 LSTM 的 70%,舍弃 30%,这是为了防止过拟合,最后全连接层的音调数就是初始定义 num_pitch 的数目,用神经网络去预测每次生成的新音调是所有音调中的哪一个,利用交叉熵和 Softmax(激活层)计算出概率最高那一个并作为输出(输出为预测音调对应的序列)。还需要在代码后面添加指定模型的损失函数和优化器设置。

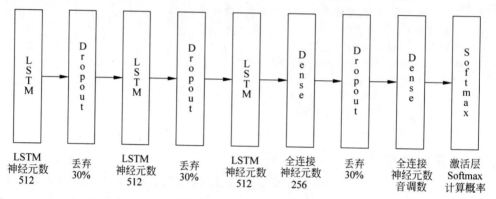

图 11-7 神经网络模型结构

```
# RNN - LSTM 循环神经网络
import tensorflow as tf
# 神经网络模型
def network_model(inputs, num_pitch, weights_file=None):
    model = tf.keras.models.Sequential()
```

首先构建一个神经网络模型(其中 Sequential 是序列的意思),在 TensorFlow 官网里可以看到基本用法,通过 add() 方法添加需要的层。Sequential 相当于一个汉堡模型,根据自己的需要按顺序填充不同层。

```
# 模型框架,第 n 层输出会成为第 n+1 层的输入,一共 9 层
    model.add(tf.keras.layers.LSTM(
        512,   # LSTM 层神经元的数目是 512,也是 LSTM 层输出的维度
        input_shape=(inputs.shape[1], inputs.shape[2]),
        # 输入的形状,对第一个 LSTM 层必须设置
        # return_sequences:控制返回类型
        # True:返回所有的输出序列
        # False:返回输出序列的最后一个输出
        # 在堆叠 LSTM 层时必须设置,最后一层 LSTM 可以不用设置
        return_sequences=True   # 返回所有的输出序列
    ))
    # 丢弃 30% 神经元,防止过拟合
    model.add(tf.keras.layers.Dropout(0.3))
    model.add(tf.keras.layers.LSTM(512, return_sequences=True))
```

```python
model.add(tf.keras.layers.Dropout(0.3))
model.add(tf.keras.layers.LSTM(512))
#return_sequences 是默认的 False,只返回输出序列的最后一个
#256 个神经元的全连接层
model.add(tf.keras.layers.Dense(256))
model.add(tf.keras.layers.Dropout(0.3))
model.add(tf.keras.layers.Dense(num_pitch))
#输出的数目等于所有不重复的音调数目:num_pitch
```

2. 优化损失函数

确定神经网络模型架构之后,需要对模型进行编译,这是回归分析问题,因此,需要用 Softmax 计算百分比概率,再用 Cross entropy(交叉熵)计算概率和对应的独热码之间的误差,使用 RMSProp 优化器优化模型参数。

```python
model.add(tf.keras.layers.Activation('softmax'))    #Softmax 激活函数算概率
    #交叉熵计算误差,使用 RMSProp 优化器
    #计算误差
    model.compile(loss = 'categorical_crossentropy', optimizer = 'rmsprop')
    #损失函数 loss,优化器 optimizer
    if weights_file is not None:    #如果是生成音乐
        #从 HDF5 文件中加载所有神经网络层的参数(Weights 权重)
        model.load_weights(weights_file)
    return model
```

11.3.4 模型训练及保存

构建完整模型后,在训练模型之前需要准备输入序列,创建一个字典,用于映射音调和整数,同样需要字典反向将整数映射成音调。除此之外,将输入序列的形状转成神经网络模型可接受的形式,输入归一化。本文前面在构建神经网络模型时定义了损失函数,是用布尔的形式计算交叉熵,所以要将期望输出转换成由 0 和 1 组成的布尔矩阵。

1. 模型训练

```python
import numpy as np
import tensorflow as tf
from utils import *
from network import *
#训练神经网络
def train():
    notes = get_notes()
    #得到所有不重复的音调数目
    num_pitch = len(set(notes))
    network_input, network_output = prepare_sequences(notes, num_pitch)
    model = network_model(network_input, num_pitch)
    filepath = "weights-{epoch:02d}-{loss:.4f}.hdf5"
```

在训练模型之前,需要定义一个检查点,其目的是在每轮结束时保存模型参数(weights),在训练过程中不会丢失模型参数,而且在对损失满意时随时停止训练。本文根据官方文件提供的示例格式设置了文件路径,不断更新保存模型参数 weights,文件的格式也提到过.hdf5。其中 checkpoint 中参数设置 save_best_only=Ture 是指监视器 monitor="loss"监视保存最好的损失,如果这次损失比上次损失小,则上次参数就会被覆盖。

```python
checkpoint = tf.keras.callbacks.ModelCheckpoint(
    filepath,           #保存的文件路径
    monitor = 'loss',   #监控的对象是损失(loss)
    verbose = 0,
    save_best_only = True,  #不替换最近数值最佳监控对象的文件
    mode = 'min'        #取损失最小的
)
callbacks_list = [checkpoint]
#用 fit()方法训练模型
model.fit(network_input, network_output, epochs = 100, batch_size = 64, callbacks = callbacks_list)
#为神经网络准备好训练的序列
def prepare_sequences(notes, num_pitch):
    sequence_length = 100    #序列长度
    #得到所有不重复音调的名字
    pitch_names = sorted(set(item for item in notes))  #sorted用于字母排序
    #创建一个字典,用于映射音调和整数
    pitch_to_int = dict((pitch,num) for num,pitch in enumerate(pitch_names))
    #enumerate 是枚举
    #创建神经网络的输入序列和输出序列
    network_input = []
    network_output = []
    for i in range(0, len(notes) - sequence_length, 1):
        #每隔一个音符就取前面的 100 个音符用来训练
        sequence_in = notes[i: i + sequence_length]
        sequence_out = notes[i + sequence_length]
        network_input.append([pitch_to_int[char] for char in sequence_in])
```

Batch size 是批次(样本)数目。它是一次迭代所用的样本数目。Iteration 是迭代,每次迭代更新一次权重(网络参数),每次权重更新需要 Batch size 个数据进行前向运算,再进行反向运算,一个 Epoch 指所有的训练样本完成一次迭代。

2. 模型保存

训练神经网络后,将参数(weight)存入 HDF5 文件。

```python
#把 sequence_in 里的每个字符转成数字后存入 network_input
        network_output.append(pitch_to_int[sequence_out])
    n_patterns = len(network_input)
    #将输入的形状转换成神经网络模型可以接受的形式
```

```python
    network_input = np.reshape(network_input,(n_patterns,sequence_length, 1))
    #将输入标准化/归一化
    #归一化可以让之后的优化器(optimizer)更快更好地找到误差最小值
    network_input = network_input / float(num_pitch)
    #将期望输出转换成{0, 1}组成的布尔矩阵,为配合误差算法使用
    network_output = tf.keras.utils.to_categorical(network_output)
    return network_input, network_output
if __name__ == '__main__':
    train()
```

11.3.5 音乐生成

该应用主要有序列准备、音符生成和音乐生成,有3种作用:①为神经网络准备好供训练的序列;②基于序列音符,用神经网络生成新的音符;③用训练好的神经网络模型参数作曲。

本文在训练模型时用fit()方法,模型预测数据时用predict()方法得到最大的维度,也就是概率最高的音符。将实际预测的整数转换成音调保存,输入序列向后移动,不断生成新的音调。

1. 序列准备

```python
def prepare_sequences(notes, pitch_names, num_pitch):
    #为神经网络准备好供训练的序列
    sequence_length = 100
    #创建一个字典,用于映射音调和整数
pitch_to_int = dict((pitch,num) for num, pitch in enumerate(pitch_names))
    #创建神经网络的输入序列和输出序列
    network_input = []
    network_output = []
    for i in range(0, len(notes) - sequence_length, 1):
        sequence_in = notes[i: i + sequence_length]
        sequence_out = notes[i + sequence_length]
        network_input.append([pitch_to_int[char] for char in sequence_in])
        network_output.append(pitch_to_int[sequence_out])
    n_patterns = len(network_input)
    #将输入的形状转换成神经网络模型可以接受的形式
normalized_input = np.reshape(network_input,(n_patterns sequence_length, 1))
    #将输入标准化/归一化
    normalized_input = normalized_input / float(num_pitch)
    return network_input, normalized_input
```

2. 音符生成

```python
def generate_notes(model, network_input, pitch_names, num_pitch):
    #基于一序列音符,用神经网络生成新的音符
```

```python
        # 从输入里随机选择一个序列，作为"预测"/生成音乐的起始点
        start = np.random.randint(0, len(network_input) - 1)
        # 创建一个字典，用于映射整数和音调
        int_to_pitch = dict((num, pitch) for num, pitch in enumerate(pitch_names))
        pattern = network_input[start]
        # 神经网络实际生成的音调
        prediction_output = []
        # 生成700个音符/音调
        for note_index in range(700):
            prediction_input = np.reshape(pattern, (1, len(pattern), 1))
            # 输入归一化
            prediction_input = prediction_input / float(num_pitch)
            # 用载入了训练所得最佳参数文件的神经网络预测/生成新的音调
            prediction = model.predict(prediction_input, verbose=0)
            # argmax 取最大的维度
            index = np.argmax(prediction)
            # 将整数转成音调
            result = int_to_pitch[index]
            prediction_output.append(result)
            # 向后移动
            pattern.append(index)
            pattern = pattern[1:len(pattern)]
        return prediction_output
if __name__ == '__main__':
    generate()
```

3. 音乐生成

```python
# 使用之前训练所得的最佳参数生成音乐
def generate():
    # 加载用于训练神经网络的音乐数据
    with open('data/notes', 'rb') as filepath:
        notes = pickle.load(filepath)
    # 得到所有音调的名字
    pitch_names = sorted(set(item for item in notes))
    # 得到所有不重复的音调数目
    num_pitch = len(set(notes))
    network_input, normalized_input = prepare_sequences(notes, pitch_names, num_pitch)
    # 载入之前训练时最好的参数文件，生成神经网络模型
    model = network_model(normalized_input, num_pitch, "best-weights.hdf5")
    # 用神经网络生成音乐数据
    prediction = generate_notes(model, network_input, pitch_names, num_pitch)
    # 用预测的音乐数据生成 MIDI 文件，再转换成 MP3
    create_music(prediction)
```

11.4 系统测试

本部分包括模型训练及测试效果。

11.4.1 模型训练

运行 python train.py 开始训练。默认训练 100 个 Epoch,可使用组合键 Ctrl+C 结束训练,测试过程如图 11-8 所示。

```
Epoch 27/400
42685/42685 [==============================]42685/42685 [==============================] - 1858s 44ms/step - loss: 4.5118
Epoch 28/400
42685/42685 [==============================]42685/42685 [==============================] - 1855s 43ms/step - loss: 4.4739
Epoch 29/400
42685/42685 [==============================]42685/42685 [==============================] - 1853s 43ms/step - loss: 4.3547
Epoch 30/400
42685/42685 [==============================]42685/42685 [==============================] - 1853s 43ms/step - loss: 4.2431
Epoch 31/400
42685/42685 [==============================]42685/42685 [==============================] - 1850s 43ms/step - loss: 4.1182
Epoch 32/400
42685/42685 [==============================]42685/42685 [==============================] - 1849s 43ms/step - loss: 3.9861
Epoch 33/400
42685/42685 [==============================]42685/42685 [==============================] - 1847s 43ms/step - loss: 3.8438
Epoch 34/400
42685/42685 [==============================]42685/42685 [==============================] - 1845s 43ms/step - loss: 3.6849
Epoch 35/400
42685/42685 [==============================]42685/42685 [==============================] - 1842s 43ms/step - loss: 3.5315
Epoch 36/400
42685/42685 [==============================]42685/42685 [==============================] - 1844s 43ms/step - loss: 3.3884
Epoch 37/400
42685/42685 [==============================]42685/42685 [==============================] - 1845s 43ms/step - loss: 3.2341
Epoch 38/400
42685/42685 [==============================]42685/42685 [==============================] - 1845s 43ms/step - loss: 3.0969
Epoch 39/400
42685/42685 [==============================]42685/42685 [==============================] - 1849s 43ms/step - loss: 2.9628
```

图 11-8 训练过程

当 Epoch 次数增加后,损失率越来越低,模型在训练数据、测试数据上的损失和准确率逐渐收敛,最终趋于稳定。

生成 MP3 音乐时,先生成 output.mid 这个 MIDI 文件,再从 output.mid 生成 output.mp3 文件——确保其位于 generate.py 同级目录下,运行 python generate.py 即可生成 MP3 音乐。

11.4.2 测试效果

生成结果如图 11-9 所示,output.mid 是直接生成的 MIDI 文件,output.mp3 是转换后的 MP3 流媒体格式文件。

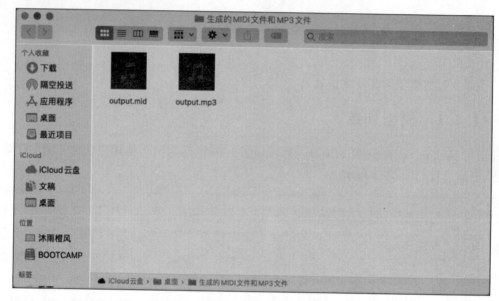

图 11-9 生成的 MIDI 文件和 MP3 文件

利用 Garage Band 尝试播放生成的音乐,如图 11-10 所示。

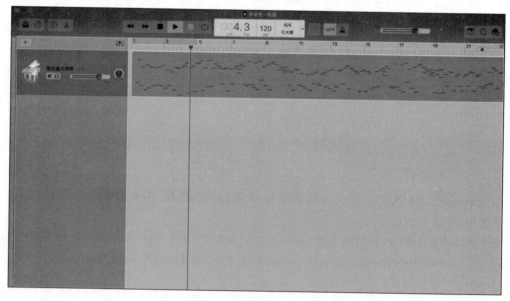

图 11-10 播放 output.mid

项目 12 智能作文打分系统

PROJECT 12

本项目基于 Kaggle 提供的 ASAP 数据集,构建 LSTM(Long Short Term Memory network)模型,实现对用户输入文章的分数预测。

12.1 总体设计

本部分包括系统整体结构图、系统流程图和前端流程图。

12.1.1 系统整体结构图

系统整体结构如图 12-1 所示。

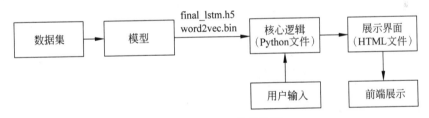

图 12-1 系统整体结构图

12.1.2 系统流程图

系统流程如图 12-2 所示。

12.1.3 前端流程图

前端流程如图 12-3 所示。

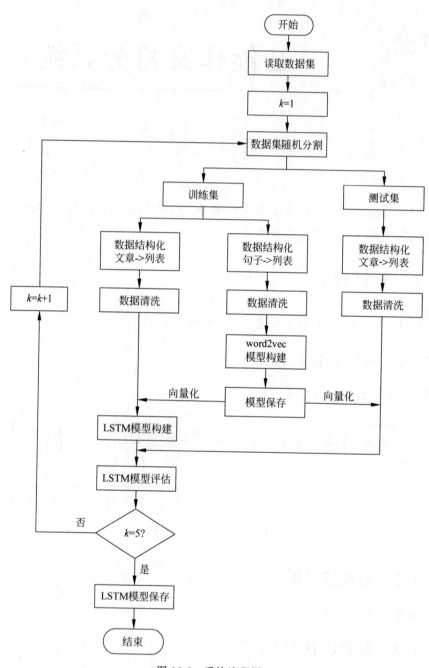

图 12-2 系统流程图

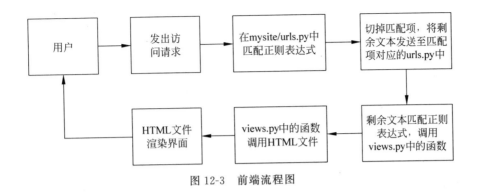

图 12-3　前端流程图

12.2　运行环境

本部分包括 Python 环境、Keras 环境和 Django 环境。

12.2.1　Python 环境

需要 Python 3.6 及以上配置，可以在 Windows 环境下载 Anaconda 完成 Python 所需的配置，下载地址为 https://docs.anaconda.com/anaconda/install/windows/。

打开 Anaconda Prompt，输入 conda list 查看已经安装的名称和版本号。若结果可以正常显示，则说明安装成功。

12.2.2　Keras 环境

创建 Python 3.6 的虚拟环境，名称为 auto_grade。打开 Anaconda Prompt，输入命令：

```
conda create -n auto_grade python=3.6
```

即可创建 auto_grade 虚拟环境。后面所有相关包的安装都依赖于该虚拟环境。在 Anaconda Prompt 中激活 TensorFlow 环境，输入命令：

```
activate auto_grade
```

安装 Keras 环境时，输入命令：

```
pip install keras
```

其他相关依赖包的安装方式和 Keras 类似，直接在虚拟环境中输入命令：

```
pip install package_name
```

安装完成后，可输入命令：

```
conda list
```

以检查是否安装成功。

12.2.3 Django 环境

Django 是基于 Python 的 Web 框架，可直接在 Anaconda Prompt 中使用 pip 安装，无需下载其他软件。

在 Anaconda Prompt 中输入命令：

```
python -m pip install Django
```

即可安装 Django 环境。若要验证 Django 是否能被 Python 识别，可以在 shell 中输入 Python，然后尝试导入 Django，输入命令：

```
>>> import django
>>> print(django.get_version())
3.0.4
```

若出现以上内容则说明 Django 成功安装，且可被 Python 识别。

12.3 模块实现

本项目包括 4 个模块：数据预处理、模型构建、模型训练及保存、模型测试。下面分别给出各模块的功能介绍及相关代码。

12.3.1 数据预处理

ASAP 数据集是 Kaggle 提供的作文评分数据集，包含 8 组不同题目的作文，共 12 978 篇，下载地址为 https://www.kaggle.com/c/asap-aes/data。其中每篇作文由两个评委打分，每组作文的评分标准不同，有不同的（最大、最小）分值。两个评委对作文的评分、文章的总得分、各组文章总得分进行归一化处理，使评分标准一致，分数为[0,10]内的整数。读取已下载的数据集并处理，生成 DataFrame 的相关代码如下：

```
#导入相应数据包
import os
import pandas as pd
import numpy as np
#文件目录
DATASET_DIR = './data/'
GLOVE_DIR = './glove.6B/'
SAVE_DIR = './'
#读取数据集
X = pd.read_csv(os.path.join(DATASET_DIR, 'training_set_rel3.tsv'), sep='\t', encoding='ISO-8859-1')    #读取文件
```

```
X = X.dropna(axis = 1)   #删除缺省的属性
X = X.drop(columns = ['rater1_domain1', 'rater2_domain1'])   #删除各评委的打分
#标签 y:文章分数(两位评委对文章的评分和)
y = X['domain1_score']
#各组文章最大分值
max_score = [12, 6, 3, 3, 4, 4, 30, 60]
#将不同组文章评分归一化到[0,10]
for i in range(r):
    for j in range(8):
        if X.iloc[i, 1] == j + 1:
            X.iloc[i, 3] = X.iloc[i, 3] /max_score[j]
```

DataFrame 前 5 行如图 12-4 所示，各属性分别为文章编号、所属题组、文章内容和综合评分。

essay_id	essay_set	essay	domain1_score
1	1	Dear local newspaper, I think effects computer...	8
2	1	Dear @CAPS1 @CAPS2, I believe that using compu...	9
3	1	Dear, @CAPS1 @CAPS2 @CAPS3 More and more peopl...	7
4	1	Dear Local Newspaper, @CAPS1 I have found that...	10
5	1	Dear @LOCATION1, I know having computers has a...	8

图 12-4 数据集 Dataframe 示意图

由于参与数据集较小，不同训练集和测试集的划分可能会使训练后模型参数产生变化，因此，使用 K 折交叉验证的方法将数据集划分为等长的 K 份，选取其中一份作为测试集，其他作为训练集，对不同的测试集进行 K 次训练，最后将 K 次训练模型的平均评价指标作为最终结果，相关代码如下：

```
#导入相关数据包
from sklearn.model_selection import KFold
#5 折交叉验证实例
cv = KFold(n_splits = 5, shuffle = True) #5 折交叉验证实例
results = []
y_pred_list = []
count = 1
#K 次划分数据集并训练
for traincv, testcv in cv.split(X): #将数据集划分成训练集和测试集,返回 5 组索引
    print("\n-------- Fold {} --------\n".format(count))
    #按索引划分训练集和测试集
    X_test, X_train, y_test, y_train = X.iloc[testcv], X.iloc[traincv], y.iloc[testcv], y.
iloc[traincv]
    train_essays = X_train['essay'] #输入 X:文章
    test_essays = X_test['essay']
```

深度学习模型通过对文章的词向量进行特征学习得到对应的评分。为了准备数据，需要对文章进行数据预处理，分为数据结构化、数据清洗、数据向量化三步。数据结构化是对每篇文章按词分隔，并存储在列表中，列表元素为词。数据清洗是去除文章中非英文字母的字符和停用词（即没有实际含义的功能词，如 but、your、this、a 等），并进行统一小写等处理。相关代码如下：

```python
#导入相应数据包
import nltk
nltk.download('stopwords') #下载停止词数据包
nltk.download('punkt') #下载分词工具
#将文章数据结构化和数据清洗存储在 clean_train_essays 列表中
clean_train_essays = []
for essay_v in train_essays:
        clean_train_essays.append(essay_to_wordlist(essay_v, remove_stopwords = True))
#清洗句子/文章，得到句子/文章的词列表
def essay_to_wordlist(essay_v, remove_stopwords):
    #去除非大小写字母以外的字符
    essay_v = re.sub("[^a-zA-Z]", " ", essay_v)
    #转化为小写，分词成词列表
    words = essay_v.lower().split()
    #去除停止符
    if remove_stopwords:
        stops = set(stopwords.words("english"))
        words = [w for w in words if not w in stops]
    return (words)
```

经过以上步骤得到了"干净"的文章数据，但要想输入 LSTM，还需将文章的词列表用数值向量表示。使用当前数据集训练 Word2Vec 模型，将文章数据输入训练好的 Word2Vec 模型中，得到词向量表示。Word2Vec 模型进行词向量化的依据是词在句子中的上下文关系，训练 Word2Vec 模型时，输入应为句子的词列表，将所有文章中的句子进行数据结构化和数据清洗，与之前操作类似，不同的是操作对象变成了句子，相关代码如下：

```python
#将句子数据结构化和数据清洗存储在 sentences 列表中
sentences = []
for essay in train_essays:
        sentences += essay_to_sentences(essay, remove_stopwords = True)
#将文章分句，并调用 essay_to_wordlist()对句子处理
def essay_to_sentences(essay_v, remove_stopwords):
    #加载英文划分句子的模型(英文句子特点:.之后有空格)
    tokenizer = nltk.data.load('tokenizers/punkt/english.pickle')
    #去除首尾的空格，得到句子列表
    raw_sentences = tokenizer.tokenize(essay_v.strip())
    sentences = []
    #调用 essay_to_wordlist()对句子进行数据结构化和数据清洗
    for raw_sentence in raw_sentences:
```

```
        if len(raw_sentence) > 0:
            sentences.append(essay_to_wordlist(raw_sentence,remove_stopwords))
    return sentences
```

Python中的Gensim工具包封装了Word2Vec模型的代码,直接将句子的词向量输入模型即可训练,Word2Vec模型训练和保存的相关代码如下:

```
#导入相应数据包
from gensim.models import Word2Vec
#设置Word2Vec模型的参数
num_features = 300 #特征向量的维度
min_word_count = 40 #最小词频,小于min_word_count的词被丢弃
num_workers = 4 #训练的并行数
context = 10 #当前词与预测词在一个句子中的最大距离
downsampling = 1e-3 #高频词汇随机降采样的配置阈值
#训练模型
print("Training Word2Vec Model...")
model = Word2Vec(sentences, workers = num_workers, size = num_features, min_count = min_word_
count, window = context, sample = downsampling)
#结束训练后锁定模型,使模型的存储更加高效
model.init_sims(replace = True)
#保存模型
model.wv.save_word2vec_format('word2vecmodel.bin', binary = True)
```

训练好Word2Vec模型后,对"干净"的文章数据进行预处理——数据向量化。将文章中的每个词输入Word2Vec模型,得到一个长度为300的向量,按位取平均,得到词向量。

```
trainDataVecs = getAvgFeatureVecs(clean_train_essays, model, num_features)
#对每个文章调用makeFeatureVec()向量化并合并文章向量
def getAvgFeatureVecs(essays, model, num_features):
    counter = 0
    #设置Numpy变量存储向量化的所有文章
    essayFeatureVecs = np.zeros((len(essays), num_features), dtype = "float32")
    #对每个文章调用makeFeatureVec()向量化
    for essay in essays:
        essayFeatureVecs[counter] = makeFeatureVec(essay,model,num_features)
        counter = counter + 1
    return essayFeatureVecs
#从文章的单词列表中制作特征向量
def makeFeatureVec(words, model, num_features):
    #设置Numpy变量存储向量化的文章
    featureVec = np.zeros((num_features,), dtype = "float32")
    num_words = 0.
    #训练集中留下的词列表
    index2word_set = set(model.wv.index2word)
    #将文章中每个词输入Word2Vec模型,得到各个词向量
    for word in words:
```

```
        if word in index2word_set:
            num_words += 1
            #将每个词向量按位相加
            featureVec = np.add(featureVec, model[word])#将每个词向量叠加
    #词向量为文章中各词向量的平均
    featureVec = np.divide(featureVec, num_words)
    return featureVec
```

最后,为确保数据结构符合模型要求,再次将数据的格式进行处理,相关代码如下:

```
#转换训练向量和测试向量为 Numpy 数组,提高运行效率
trainDataVecs = np.array(trainDataVecs)
#将训练向量和测试向量重塑为 3 维(1 代表一个时间步长)
trainDataVecs = np.reshape(trainDataVecs, (trainDataVecs.shape[0], 1, trainDataVecs.shape[1]))
```

对测试集进行同样的数据预处理操作(不参与训练 Word2Vec 模型),得到输入测试集 testDataVecs。

12.3.2 模型构建

数据加载到模型之后,定义模型结构,并优化损失函数和性能指标。模型的架构如下:2 个 LSTM 层,提取文章的特征;在其后连接丢弃层进行正则化,以防止模型过拟合;最后加一个全连接层,激活函数为 Relu,总结最终的评分结果。

文章评分是预测问题,使用均方误差作为损失函数。由于不同标签的样本数量不同,如 5～8 分的文章数较多,而 0～2 分的数量较少,因此,使用二次加权 Kappa 系数作为性能指标,表征分类结果与随机选取结果的差异程度,并且与样本数量无关,RMSProp 算法采用梯度下降的方法优化模型参数。

Keras 是 TensorFlow 高阶 API,其完全模块化和可扩展性使神经网络的代码更加简洁,因此,使用 Keras 创建模型。相关代码如下:

```
#引用相关的数据包
from keras.layers import Embedding, LSTM, Dense, Dropout, Lambda, Flatten
from keras.models import Sequential, load_model, model_from_config
import keras.backend as K
#构建 RNN 模型
def get_model():
    #定义顺序结构的模型
    model = Sequential()
    #第一层 LSTM 层
    model.add(LSTM(300, dropout = 0.4, recurrent_dropout = 0.4, input_shape = [1, 300], return_sequences = True))
    #第二层 LSTM 层
    model.add(LSTM(64, recurrent_dropout = 0.4))
    #丢弃层
```

```python
model.add(Dropout(0.5))
# 全连接层
model.add(Dense(1, activation = 'relu'))
# 对网络的学习过程进行配置,损失函数为均方误差,评价参数为平均绝对误差
model.compile(loss = 'mean_squared_error', optimizer = 'rmsprop', metrics = ['mae'])
model.summary()  # 输出模型各层的参数状况
return model
```

12.3.3 模型训练及保存

在定义模型架构和编译之后,通过训练集训练模型,使模型对文章打分。这里,使用训练集来拟合模型,并用测试集观察效果,最后保存模型。

1. 模型训练

```python
# 开始模型生成
lstm_model = get_model()
# 训练 LSTM 模型
lstm_model.fit(trainDataVecs, y_train, batch_size = 64, epochs = 25)
# 使用测试集预测模型输出
y_pred = lstm_model.predict(testDataVecs)
# 将预测值 y_pred 舍入到最接近的整数
y_pred = np.around(y_pred)
# 评估测试结果
result = cohen_kappa_score(y_test.values, y_pred, weights = 'quadratic')
```

将 10 381 个文章训练 40 次,并使用 5 折交叉验证训练划分数据集,将上述过程重复 5 次。每折训练结束后,用测试集的二次加权 Kappa 系数作为评价指标,如图 12-5 所示。从训练结果中看出,各折训练得到的评价指标都在 0.95 左右,波动不大,因此,判定训练集与测试集的划分对模型的参数影响极小。

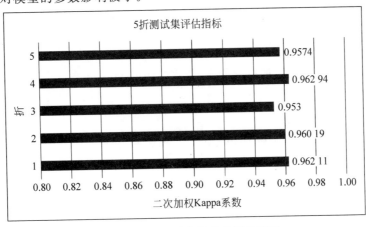

图 12-5 5 折测试集评估指标条形图

测试集上的评估指标为 0.9591。若 Kappa 系数大于 0.8，则模型预测较为准确。

2. 模型保存

使用 Keras 的 model.save() 函数直接保存模型，既保持了图的结构，又保存了参数。本文保存最后一折，即第 5 折的模型，保存后可以被重用，也可以移植到其他环境中使用。

```
#存储5个模型中最后一个(5折交叉验证训练)
if count == 5:
    lstm_model.save('./model_weights/final_lstm.h5')
```

12.3.4 模型测试

完成模型训练后，用 Web 前端展示训练结果，主要分为创建工程文件、应用程序交互界面设计和应用程序核心逻辑设计。

1. 创建工程文件

建立前端所需的工程文件，具体步骤如下。

1) 创建项目

在 Pycharm 中，进入存放项目代码的目录，运行以下命令：

```
django-admin startproject mysite
```

此命令创建名为 mysite 的项目，会自动创建一些文件，目录结构如下：

```
mysite/
└── manage.py
└── mysite/
    └── __init__.py
    └── settings.py
    └── urls.py
    └── asgi.py
    └── wsgi.py
```

其中：manage.py 为整个项目的控制程序，与它同级的 mysite 文件夹中存放了项目的具体设置。

2) 创建应用

一个项目内可以有多个应用，使用如下命令创建 grader：

```
py manage.py startapp grader
```

目录结构如下：

```
grader/
└── __init__.py
```

```
└──admin.py
└──apps.py
└──migrations/
    └──__init__.py
└──models.py
└──tests.py
└──views.py
```

其中,admin.py 为后台相关设置;apps.py 为 App 相关的设置;apps.py 为一个表单类,用来存放填入文章信息;migrations 为文件夹存放数据库内容,其中定义了数据库格式;models.py 定义了数据库模型;tests.py 为单元测试文件;views.py 为整个 App 的视图逻辑。

3) 创建文件或文件夹

(1) 创建 urls.py 文件,定义 App 内不同网页的 URL。

(2) 创建 deep_learning_files 文件夹,存放训练好的模型参数。

(3) 创建 static 文件夹,存放 css 和 js 静态文件。

(4) 创建 templates 文件夹,并在其中创建 3 个 HTML 文件,分别为 index.html、question.html 和 essay.html,描述网站主界面、写作界面及得分界面的展示形式。

(5) 创建 utils 文件夹,存放深度学习模型。在其中创建两个文件,分别为 model.py 和 helper.py,前者与训练时用的模型相同,后者描述数据处理中的一些参数。

(6) 在外层的 mysite 文件夹中创建 templates 的文件夹,其中创建 base.html 的 HTML 文件,此文件作为基础模板被 index.html、question.html 和 essay.html 扩展。

最终,前端部分的文件结构如下:

```
mysite
├── mysite
│   └── setting.py
│   └── urls.py
│   └── wsgi.py
├── templates
│   └── base.html
├── manage.py
├── grader
│   └── deep_learning_files
│   └── migrations
│       └── __init__.py
│   └── templates
│   └── static
```

```
|          └── utils
|                   └── model.py
|                   └── helper.py
|          └── admin.py
|          └── apps.py
|          └── forms.py
|          └── models.py
|          └── tests.py
|          └── urls.py
|          └── views.py
```

2. 应用程序交互界面设计

（1）在 base.html 文件中写入网站的通用格式。将这个文件作为基础模板，扩展到网站的每一页。也可以不使用 base.html，而是将此文件中的内容复制到每个 HTML 文件中，扩展的方式可以使代码看起来更加整洁，容易维护。

```html
<!-- 定义页面最上方导航链接 -->
<nav class="navbar navbar-expand-lg navbar-dark bg-dark">
    <div class="container">
        <a class="navbar-brand" href="{% url 'index' %}">Essays</a>
    </div>
</nav>
<!-- 模板扩展 -->
<div class="container">
    <ol class="breadcrumb my-4">
        {% block breadcrumb %}
        {% endblock %}
    </ol>
    {% block content %}
    {% endblock %}
</div>
```

（2）在 index.html 创建网站主页。主页将显示 6 个可选的题目，单击文章标题进入每个题目。当光标放在题目上时，该题目的背景颜色变为紫色，index.html 扩展 base.html 中的内容。

```html
<!-- 扩展 base 模板 -->
{% extends 'base.html' %}
<!-- 使用扩展的方式加载导航 -->
{% block breadcrumb %}
    <li class="breadcrumb-item active">Question Sets</li>
{% endblock %}
<!-- 加载题目列表 -->
```

```
{% block content %}
    {% if questions_list %}
        <p class="h3">Alright! Let's select a Question Set to start writing!</p>
        <table class="table">
            <thead>
                <tr>
                    <th scope="col">#</th><!--列标题-->
                    <th scope="col">Question</th>
                    <th scope="col">Min Score</th>
                    <th scope="col">Max Score</th>
                </tr>
            </thead>
            {% for question in questions_list %}
                <tr class="clickable-row" data-href='/{{question.set}}'>
                    <th scope="row">{{ question.set }}</th>
                    <td>{{ question.question_title|truncatewords:15 }}</td><!--只
截断显示前15个词-->
                    <td>{{ question.min_score }}</td><!--显示最小得分-->
                    <td>{{ question.max_score }}</td><!--显示最大得分-->
                </tr>
            {% endfor %}
            <tbody>
            </tbody>
        </table>
<!--放置光标的背景色和光标形状-->
        <style type="text/css">
            tr:hover {
                background-color: #cc99ff;
                cursor: pointer;
            }
        </style>
{% endblock %}
```

（3）在 question.html 文件中创建写作界面。此页面展示用户所选文章的题干，可输入文章并单击"提交作文"按钮，question.html 扩展了 base.html 中的内容。

```
<!--扩展base模板-->
{% extends 'base.html' %}
{% block content %}
<!--展示题目和题干-->
{% if question %}
    <h1>Question Set {{ question.set }}</h1><!--题目-->
    <p class="text-justify">{{ question.question_title }}</p><!--题干-->
    <form method="post" novalidate><!--提交时不验证-->
        {% csrf_token %}
        {% include 'includes/form.html' %}
        <button type="submit" class="btn btn-success">Grade Me!</button><!--提交按钮-->
```

```
</form>
```

（4）在 essay.html 中创建最终得分和用户输入文章的展示界面。essay.html 扩展了 base.html 中的内容。

```
<!-- 扩展 base 模板 -->
{% extends 'base.html' %}
{% block content %}
<!-- jumbotron 块用来展示文章得分 -->
    <div class="jumbotron">
        <h1 class="display-4">Grade {{ essay.score }}</h1>
        <p class="lead">Congratulations! Maximum Possible Score on this question is {{ essay.question.max_score }}</p>
        <hr class="my-4">
     <p>Your essay is graded by the magical power of neural networks.</p>
        <a class="btn btn-primary btn-lg" href="#" role="button">Learn more</a>
    </div>
<!-- container 块用来展示文章 -->
    <div class="container">
        <h2 class="display-4">Your Submission</h2>
        <p class="text-justify">{{ essay.content }}</p>
    </div>
{% endblock %}
```

3. 应用程序核心逻辑设计

Django 框架的核心逻辑是当用户请求网站的某个页面时，Django 将会载入 mysite/urls.py 模块。寻找 urlpatterns 的变量并且按序匹配正则表达式。找到匹配项后，切掉匹配的文本，将剩余文本发送至匹配项对应的 urls.py 文件，做进一步匹配。根据剩余文本匹配的内容，调用 views.py 中的函数。完成相应操作后，调用 templates 中的 HTML 文件渲染前端界面展示给用户。

（1）在 mysite/urls.py 中建立 urlpatterns 变量，并为其赋予正则表达式，使 URL 能通过它进入对应的 urls.py 文件。

```
urlpatterns = [
    path('', include('grader.urls')),  # 指向 app grader 中的 urls
    path('admin/', admin.site.urls),]
```

本项目只有一个 App，因此，每个 URL 都会被指向 grader/urls.py。

（2）在 grader/urls.py 中建立 urlpatterns 变量，并创建 3 条正则表达式，分别指向主页、写作页、得分页，3 个页面对应 views.py 中的函数。当用户请求的 URL 与其中一项匹配时，读取 views.py 中的对应函数，进行下一步的操作。

```
urlpatterns = [
# 匹配 url,对应视图函数
```

```python
path('', views.index, name = 'index'),
path('< int:question_id >/', views.question, name = 'question'),
path('< int:question_id >/essay< int:essay_id >/', views.essay, name = 'essay'),]
```

（3）在 views.py 中分别创建对应主页、写作页、得分页的函数，这三个函数最终的返回结果即为页面所对应的 HTML 文件。

（4）主页函数 index(request) 通过从数据库中提取题目数据，调用并将数据传给 index.html 文件后渲染页面。

```python
def index(request):
    #从 sql 中提取 questions 数据
    questions_list = Question.objects.order_by('set')
    context = {
        'questions_list': questions_list,
    }
    return render(request, 'grader/index.html', context)      #返回页面 index
```

（5）写作页函数 question(request，question_id) 从数据库中提取题目的数据，若用户没有在输入框中输入文章，则调用 question.html 显示写作界面，对应 else 部分：

```python
        else: #没有填写文章
            form = AnswerForm() #创建空表单实例
        context = {
            "question": question,
            "form": form,
        }
    #仍返回此页面
    return render(request, 'grader/question.html', context)
```

首先，单击 grade me 按钮，触发 post 请求，读取输入内容并进行处理。限定文章最低字数为 20，若字数不足，则判为 0 分；其次，构建 LSTM 模型并加载训练好的数据，预测文章得分。预测打分可能会大于满分、小于 0 分，将大于满分的数值都置为满分，将小于 0 分的数值都置为 0 分；最后，将文章和得分写入数据库中，URL 重定向到 essay 页面并调用 views.py 内的 essay() 函数渲染最终得分页面。

```python
def question(request, question_id):
    question = get_object_or_404(Question, pk = question_id)
    #提取 question_ID 的数据
    if request.method == 'POST':
        #创建一个表单实例，并使用请求中的数据填充它
        form = AnswerForm(request.POST)
        if form.is_valid():
            content = form.cleaned_data.get('answer')
            #读取 name 为 'answer' 的表单提交值
            if len(content) > 20: #文章长度大于 20
                num_features = 300
```

```python
        # 加载训练好的 word2vec 模型
        model = word2vec.KeyedVectors.load_word2vec_format(os.path.join(current_path,"deep_learning_files/word2vec.bin"),binary = True)
        # 处理 content,即输入的文章
        clean_test_essays = []
        clean_test_essays.append(essay_to_wordlist(content, remove_stopwords = True ))
        testDataVecs = getAvgFeatureVecs(clean_test_essays, model, num_features )
        testDataVecs = np.array(testDataVecs)
        testDataVecs = np.reshape(testDataVecs, (testDataVecs.shape[0], 1, testDataVecs.shape[1]))
        # 构建 LSTM 模型,并加载训练好的数据
        lstm_model = get_model()
        lstm_model.load_weights(os.path.join(current_path, "deep_learning_files/final_lstm.h5"))
        preds = lstm_model.predict(testDataVecs)  # 分数预测值
        if math.isnan(preds):  # 判断预测值是否有效
            preds = 0
        else:
            preds = np.around(preds)
        # 限定分数的最大最小值
        if preds < 0:
            preds = 0
        if preds > 10:
            preds = 10
    else:  # 若文章长度小于 20,分数判为 0
        preds = 0
    K.clear_session()
    # 将此文章数据写入
    essay = Essay.objects.create(
        content = content,
        question = question,
        score = preds
    )
    return redirect('essay', question_id = question.set, essay_id = essay.id)
```

(6) 得分页函数 essay(request、question_id、essay_id)通过从数据库中提取题目和用户所写文章的数据,调用并将数据传给 essay.html 文件后渲染页面。

```python
def essay(request, question_id, essay_id):
    # 提取 essay_id 的数据
    essay = get_object_or_404(Essay, pk = essay_id)
    context = {
        "essay": essay,
    }
    return render(request, 'grader/essay.html', context)  # 返回页面 essay
```

12.4 系统测试

本部分包括训练准确率、模型应用及测试效果。

12.4.1 训练准确率

测试二次加权 Kappa 值达到 0.95 及以上,这意味着预测模型训练比较成功。如果查看整个训练日志,会发现随着 epoch 次数的增多,模型在训练数据、测试数据上的损失和平均绝对误差逐渐收敛,最终趋于稳定,以第 5 折训练过程为例,图 12-6 和图 12-7 分别为模型训练损失图和模型测试集评价指标图。

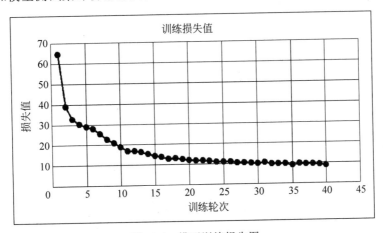

图 12-6 模型训练损失图

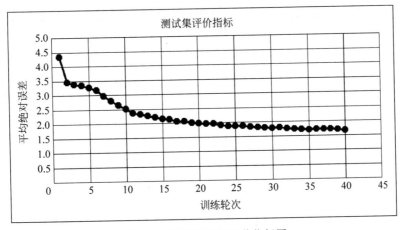

图 12-7 模型测试集评价指标图

12.4.2 模型应用

在 Anaconda Prompt 或 pycharm 的 terminal 中,进入 mysite 文件夹,依次运行:

python manage.py migrate
python manage.py runserver

在 http://127.0.0.1:8000/中运行 Web 端,如图 12-8 所示,展示了每个题目要求的前 15 词及每题分数的最大最小值,单击任意题目即可进入写作界面。

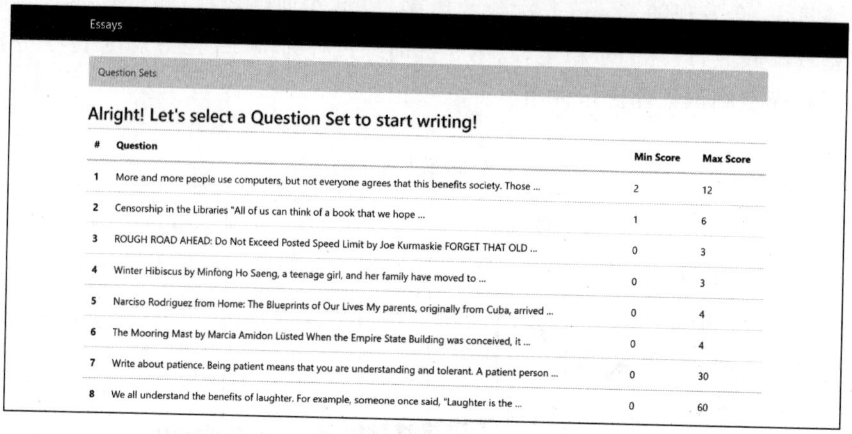

图 12-8 主页界面

写作界面如图 12-9 所示。其中,展示了作文题目要求,下方有文本输入框,用户可在输入框中写入作文,单击 Grade Me 按钮即可进入得分页查看分数。页面上方有导航栏,可单击 Home 按钮返回主界面。

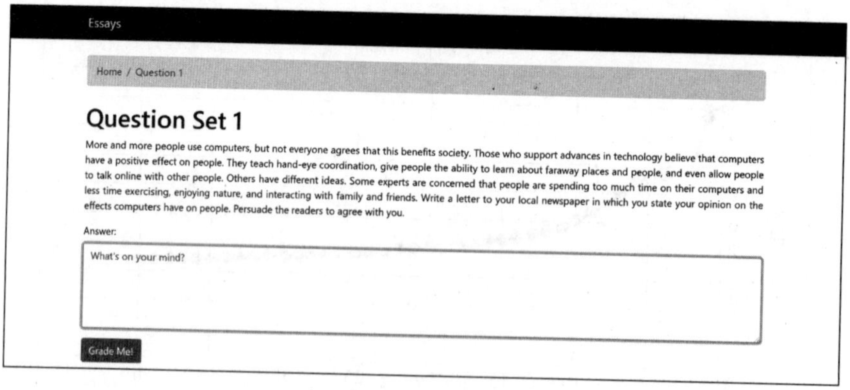

图 12-9 写作界面

得分界面如图 12-10 所示,展示用户得分及提交的文章。Learn more 按钮没有指定新的界面,将实现方法补充到网站中,帮助用户理解评分原理。

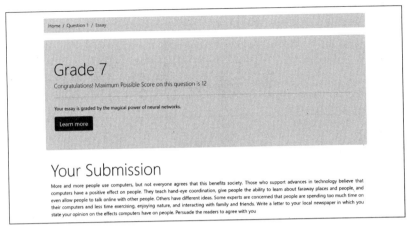

图 12-10　得分界面

12.4.3　测试效果

从数据集中取一篇文章,在 Web 端的测试结果,将预测得分与真实得分做对比,经过验证,模型对得分的预测比较准确,如图 12-11 所示,这篇文章的真实得分为 8 分。

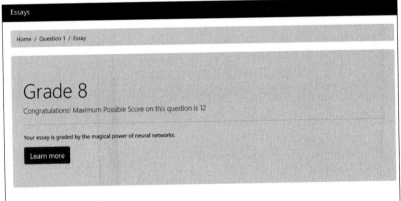

图 12-11　模型训练效果

项目 13　新冠疫情舆情监督

PROJECT 13

本项目基于循环神经网络的 LSTM 检测模型,实现微博谣言检测功能。

13.1　总体设计

本部分包括系统整体结构图和系统流程图。

13.1.1　系统整体结构图

系统整体结构如图 13-1 所示。

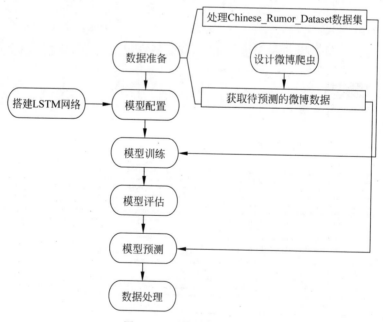

图 13-1　系统整体结构图

13.1.2 系统流程图

系统流程如图 13-2 所示。

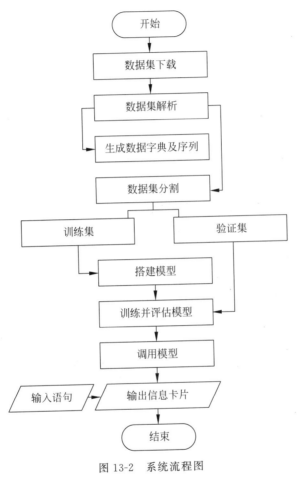

图 13-2 系统流程图

13.2 运行环境

本部分包括 Python 环境和 PaddlePaddle 环境。

13.2.1 Python 环境

需要 Python 3.6 及以上配置,在 Anaconda Python 3.7.0 环境下运行。

13.2.2 PaddlePaddle 环境

使用 Conda 安装的步骤如下

1. 创建虚拟环境,输入命令:

```
conda create -n paddle_env python=3.7
```

2. 添加清华源,输入命令:

```
conda config --add channels https://mirrors.tuna.tsinghua.edu.cn/anaconda/pkgs/free/
conda config --add channels https://mirrors.tuna.tsinghua.edu.cn/anaconda/pkgs/main/
conda config --add channels https://mirrors.tuna.tsinghua.edu.cn/anaconda/cloud/Paddle/
conda config --set show_channel_urls yes
```

3. 安装 PaddlePaddle(CPU 版本),输入命令:

```
pip install paddlepaddle
```

13.3 模块实现

本项目包括 5 个模块:准备预处理、模型构建、模型训练、模型评估和模型预测。下面分别给出各模块的功能介绍及相关代码。

13.3.1 准备预处理

本部分包括数据获取、数据预处理、定义数据和生成数据。

1. 数据获取

本部分包括获取已有数据和数据爬取。

1) 获取已有数据

GitHub 是开源数据集,下载地址为 https://github.com/thunlp/Chinese_Rumor_Dataset。

```
# 引用数据包含与微博原文相关的转发与评论信息,数据集中共包含谣言 1538 条和非谣言 1849 条
# @article{song2018ced,
# title = {CED: Credible Early Detection of Social Media Rumors},
# author = {Song, Changhe and Tu, Cunchao and Yang, Cheng and Liu, Zhiyuan and Sun, Maosong},
# journal = {arXiv preprint arXiv:1811.04175},
# year = {2018}
```

```python
#}
import zipfile
import os
import random
import json
src_path = "D:/_Projects/Rumor_Prediction/Chinese_Rumor_Dataset.zip"
#所下载数据集位置
target_path = "D:/_Projects/Rumor_Prediction/Chinese_Rumor_Dataset-master" #欲存储数据位置
if(not os.path.isdir(target_path)):
    z = zipfile.ZipFile(src_path, 'r')
    z.extractall(path = target_path) #对下载数据集进行解压
    z.close()
#保存路径
rumor_class_dirs = os.listdir(target_path + "/CED_Dataset/rumor-repost/")
non_rumor_class_dirs = os.listdir(target_path + "/CED_Dataset/non-rumor-repost/")
original_microblog = target_path + "/CED_Dataset/original-microblog/"
#谣言/非谣言标签
rumor_label = "0"
non_rumor_label = "1"
#谣言/非谣言总数
rumor_num = 0
non_rumor_num = 0
all_rumor_list = []
all_non_rumor_list = []
```

2) 数据爬取

数据爬取相关代码如下：

```python
import requests                                    #导入需要的模块
import codecs
from pyquery import PyQuery as pq
import time
#from pymongo import MongoClient
from urllib.parse import quote
headers = {
    'Host': 'm.weibo.cn',
    'User-Agent': 'Mozilla/5.0 (Macintosh; Intel Mac OS X 10_12_3) AppleWebKit/537.36 (KHTML, like Gecko) Chrome/58.0.3029.110 Safari/537.36',
    'X-Requested-With': 'XMLHttpRequest',
}
m = input('你想查找的内容:')                        #控制检索关键词
def get_page(page):                                #获取页面
    url = 'https://m.weibo.cn/api/container/getIndex?containerid=100103type%3D1%26q%3D' + quote(m) + '&page_type=searchall&page=' + str(page)
    try:
        response = requests.get(url, headers = headers)
```

```python
            if response.status_code == 200:
                return response.json()
        except requests.ConnectionError as e:    #异常处理
            print('Error', e.args)
def parse_page(json):                             #解析页面
    if json:
        items = json.get('data').get('cards')
        for i in items:
            groups = i.get('card_group')
            if groups == None:
                continue
            for item in groups:
                item = item.get('mblog')
                if item == None:
                    continue
                weibo = {}
                weibo['id'] = item.get('id')
                weibo['text'] = pq(item.get('text')).text()
                weibo['name'] = item.get('user').get('screen_name')
                if item.get('longText') != None :
                #微博分长文本与文本,较长的文本会显示不全,故要判断并抓取
                    weibo['longText'] = item.get('longText').get('longTextContent')
                else:
                    weibo['longText'] = None
                    print(weibo['text'])
                print(weibo['name'])
                if weibo['longText'] != None:
                    print(weibo['longText'])  #判断长文本是否为 None,如果是,不输出
                weibo['attitudes'] = item.get('attitudes_count')
                weibo['comments'] = item.get('comments_count')
                weibo['reposts'] = item.get('reposts_count')
                weibo['time'] = item.get('created_at')
                yield weibo
if __name__ == '__main__':                        #主函数
    for page in range(1,10):                      #循环页面
        json2 = get_page(page)
        results = parse_page(json2)
        for result in results:
            print(result)
            with codecs.open('d:\\weibodata.txt', mode = 'a', encoding = 'utf-8') as file_txt:
                file_txt.write(json.dumps(result))    #存储文件
```

以新冠肺炎为例,爬虫爬取结果如图 13-3 所示。

从网页获取的文本数据储存在 weibodata.txt 中,用于 RNN 网络的评估及预测。

2. 数据预处理

由于未将两组数据连接起来,以下对已有数据集以 CED_Dataset 处理为主。对预先分

```
你想查找的内容：新冠肺炎
你想查找多少页：10
新华视点
【定了！#中国将首次完全以网络形式举办广交会#】7日召开的国务院常务会议决定，第127
届广交会于6月中下旬在网上举办。这将是中国历史最为悠久的贸易盛会首次完全以网络形
式举办，实现中外客商足不出户下订单、做生意。
    当前，新冠肺炎疫情在全球蔓延，形势严峻。会议决定，广邀海内外客商在线展示产
品，运用先进信息技术，提供全天候网上推介、供采对接、在线洽谈等服务，打造优质特色
商品的线上外贸平台。（记者刘红霞、王攀）
11小时前
四平日报V
#扫黑除恶# #吉林新闻# 【双辽市郑佰文涉黑案一审获刑25年】3月31日，受新冠肺炎疫情
影响，铁东区人民法院通过远程视频一审公开宣判郑佰文、郑佰战、郑佰武、郑佰勇、郑龙
等26名被告人犯组织、领导、参加黑社会性质组织罪、寻衅滋事罪、聚众斗殴罪、妨害公务
罪等一案。法庭通过云审判信息平台与看守所远程连线，25名被告人在监所接受远程宣判，
1名取保候审涉黑人员在法庭接受宣判。主要涉黑成员郑佰文被判处有期徒刑25年，剥夺政
治权利5年，并处没收个人全部财产。
```

图 13-3 微博信息爬取结果

好类的谣言和非谣言数据分别进行解析，并计数统计是否有缺漏的情况。

```python
for rumor_class_dir in rumor_class_dirs:                         #解析谣言数据
    if(rumor_class_dir != '.DS_Store'):
        with open(original_microblog + rumor_class_dir, 'r', encoding = 'utf-8') as f:
            try:
                rumor_content = f.read()                         #打开文件
            except UnicodeDecodeError:                           #异常处理
                continue
            else:
                rumor_dict = json.loads(rumor_content)           #加载数据
                all_rumor_list.append(rumor_label + "\t" + rumor_dict["text"] + "\n")
                rumor_num += 1
for non_rumor_class_dir in non_rumor_class_dirs:                 #解析非谣言数据
    if(non_rumor_class_dir != '.DS_Store'):
        with open(original_microblog + non_rumor_class_dir, 'r', encoding = 'utf-8') as f2:
            try:
                non_rumor_content = f2.read()                    #打开文件
            except UnicodeDecodeError:                           #异常处理
                continue
            else:
                non_rumor_dict = json.loads(non_rumor_content)
                all_non_rumor_list.append(non_rumor_label + "\t" + non_rumor_dict["text"] + "\n")
                non_rumor_num += 1
print("谣言数据总量为：" + str(rumor_num))
print("非谣言数据总量为：" + str(non_rumor_num))
```

该数据集中所有的数据都得到了处理和统计。在上述代码中，为了正确读取文件，必须设置 encoding='utf-8'参数，否则会出现报错 UnicodeDecodeError。因此，做双重保险，用 try except else 语句进行异常处理。

经过整理后,文段被区分为谣言或非谣言,不利于后面的训练与验证,所以将所有的文段打乱顺序后作为最终的文字文件。

```python
data_list_path = "D:/_Projects/Rumor_Prediction/"        #乱序保存
all_data_path = data_list_path + "all_data.txt"
all_data_list = all_rumor_list + all_non_rumor_list
random.shuffle(all_data_list)
with open(all_data_path, 'w', encoding = 'utf-8') as f:   #打开文件
    f.seek(0)
    f.truncate()
with open(all_data_path, 'a', encoding = 'utf-8') as f:   #数据写入
    for data in all_data_list:
        f.write(data)
```

最终生成的文字文件如图 13-4 所示;其中 0 与 1 分别代表了谣言和非谣言,标注在每一段文字前。

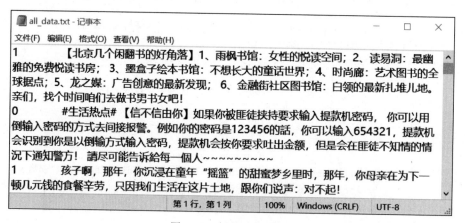

图 13-4 标签乱序结果

3. 定义数据

首先,定义数据字典的生成方式。剔除 all_data.txt 中对每段文字的标签,并整合所有文字;其次,分割每一个文字,并使其与特定的数字相对应。

```python
#数据字典
def create_dict(data_path, dict_path):
    dict_set = set()
    with open(data_path, 'r', encoding = 'utf-8') as f:
        lines = f.readlines()      #读取所有数据
    for line in lines:
        content = line.split('\t')[-1].replace('\n','')   #整合所有文字信息统一处理
        for s in content:
            dict_set.add(s)
    dict_list = []
```

```
    i = 0
    for s in dict_set:
        dict_list.append([s, i])  #使单字与数字相对应
        i += 1
    dict_txt = dict(dict_list)
    end_dict = {"<unk>": i}       #添加未知字符
    dict_txt.update(end_dict)
    with open(dict_path, 'w', encoding = 'utf-8') as f:
        f.write(str(dict_txt))
    print("数据字典生成完成!")
```

数据字典 dict.txt 的内容如图 13-5 所示。

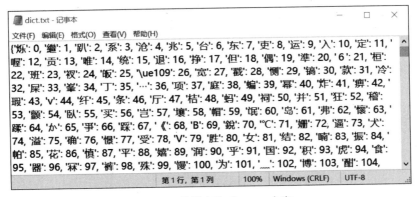

图 13-5　数据字典 dict.txt 内容

定义字典长度,为后续计算提供数据信息。

```
#字典长度
def get_dict_len(dict_path):
    with open(dict_path, 'r', encoding = 'utf-8') as f:
        line = eval(f.readlines()[0])
    return len(line.keys())
```

定义数据列表。当所有文字按照数据字典替换为数字后,按照 7∶1 的比例将数据集分为训练集和验证集。

```
#序列化表示数据
def create_data_list(data_list_path):
    with open(os.path.join(data_list_path, 'eval_list.txt'), 'w', encoding = 'utf-8') as f_eval:
        f_eval.seek(0)
        f_eval.truncate()  #清空 eval_list.txt
    with open(os.path.join(data_list_path, 'train_list.txt'), 'w', encoding = 'utf-8') as f_train:
        f_train.seek(0)
        f_train.truncate()  #清空 train_list.txt
```

```python
        with open(os.path.join(data_list_path, 'dict.txt'), 'r', encoding = 'utf-8') as f_data:
            dict_txt = eval(f_data.readlines()[0])  # 验证集词典
        with open(os.path.join(data_list_path, 'all_data.txt'), 'r', encoding = 'utf-8') as f_data:
            lines = f_data.readlines()  # 所有数据
        i = 0
        with open(
            os.path.join(data_list_path, 'eval_list.txt'), 'a', encoding = 'utf-8') as f_eval, \
open(  # 打开验证集数据
            os.path.join(data_list_path, 'train_list.txt'), 'a', encoding = 'utf-8') as f_train:
# 打开训练集数据
            for line in lines:
                words = line.split('\t')[-1].replace('\n', '')  # 分割
                label = line.split('\t')[0]
                labs = ""
                if i % 8 == 0:
                    for s in words:
                        lab = str(dict_txt[s])
                        labs = labs + lab + ','
                    labs = labs[:-1]
                    labs = labs + '\t' + label + '\n'
                    f_eval.write(labs)  # 作为验证集
                else:
                    for s in words:
                        lab = str(dict_txt[s])
                        labs = labs + lab + ','
                    labs = labs[:-1]
                    labs = labs + '\t' + label + '\n'
                    f_train.write(labs)  # 作为训练集
                i += 1
    print("数据列表生成完成!")
```

数据列表 train_list.txt 内容如图 13-6 所示, eval_list.txt 内容类似。

4. 生成数据

完成定义后, 生成数据字典和数据列表。

```
dict_path = data_list_path + "dict.txt"              # 词典路径
with open(dict_path, 'w') as f:
    f.seek(0)
    f.truncate()
create_dict(all_data_path, dict_path)                # 生成数据词典和列表
create_data_list(data_list_path)
```

13.3.2 模型构建

数据加载进模型之后, 需要定义数据读取工具、搭建模型和定义函数。

图 13-6　数据列表 train_list.txt 内容

1. 定义数据读取工具

使用 PaddlePaddle 框架中的 paddle.fluid.io.xmap_readers(mapper、reader、process_num、buffer_size、order=False)函数,其功能为多线程下,使用自定义映射器返回样本到输出队列。

```
#数据映射关系
def data_mapper(sample):
    data, label = sample
    data = [int(data) for data in data.split(',')]
    return data, int(label)
#数据读取器
def data_reader(data_path):
    def reader():
        with open(data_path, 'r', encoding = 'utf-8') as f:
            lines = f.readlines()
            for line in lines:
                data, label = line.split('\t')
                yield data, label
    return paddle.reader.xmap_readers(data_mapper, reader, cpu_count(), 1024) #多线程下,使用自定义映射器 reader 返回样本到输出队列
```

完成数据读取器的定义后,根据训练集和验证集数据分别生成对应的数据读取器。同时设置了一次训练所选取的样本数 BATCH_SIZE 为 128。

```python
# 获取训练数据读取器和测试数据读取器
BATCH_SIZE = 128
train_list_path = data_list_path + 'train_list.txt'
eval_list_path = data_list_path + 'eval_list.txt'
train_reader = paddle.batch(   # 读取训练集
        reader = data_reader(train_list_path),
        batch_size = BATCH_SIZE)
eval_reader = paddle.batch(    # 读取验证集
        reader = data_reader(eval_list_path),
        batch_size = BATCH_SIZE)
```

2. 搭建模型

选择 RNN 中的 LSTM 作为计算网络。PapplePaddle 框架中为 LSTM 网络提供了 fluid.layers.fc() 和 fluid.layers.dynamic_lstm() 两种函数，分别构建全连接层和实现 LSTM。

```python
# 定义长短期记忆网络
def lstm_net(ipt, input_dim):
    emb = fluid.layers.embedding(input = ipt, size = [input_dim, 128], is_sparse = True)
    fc1 = fluid.layers.fc(input = emb, size = 128)
    lstm1, cell = fluid.layers.dynamic_lstm(input = fc1,
            # 返回:隐藏状态 LSTM 的神经元状态
                                            size = 128)  # size = 4 * hidden_size
    fc2 = fluid.layers.sequence_pool(input = fc1, pool_type = 'max')
    lstm2 = fluid.layers.sequence_pool(input = lstm1, pool_type = 'max')
    out = fluid.layers.fc(input = [fc2, lstm2], size = 2, act = 'softmax')
    return out
```

输入网络的数据运用 fluid.data() 进行处理，其中 PaddlePaddle 框架中独具特色的部分 LoDTensor 也有所体现，即 lod_level 参数的设置，默认值 0 代表该数据为非序列数据，1 代表该数据为序列数据。

```python
# 定义数据
words = fluid.data(name = 'words', shape = [None,1], dtype = 'int64', lod_level = 1) # 指定输入数据为序列数据
label = fluid.data(name = 'label', shape = [None,1], dtype = 'int64')
# 默认非序列数据,有了上述的模型与数据,则可以得到 LSTM 的分类器。
# 获取字典长度,即类别数量
dict_dim = get_dict_len(dict_path)
# 获取 LSTM 的分类器
model = lstm_net(words, dict_dim)
```

3. 定义函数

判断所得到分类器实现效果的依据是损失函数和准确率。首先,运用 PaddlePaddle 框架中的 fluid.layers.cross_entropy() 和 fluid.layers.accuracy() 计算；其次,需要对模型进

行不断优化,初步设置学习率为 0.001,运用自适应梯度优化器即函数 fluid.optimizer.AdagradOptimizer()进行优化。

```
#获取性质函数:损失函数 & 准确率
cost = fluid.layers.cross_entropy(input = model, label = label)    #定义损失函数
avg_cost = fluid.layers.mean(cost)                                  #定义损失平均值
acc = fluid.layers.accuracy(input = model, label = label)           #定义准确率
#获取预测程序
test_program = fluid.default_main_program().clone(for_test = True) #复制一个主程序
#定义优化方法
optimizer = fluid.optimizer.AdagradOptimizer(learning_rate = 0.001)
opt = optimizer.minimize(avg_cost)                                  #定义优化方法
```

13.3.3 模型训练

定义模型架构和编译之后,使用训练集训练模型,使模型可以分辨言论是否为谣言。这里,使用训练集和测试集来拟合并保存模型。

1. 初始化参数设置

初始化分为执行初始化和数据初始化。在 PaddlePaddle 框架中,执行模型需使用执行器 fluid.Executor 来执行各类操作;且用 fluid.CUDAPlace 和 fluid.CPUPlace 指定执行地点。进行数据初始化时,需要考虑到训练集和验证集两类数据集,其中包括损失率和准确率的计算。

```
#执行初始化
use_cuda = False
place = fluid.CUDAPlace(0) if use_cuda else fluid.CPUPlace()
#定义执行地点,此处为混合设备运行
exe = fluid.Executor(place)  #创建执行器
exe.run(fluid.default_startup_program())  #进行初始化
#定义数据映射器
feeder = fluid.DataFeeder(place = place, feed_list = [words, label])
#数据初始化
all_train_iter = 0
all_train_iters = []
all_train_costs = []
all_train_accs = []
all_eval_iter = 0
all_eval_iters = []
all_eval_costs = []
all_eval_accs = []
```

2. 定义数据可视化

为了能够对模型性质进行分析,需要提前定义模型训练产生的参数与可视化图表。

```python
def draw_process(title,iters,costs,accs,label_cost,lable_acc):   # 画图参数
    plt.title(title, fontsize=24)
    plt.xlabel("iter", fontsize=20)
    plt.ylabel("cost/acc", fontsize=20)
    plt.plot(iters, costs,color='red',label=label_cost)
    plt.plot(iters, accs,color='green',label=lable_acc)
    plt.legend()
    plt.grid()
    plt.show()
```

3. 完成训练与验证模型

训练时运用已划分好的训练集和验证集分别对模型进行训练和验证。设置 EPOCH 为 10 轮，每训练一次就会对现有模型验证一次，并对每轮训练或验证计算损失率和准确率。

```python
EPOCH_NUM = 10  # 轮次及模型保存
model_save_dir = 'D:/_Projects/Rumor_Prediction/work/infer_model/'
for pass_id in range(EPOCH_NUM):                          # 训练
    for batch_id, data in enumerate(train_reader()):
        train_cost,train_acc = exe.run(program = fluid.default_main_program(),
                                       feed = feeder.feed(data),
                                       fetch_list = [avg_cost, acc])
        all_train_iter = all_train_iter + BATCH_SIZE
        all_train_iters.append(all_train_iter)
        all_train_costs.append(train_cost[0])    # 代价
        all_train_accs.append(train_acc[0])      # 准确度
        if batch_id % 100 == 0:
            print('Pass:%d, Batch:%d, Cost:%0.5f, Acc:%0.5f' % (pass_id, batch_id,
train_cost[0], train_acc[0]))
    eval_costs = []                                       # 验证
    eval_accs = []
    for batch_id, data in enumerate(eval_reader()):
        eval_cost, eval_acc = exe.run(program = test_program,
                                      feed = feeder.feed(data),
                                      fetch_list = [avg_cost, acc])
        eval_costs.append(eval_cost[0])
        eval_accs.append(eval_acc[0])
        all_eval_iter = all_eval_iter + BATCH_SIZE
        all_eval_iters.append(all_eval_iter)
        all_eval_costs.append(eval_cost[0])
        all_eval_accs.append(eval_acc[0])
    # 平均验证损失率和准确率
    eval_cost = (sum(eval_costs) / len(eval_costs))
    eval_acc = (sum(eval_accs) / len(eval_accs))
    print('Test:%d, Cost:%0.5f, ACC:%0.5f' % (pass_id, eval_cost, eval_acc))
if not os.path.exists(model_save_dir):                    # 保存模型
    os.makedirs(model_save_dir)
```

```
fluid.io.save_inference_model(model_save_dir,
                              feeded_var_names = [words.name],
                              target_vars = [model],
                              executor = exe)
print('训练模型保存完成!')
draw_process("train",all_train_iters,all_train_costs,all_train_accs,"trainning cost",
"trainning acc")
draw_process("eval",all_eval_iters,all_eval_costs,all_eval_accs,"evaling cost","evaling
acc")
```

13.3.4 模型评估

完成模型训练后绘制出训练结果,供模型性质分析。

```
draw_process("train",all_train_iters,all_train_costs,all_train_accs,"trainning cost",
"trainning acc")
draw_process("eval",all_eval_iters,all_eval_costs,all_eval_accs,"evaling cost","evaling
acc")
```

13.3.5 模型预测

将上述步骤中已经搭建好的模型运用于预测中。

1. 调用生成模型

调用生成模型时会使用到和模型训练验证时类似的程序。除此之外,还需结合数据准备实现序列的部分程序,将用户输入的文字按照数据字典转化为序列。

```
#创建执行器
place = fluid.CPUPlace()
infer_exe = fluid.Executor(place)
infer_exe.run(fluid.default_startup_program())
save_path = 'D:/_Projects/Rumor_Prediction/work/infer_model/'
#从模型中获取预测程序、输入数据名称列表、分类器
[infer_program, feeded_var_names, target_var] = fluid.io.load_inference_model(dirname =
save_path, executor = infer_exe)
#获取数据
def get_data(sentence):
    #读取数据字典
    with open('D:/_Projects/Rumor_Prediction/dict.txt', 'r', encoding = 'utf-8') as f_data:
        dict_txt = eval(f_data.readlines()[0])
    dict_txt = dict(dict_txt)
    #把字符串数据转换成列表数据
    keys = dict_txt.keys()
    data = []
    for s in sentence:
```

```
        # 判断是否存在未知字符
        if not s in keys:
            s = '<unk>'
        data.append(int(dict_txt[s]))
    return data
```

2. 输入需验证的信息

使用 input 语句与用户实现互动，用户可通过下方窗口输入需验证的信息。

```
data = []
question = input("您想验证的语句为:")
data_1 = np.int64(get_data(question)) # 创建执行器
place = fluid.CPUPlace()
infer_exe = fluid.Executor(place)
infer_exe.run(fluid.default_startup_program())
save_path = 'D:/_Projects/Rumor_Prediction/work/infer_model/'
# 从模型中获取预测程序、输入数据名称列表、分类器
[infer_program, feeded_var_names, target_var] = fluid.io.load_inference_model(dirname =
save_path, executor = infer_exe)
# 获取数据
def get_data(sentence):
    # 读取数据字典
    with open('D:/_Projects/Rumor_Prediction/dict.txt', 'r', encoding = 'utf-8') as f_data:
        dict_txt = eval(f_data.readlines()[0])
    dict_txt = dict(dict_txt)
    # 把字符串数据转换成列表数据
    keys = dict_txt.keys()
    data = []
    for s in sentence:
        # 判断是否存在未知字符
        if not s in keys:
            s = '<unk>'
        data.append(int(dict_txt[s]))
    return data
data.append(data_1)
# 获取每句话的单词数量
base_shape = [[len(c) for c in data]]
```

3. 定义输出格式

调用 OpenCV 库处理信息卡片原始模板，将用户欲验证的信息和计算结果输出至最终的信息卡片中，运用 PhotoShop 软件计算输出文字的位置信息。

```
import cv2
from PIL import Image, ImageDraw, ImageFont
def out_img(results):                    # 输出的参数定义
    img_OpenCV = cv2.imread("D:/_Projects/Rumor_Prediction/original.png")
```

```
img_PIL = Image.fromarray(cv2.cvtColor(img_OpenCV, cv2.COLOR_BGR2RGB))
font = ImageFont.truetype('simhei.ttf', 40)
fillColor_1 = (255, 255, 255)          #填充颜色
fillColor_2 = (64, 79, 105)
position_1 = (64, 46)                  #位置
position_2 = (64, 405)
draw = ImageDraw.Draw(img_PIL)
draw.text(position_1, question, font = font, fill = fillColor_1)
draw.text(position_2, results, font = font, fill = fillColor_2)
img_PIL.save('C:/Users/iris - /Desktop/prediction_results.jpg', 'jpeg')
```

4. 生成信息卡片

通过运行之前生成的数据信息得到计算结果。

```
#生成预测数据
tensor_words = fluid.create_lod_tensor(data, base_shape, place)
#执行预测
result = exe.run(program = infer_program,
                feed = {feeded_var_names[0]: tensor_words},
                fetch_list = target_var)
#分类名称
names = [ '谣言', '非谣言']
#获取结果概率最大的label
for i in range(len(data)):
    lab = np.argsort(result)[0][i][-1]
    prediction = '\n预测结果标签:' + str(lab) + ', 分类:' + str(names[lab]) + '\n\n概率:' + str(result[0][i][lab])
    out_img(prediction)
```

13.4 系统测试

本部分包括训练准确率、测试效果及模型应用。

13.4.1 训练准确率

训练准确率接近90%则意味着这个预测模型训练比较成功。通过查看训练日志,随着epoch次数的增多,模型在训练数据、测试数据上的损失和准确率逐渐收敛,最终趋于稳定,如图13-7所示。

13.4.2 测试效果

将测试集的数据代入模型进行测试,对分类的标签与原始数据进行显示和比较,测试准确率接近80%,如图13-8所示,虽然比训练准确率低,但足够判断大部分数据。

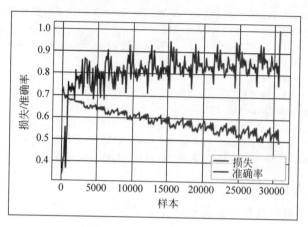

图 13-7　模型训练准确率

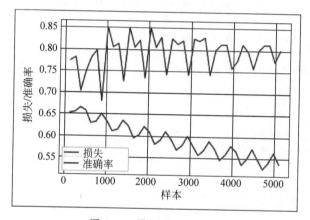

图 13-8　模型验证准确率

输入需运用模型进行判断的语句,这里以来自腾讯较真查证平台的谣言语句为例,选取非谣言语句如图 13-9 所示,非谣言语句结果如图 13-10 所示,选取谣言语句如图 13-11 所示,谣言语句结果如图 13-12 所示。

图 13-9　选取一条非谣言语句

图 13-10　模型测试非谣言语句结果

项目13　新冠疫情舆情监督

图13-11　选取一条谣言语句

图13-12　模型测试谣言语句结果

综上可知,模型可以完成新冠疫情谣言检测的功能。

13.4.3　模型应用

本部分包括程序使用说明和测试结果。

1. 程序使用说明

模型编译成功后,将已完成的模型与原始图放置于同一文件夹中。运行文件前,需修改模型 Rumor_Prediction 文件夹所在位置和希望生成的信息卡片保存的位置,如图13-13所示。

```
path = 'D:/_Projects/Rumor_Prediction/'   ## 需修改单引号中的地址为目前文件保存位置
path_pics = 'C:/Users/iris-/Desktop/'     ## 需修改单引号中的地址为,希望信息卡片所保存到的位置
```

图13-13　运行程序前需修改的两个地址参数

完成修改后,运行 Rumor_user.ipynb 文件,按照提示输入希望验证的语句信息,并得到生成的信息卡片。

2. 测试结果

本谣言检测可以提供基本判断,最终生成的信息卡片不仅呈现出待验证语句是否为谣言,还会提供可信概率供用户参考。图13-14中虽然为"非谣言",但是可信概率为0.51,较为中立,这是由于具体的判断时间不同导致的。

图13-14　测试结果示例

项目 14　语音识别——视频添加字幕

PROJECT 14

本项目通过 THCHS30 数据集进行 B-RNN 网络模型训练,实现对音频信息进行语音识别,并生成字幕文本。

14.1　总体设计

本部分包括系统整体结构图和系统流程图。

14.1.1　系统整体结构图

系统整体结构如图 14-1 所示。

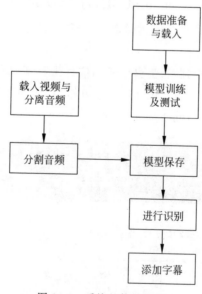

图 14-1　系统整体结构图

14.1.2 系统流程图

系统流程如图 14-2 所示。

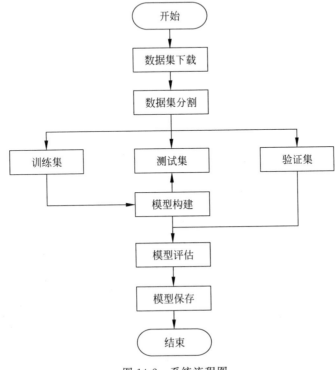

图 14-2 系统流程图

14.2 运行环境

硬件平台为 PC；操作系统为 Windows 10；编程平台为 Anaconda；开发环境为 TensorFlow；开发语言为 Python；数据集为公开的语料库样本 THCHS30。数据集下载地址为 http://www.openslr.org/18/。

14.3 模块实现

本项目包括 7 个模块：分离音频、分割音频、提取音频、模型构建、识别音频、添加字幕、GUI 界面。下面分别给出各模块的功能介绍及相关代码。

14.3.1 分离音频

用 MoviePy 库将音频从视频中分离,它是 Python 视频编辑库,可裁剪、拼接、标题插入、视频合成、视频处理和自定义效果。不安装 moviepy 视频编辑库可以直接使用 ffmpeg-python 库。相关代码如下:

```python
from moviepy.editor import *              #导入模块
def MovieToVoice(input,output):           #视频转为音频
    video = VideoFileClip(input)
    video.audio.write_audiofile(output)   #输出音频
input_str = r'视频地址'
output_str = r'音频地址'
MovieToVoice(input_str,output_str)
```

运行结果如图 14-3 所示。

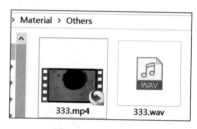

图 14-3　运行结果

14.3.2 分割音频

用静音检测方法切分音频,为分类奠定基础。音频打点切分完成后,将各音频片段保存。使用 pydub 库中的函数如下:

```
split_on_silence(soun,min_silence_len, silence_thresh, keep_silence=400)
```

第一个参数为待分割音频,第二个为多少秒"没声"代表沉默,第三个为分贝小于多少 dBFS 时代表沉默,第四个为截出的每个音频添加多少 ms 无声。相关代码如下:

```python
from pydub import AudioSegment            #导入模块
from pydub.silence import split_on_silence
def Split(str1):  #分割
    sound = AudioSegment.from_mp3(str1)
    loudness = sound.dBFS
    #print(loudness)
    chunks = split_on_silence(sound,
        #沉默半秒
        min_silence_len=10,
                #声音小于-16dBFS 认为沉默
        silence_thresh=-52,
        keep_silence=0
        )
    print('总分段:', len(chunks))
    #根据后续的情况选择
    '''
        # 放弃长度小于2秒的录音片段
        for i in list(range(len(chunks)))[::-1]:
```

```
            if len(chunks[i]) <= 2000 or len(chunks[i]) >= 10000:
                chunks.pop(i)
    print('取有效分段(大于2s小于10s):', len(chunks))
    '''
    #将分割的音频保存
    for i, chunk in enumerate(chunks):
        str = "chunk{0}.wav".format(i)
        chunk.export(str, format = "wav")
```

结果如图14-4所示。

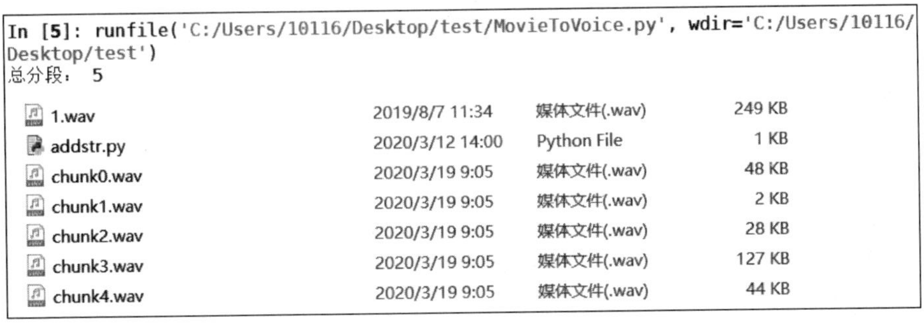

```
In [5]: runfile('C:/Users/10116/Desktop/test/MovieToVoice.py', wdir='C:/Users/10116/
Desktop/test')
总分段： 5
 1.wav          2019/8/7 11:34    媒体文件(.wav)   249 KB
 addstr.py      2020/3/12 14:00   Python File     1 KB
 chunk0.wav     2020/3/19 9:05    媒体文件(.wav)   48 KB
 chunk1.wav     2020/3/19 9:05    媒体文件(.wav)   2 KB
 chunk2.wav     2020/3/19 9:05    媒体文件(.wav)   28 KB
 chunk3.wav     2020/3/19 9:05    媒体文件(.wav)   127 KB
 chunk4.wav     2020/3/19 9:05    媒体文件(.wav)   44 KB
```

图14-4 代码运行成功示意图

14.3.3 提取音频

MFCC提取过程如下：对语音进行预减轻、分帧和加窗；每个短时分析窗通过FFT去掉对应的频谱；将上面的频谱通过梅尔滤波器组去掉梅尔频谱；在梅尔频谱上面进行倒谱分析(先取对数，然后做逆变换，逆变换通过DCT离散余弦变换来实现，取DCT后的2～13作为MFCC系数)，取得梅尔频率倒谱系数MFCC，就是这帧语音的特征。

将语音数据转换为需要计算13位或26位不同倒谱特征的MFCC，作为模型的输入。经过转换，数据存储在一个频率特征系数(行)和时间(列)的矩阵中。相关代码如下：

```
#将音频信息转成MFCC特征
#参数说明 --- audio_filename:音频文件,numcep:梅尔倒谱系数个数
#numcontext:对于每个时间段,要包含上下文样本个数
def audiofile_to_input_vector(audio_filename, numcep, numcontext):
    #加载音频文件
    fs, audio = wav.read(audio_filename)
    #获取MFCC系数
    orig_inputs = mfcc(audio, samplerate = fs, numcep = numcep)
    #打印MFCC系数的形状,得到(955, 26)的形状
    #955表示时间序列,26表示每个序列MFCC的特征值
    #形状因文件而异,不同文件可能有不同长度的时间序列,但是,每个序列的特征值数量都相同
    print(np.shape(orig_inputs))
```

使用双向循环神经网络训练,输出包含正、反向的结果,相当于每一个时间序列扩大一倍,为了保证总时序不变,使用 orig_inputs = orig_inputs[::2]对 orig_inputs 每隔一行进行一次取样。这样被忽略的序列可以用后文中反向 RNN 生成的输出代替,维持总的序列长度。

```python
orig_inputs = orig_inputs[::2] #(478, 26)
print(np.shape(orig_inputs))
#实际使用 numcontext = 9
#返回数据,考虑前9个和后9个时间序列,每个时间序列组合 19 * 26 = 494 个 MFCC 特征数
train_inputs = np.array([], np.float32)
train_inputs.resize((orig_inputs.shape[0], numcep + 2 * numcep * numcontext))
print(np.shape(train_inputs)) #(478, 494)
#准备修复前修复后上下文
empty_mfcc = np.array([])
empty_mfcc.resize((numcep))
#使用过去和将来的上下文准备 train_inputs
#time_slices 保存的是时间切片,也就是有多少个时间序列
time_slices = range(train_inputs.shape[0])
#context_past_min 和 context_future_max 用来计算哪些序列需要补零
context_past_min = time_slices[0] + numcontext
context_future_max = time_slices[-1] - numcontext
#开始遍历所有序列
for time_slice in time_slices:
    #对前9个时间序列的 MFCC 特征补 0,不需要补零的,则直接获取前9个时间序列的特征
    need_empty_past = max(0, (context_past_min - time_slice))
    empty_source_past = list(empty_mfcc for empty_slots in range(need_empty_past))
    data_source_past = orig_inputs[max(0, time_slice - numcontext):time_slice]
    assert(len(empty_source_past) + len(data_source_past) == numcontext)
    #对后9个时间序列的 MFCC 特征补 0,不需要补零的,则直接获取后9个时间序列的特征
    need_empty_future = max(0, (time_slice - context_future_max))
    empty_source_future = list(empty_mfcc for empty_slots in range(need_empty_future))
    data_source_future = orig_inputs[time_slice + 1:time_slice + numcontext + 1]
    assert(len(empty_source_future) + len(data_source_future) == numcontext)
    #前9个时间序列的特征
    if need_empty_past:
        past = np.concatenate((empty_source_past, data_source_past))
    else:
        past = data_source_past
    #后9个时间序列的特征
    if need_empty_future:
        future = np.concatenate((data_source_future, empty_source_future))
    else:
        future = data_source_future
    #将前9个时间序列、当前时间序列以及后9个时间序列组合
    past = np.reshape(past, numcontext * numcep)
    now = orig_inputs[time_slice]
```

```
        future = np.reshape(future, numcontext * numcep)
        train_inputs[time_slice] = np.concatenate((past, now, future))
        assert(len(train_inputs[time_slice]) == numcep + 2 * numcep * numcontext)
    # 将数据使用正太分布标准化,减去均值再除以方差
    train_inputs = (train_inputs - np.mean(train_inputs)) / np.std(train_inputs)
    return train_inputs
```

14.3.4 模型构建

本部分包括定义模型结构、优化损失函数、模型训练。

1. 定义模型结构

网络模型使用 3 个 1024 节点的全连接层网络,经过 Bi-RNN 网络,最后再连接两个全连接层,且都带有 dropout 层。激活函数使用带截断的 Relu,截断值设置为 20。

模型的结构变换如下:由于输入数据结构是 3 维,首先,将它变成 2 维,传入全连接层;其次,全连接层到 Bi-RNN 网络时,转成 3 维;再次,转成 2 维,传入全连接层;最后,将 2 维转成 3 维的输出。具体过程如下:

```
[batch_size, amax_stepsize, n_input + (2 * n_input * n_context)]
[amax_stepsize * batch_size, n_input + 2 * n_input * n_context]
[amax_stepsize, batch_size, 2 * n_cell_dim]
[amax_stepsize * batch_size, 2 * n_cell_dim]
```

相关代码如下:

```
def BiRNN_model(batch_x, seq_length, n_input, n_context, n_character, keep_dropout):
    # batch_x_shape: [batch_size, amax_stepsize, n_input + 2 * n_input * n_context]调试代码
    batch_x_shape = tf.shape(batch_x)
    # 将输入转成时间序列优先
    batch_x = tf.transpose(batch_x, [1, 0, 2])
    # 再转成 2 维传入第一层
    # [amax_stepsize * batch_size, n_input + 2 * n_input * n_context]
    batch_x = tf.reshape(batch_x, [-1, n_input + 2 * n_input * n_context])
    # 使用 RELU 激活和退出
    # 第 1 层
    with tf.name_scope('fc1'):
        b1 = variable_on_cpu('b1', [n_hidden_1], tf.random_normal_initializer(stddev = b_stddev))
        h1 = variable_on_cpu('h1', [n_input + 2 * n_input * n_context, n_hidden_1],
                             tf.random_normal_initializer(stddev = h_stddev))
        layer_1 = tf.minimum(tf.nn.relu(tf.add(tf.matmul(batch_x, h1), b1)), relu_clip)
        layer_1 = tf.nn.dropout(layer_1, keep_dropout)
    # 第 2 层
    with tf.name_scope('fc2'):
        b2 = variable_on_cpu('b2', [n_hidden_2], tf.random_normal_initializer(stddev = b_
```

```python
                stddev))
            h2 = variable_on_cpu('h2', [n_hidden_1, n_hidden_2], tf.random_normal_initializer
(stddev = h_stddev))
            layer_2 = tf.minimum(tf.nn.relu(tf.add(tf.matmul(layer_1, h2), b2)), relu_clip)
            layer_2 = tf.nn.dropout(layer_2, keep_dropout)
        # 第 3 层
        with tf.name_scope('fc3'):
            b3 = variable_on_cpu('b3', [n_hidden_3], tf.random_normal_initializer(stddev = b_
stddev))
            h3 = variable_on_cpu('h3', [n_hidden_2, n_hidden_3], tf.random_normal_initializer
(stddev = h_stddev))
            layer_3 = tf.minimum(tf.nn.relu(tf.add(tf.matmul(layer_2, h3), b3)), relu_clip)
            layer_3 = tf.nn.dropout(layer_3, keep_dropout)
        # 双向 RNN
        with tf.name_scope('lstm'):
            # 前向
            lstm_fw_cell = tf.contrib.rnn.BasicLSTMCell(n_cell_dim, forget_bias = 1.0, state_is_
tuple = True)
            lstm_fw_cell = tf.contrib.rnn.DropoutWrapper(lstm_fw_cell,
                                                        input_keep_prob = keep_dropout)
            # 反向
            lstm_bw_cell = tf.contrib.rnn.BasicLSTMCell(n_cell_dim, forget_bias = 1.0, state_is_
tuple = True)
            lstm_bw_cell = tf.contrib.rnn.DropoutWrapper(lstm_bw_cell,
                                                        input_keep_prob = keep_dropout)
            # 第 3 层[amax_stepsize, batch_size, 2 * n_cell_dim]
            layer_3 = tf.reshape(layer_3, [-1, batch_x_shape[0], n_hidden_3])
            outputs, output_states = tf.nn.bidirectional_dynamic_rnn(cell_fw = lstm_fw_cell,
                                                                    cell_bw = lstm_bw_cell,
                                                                    inputs = layer_3,
                                                                    dtype = tf.float32,
                                                                    time_major = True,
                                                                    sequence_length = seq_length)
            # 连接正反向结果[amax_stepsize, batch_size, 2 * n_cell_dim]
            outputs = tf.concat(outputs, 2)
            # 单个张量[amax_stepsize * batch_size, 2 * n_cell_dim]
            outputs = tf.reshape(outputs, [-1, 2 * n_cell_dim])
        with tf.name_scope('fc5'):
            b5 = variable_on_cpu('b5', [n_hidden_5], tf.random_normal_initializer(stddev = b_
stddev))
            h5 = variable_on_cpu('h5', [(2 * n_cell_dim), n_hidden_5], tf.random_normal_
initializer(stddev = h_stddev))
            layer_5 = tf.minimum(tf.nn.relu(tf.add(tf.matmul(outputs, h5), b5)), relu_clip)
            layer_5 = tf.nn.dropout(layer_5, keep_dropout)
        with tf.name_scope('fc6'):
            # 全连接层用于 softmax 分类
            b6 = variable_on_cpu('b6', [n_character], tf.random_normal_initializer(stddev = b_
```

```
            stddev))
        h6 = variable_on_cpu('h6', [n_hidden_5, n_character], tf.random_normal_initializer
(stddev = h_stddev))
        layer_6 = tf.add(tf.matmul(layer_5, h6), b6)
        # 将 2 维[amax_stepsize * batch_size, n_character]转成 3 维 time - major [amax_
stepsize, batch_size, n_character].
        layer_6 = tf.reshape(layer_6, [-1, batch_x_shape[0], n_character])
        print('n_character:' + str(n_character))
        # 输出维度[amax_stepsize, batch_size, n_character]
        return layer_6
```

2. 优化损失函数

语音识别属于时序分类任务，使用 ctc_loss 函数计算损失。而优化器使用梯度下降法 AdamOptimizer，设置学习率为 0.001。相关代码如下：

```
# 使用 ctc_loss 计算损失
avg_loss = tf.reduce_mean(ctc_ops.ctc_loss(targets, logits, seq_length))
# 优化器
learning_rate = 0.001
optimizer = tf.train.AdamOptimizer(learning_rate = learning_rate).minimize(avg_loss)
```

3. 模型训练

音频文件直接使用 train 文件夹下的数据，翻译用 data 文件夹下的数据，音频文件是 XX.wav，对应的翻译文件则是 XX.wav.trn，先找出所有 train 文件夹下的音频文件，再找 data 文件夹下音频文件名.trn 后缀的翻译文件，取第一行作为翻译内容，将音频文件和翻译的内容一一对应，加载到内存中，方便使用。相关代码如下：

```
# 迭代次数
epochs = 100
# 模型保存地址
savedir = "saver/"
# 如果该目录不存在,则新建
if os.path.exists(savedir) == False:
    os.mkdir(savedir)
# 生成 saver
saver = tf.train.Saver(max_to_keep = 1)
# 创建 session
with tf.Session() as sess:
    # 初始化
    sess.run(tf.global_variables_initializer())
    # 如果没有模型,重新初始化
    kpt = tf.train.latest_checkpoint(savedir)
    print("kpt:", kpt)
    startepo = 0
    if kpt != None:
```

```python
        saver.restore(sess, kpt)
        ind = kpt.find("-")
        #读取上次运行到哪一次epoch,这次直接跳过
        startepo = int(kpt[ind + 1:])
        print(startepo)
#准备运行训练步骤
section = '\n{0:=^40}\n'
n_batches_per_epoch = int(np.ceil(len(labels) / batch_size))
print(section.format('Run training epoch'))
    train_start = time.time()
for epoch in range(epochs): #样本集迭代次数
    epoch_start = time.time()
    #跳过之前运行过的epoch
    if epoch < startepo:
        continue
    print("epoch start:", epoch, "total epochs = ", epochs)
    print("total loop ", n_batches_per_epoch, "in one epoch,", batch_size, "items in one loop")
    train_cost = 0
    train_ler = 0
    next_idx = 0
            #读取上次运行到了哪个batch值,这次直接跳过
    last_batch_file = open(r'E:\文件夹\大三下\信息系统设计\视频加字幕\batch.txt', 'r')
    last_batch = last_batch_file.read()
    print('Last batch:' + last_batch)
    last_batch = int(last_batch)
    #判断是否已经执行完一个epoch,是则需要手动修改last batch值
    if last_batch == 1249:
        last_batch = -1
    last_batch_file.close()
    for batch in range(n_batches_per_epoch): #每次batch_size,取多少
            #跳过之前运行过的batch
            if batch <= last_batch:
                continue
        #取数据
        print('开始获取数据:' + str(batch))
        next_idx, source, source_lengths, sparse_labels = next_batch(wav_files, labels, next_idx, batch_size)
        print('数据获取结束')
        feed = {input_tensor: source, targets: sparse_labels, seq_length: source_lengths,
                keep_dropout: keep_dropout_rate}
        #计算avg_loss
        print('开始训练模型')
        batch_cost, _ = sess.run([avg_loss, optimizer], feed_dict=feed)
        print('模型训练结束')
        train_cost += batch_cost
        #模型保存,每batch都保存
        saver.save(sess, savedir + "saver.cpkt", global_step=epoch)
```

```
                last_batch_file = open(r'E:\文件夹\大三下\信息系统设计\视频加字幕\batch.txt','w')
                last_batch_file.write(str(batch))
                last_batch_file.close()
        epoch_duration = time.time() - epoch_start
        log = 'Epoch {}/{}, train_cost: {:.3f}, train_ler: {:.3f}, time: {:.2f} sec'
        print(log.format(epoch, epochs, train_cost, train_ler, epoch_duration))
    train_duration = time.time() - train_start
    print('Training complete, total duration: {:.2f} min'.format(train_duration / 60))
```

14.3.5 识别音频

使用训练好的模型,对分割后的音频文件依次识别,相关代码如下:

```
# 识别模块
decoded_str = []
with tf.Session() as sess:
    # 初始化
    sess.run(tf.global_variables_initializer())
    # 如没有模型,则重新初始化
    kpt = tf.train.latest_checkpoint(savedir)
    startepo = 0
    if kpt != None:
        saver = tf.train.import_meta_graph(meta_path)
        saver.restore(sess, kpt)
        ind = kpt.find("-")
        startepo = int(kpt[ind + 1:])
    for file in get_wav_files(split_dir):
        source, source_lengths, sparse_labels = get_speech_file(file, labels)
        feed2 = {input_tensor: source, targets: sparse_labels, seq_length: source_lengths,
keep_dropout: 1.0}
        d, train_ler = sess.run([decoded[0], ler], feed_dict = feed2)
        dense_decoded = tf.sparse_tensor_to_dense(d, default_value = -1).eval(session = sess)
        if (len(dense_decoded) > 0):
            decoded_str.append(ndarray_to_text_ch(dense_decoded[0], words))
```

14.3.6 添加字幕

需要参数如下:开始时间、结束时间、字幕文本、序号。按 srt 文件格式编辑内容,写入 srt 文件,然后关闭文件,将 srt 文件和同名的视频文件放在同一个文件夹,用迅雷播放器进行播放,可以显示字幕。同时生成一个字幕文本内容的 txt 文本,结果如图 14-5 所示。相关代码如下:

```
# 生成字幕文件模块
# import csv
```

```
subtitle_outcome = ""
txt_out = ""
for i in range(0,len(BeginTime)):
    subtitle_outcome += str(i) + "\n" + str(BeginTime[i]) + " --> " + str(EndTime[i]) + "\n" + str(decoded_str[i][0:20]) + "\n\n"
    txt_out += str(decoded_str[i][0:20]) + "\n"
with open(subtitle_path, 'w') as f:
    f.write(subtitle_outcome)
f.close()
with open(txt_path, 'w') as t:
    t.write(txt_out)
```

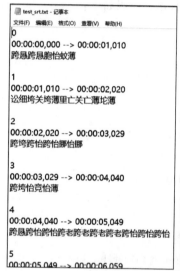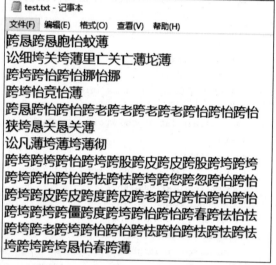

图 14-5　代码运行结果示意图

14.3.7　GUI 界面

加载文件路径，实现添加字幕功能，最后提示生成.str 文件。在 GUI 的.py 文件中，程序提取用户输入的文件路径，调用语音识别的.py 文件，将生成的.str 文件加载到视频中，最后用户得到添加好字幕的视频文件。GUI 界面如图 14-6 所示，运行结果如图 14-7 所示。

相关代码如下：

```
#!/usr/bin/env python
# coding: utf-8
import tkinter as tk  #导入模块
from tkinter import filedialog
from tkinter import *
import os
from PIL import ImageTk, Image
```

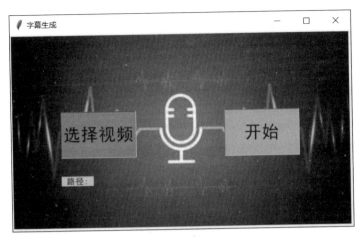

图 14-6　GUI 界面

图 14-7　运行结果

```
root = Tk()
root.title('字幕生成')
# 背景
canvas = tk.Canvas(root, width = 541, height = 300, bd = 0, cursor = 'circle')
imgpath = 'C:/Users/Administrator/Pictures/Camera Roll/test.jpg'
img = Image.open(imgpath)
photo = ImageTk.PhotoImage(img)
canvas.create_image(0, 0, anchor = NW, image = photo)
canvas.pack()
# 获取视频路径
def get_vedio_path():
    # 设置文件对话框会显示的文件类型
    my_filetypes = [('all files', '.*'), ('text files', '.txt')]
    # 请求选择文件
```

```python
            global vedio_path
            vedio_path = filedialog.askopenfilename(parent = root,
                                                    initialdir = os.getcwd(),
                                                    title = "Please select a file:",
                                                    filetypes = my_filetypes)
        #vedio路径显示
        r_path_show = tk.Label(root,text = vedio_path,fg = 'black',font = ("黑体",10))
        r_path_show.config(bg = 'lightcyan',bd = 0,height = 1,width = 55)
        canvas.create_window(130, 220,anchor = NW,window = r_path_show)
    #生成字幕
    def get_srt():
        #调用语音识别程序,生成字幕文件
        os.system("python vedio_subtitle.py %s" % (vedio_path))
        #srt生成成功显示
        global srt_path_show
        srt_path_show = tk.Label(root,text = '字幕文件生成成功',fg = 'black',font = ("黑体",15))
        srt_path_show.config(bg = 'lime',bd = 0,height = 2,width = 16)
        canvas.create_window(200, 250,anchor = NW,window = srt_path_show)
    #选择视频
    vedio_select = tk.Button(text = "选择视频",command = get_vedio_path,fg = 'black',font = ("黑体",20))
    vedio_select.config(bg = 'cornflowerblue',bd = 1,relief = 'groove',height = 2,width = 8,cursor = 'circle',activebackground = 'cornflowerblue', activeforeground = 'red')
    #显示视频路径
    vedio_show = tk.Label(root,text = '路径:',fg = 'black',font = ("黑体",10))
    vedio_show.config(bg = 'lightblue',bd = 0,height = 1,width = 7)
    #开始生成字幕文件
    subtitle_generate = tk.Button(text = "开始",command = get_srt,fg = 'black',font = ("黑体",20))
    subtitle_generate.config(bg = 'turquoise',bd = 1,relief = 'groove',height = 2,width = 8,cursor = 'circle',activebackground = 'turquoise', activeforeground = 'red')
    #将各部件放置指定位置
    canvas.create_window(80, 120,anchor = NW, window = vedio_select)
    canvas.create_window(80, 220,anchor = NW, window = vedio_show)
    canvas.create_window(340, 120,anchor = NW,window = subtitle_generate)
    #窗口运行
    root.mainloop()
```

14.4 系统测试

全数据集迭代 12 次模型时的错误率为 0.872,测试结果如图 14-8 所示;视频分离音频文件如图 14-9 所示;音频分离文件如图 14-10 所示;字幕文件如图 14-11 所示,字幕内容如图 14-12 所示;字幕文本内容如图 14-13 所示。

```
INFO:tensorflow:Restoring parameters from ./saver\saver.cpkt-12
Decoded: 这 月 恳 员 员 蓬 的
0.8367347
Decoded: 这 国  队 用 礼
0.877551
Decoded: 他 闹   生
0.8979592
Decoded: 这 年 唧  雀 嶂 培 惯 了
0.8367347
Decoded: 他 玩  子 惯 况
0.877551
Decoded: 这 韵  酷 渠 了
0.8979592
Decoded: 这 年   蓬 子 击 了
0.8367347
Decoded: 这 霞    礼
0.877551
Decoded: 他 玩   等
0.8979592
Decoded: 这 日  尼 营 礼
0.877551
Decoded: 这 启 韵 业 训  了
0.877551
error rate:0.8719851591370322
```

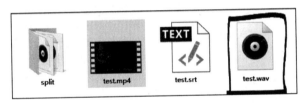

图 14-8　数据集迭代 12 次测试结果

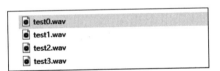

图 14-9　视频分离音频文件

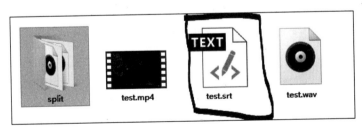

图 14-10　音频分离文件

图 14-11　字幕文件

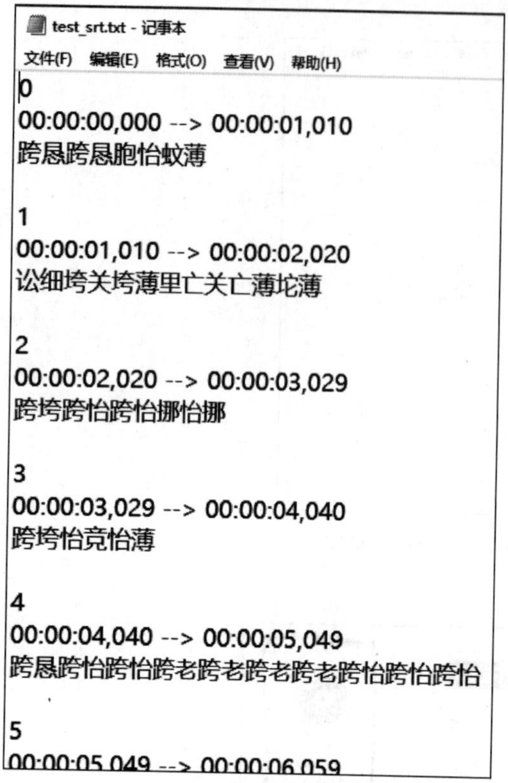

```
test_srt.txt - 记事本
文件(F) 编辑(E) 格式(O) 查看(V) 帮助(H)
0
00:00:00,000 --> 00:00:01,010
跨恳跨恳胞怡蚊薄

1
00:00:01,010 --> 00:00:02,020
讼细垮关垮薄里亡关亡薄坨薄

2
00:00:02,020 --> 00:00:03,029
跨垮跨怡跨怡挪怡挪

3
00:00:03,029 --> 00:00:04,040
跨垮怡竟怡薄

4
00:00:04,040 --> 00:00:05,049
跨恳跨怡跨怡跨老跨老跨老跨怡跨怡

5
00:00:05,049 --> 00:00:06,059
```

图 14-12 字幕内容

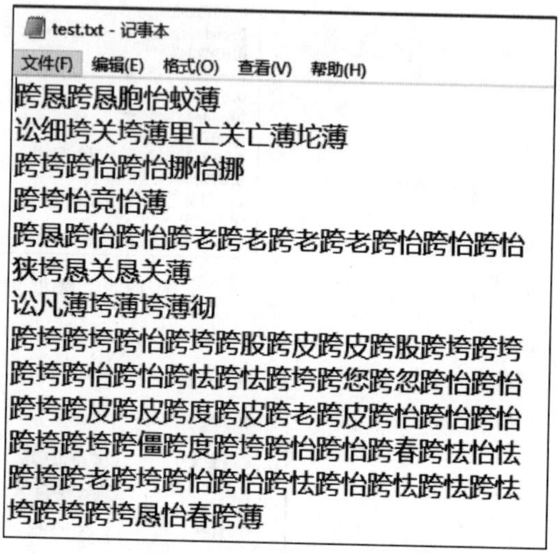

```
test.txt - 记事本
文件(F) 编辑(E) 格式(O) 查看(V) 帮助(H)
跨恳跨恳胞怡蚊薄
讼细垮关垮薄里亡关亡薄坨薄
跨垮跨怡跨怡挪怡挪
跨垮怡竟怡薄
跨恳跨怡跨怡跨老跨老跨老跨怡跨怡跨怡
狭垮恳关恳关薄
讼凡薄垮薄垮薄彻
跨垮跨垮跨怡跨薄跨股跨皮跨皮股跨垮跨垮
跨垮跨怡跨怡跨怯跨怯跨垮您跨忽跨怡跨怡
跨垮跨皮跨皮跨度跨皮跨老跨皮跨怡跨怡跨怡
跨垮跨垮跨僵跨度跨垮跨怡跨春跨怯怡怯
跨垮跨老跨垮跨怡跨垮跨怯跨怡跨怯跨怯跨怯
垮跨垮跨垮恳怡春跨薄
```

图 14-13 字幕文本内容

项目15 人脸识别与机器翻译小程序
PROJECT 15

本项目通过带有注意力机制的Seq2Seq架构及Tranformer训练机器翻译模型,实现多项人脸业务以及机器翻译的功能。

15.1 总体设计

本部分主要包括系统整体结构图和系统流程图。

15.1.1 系统整体结构图

系统整体结构如图15-1所示。

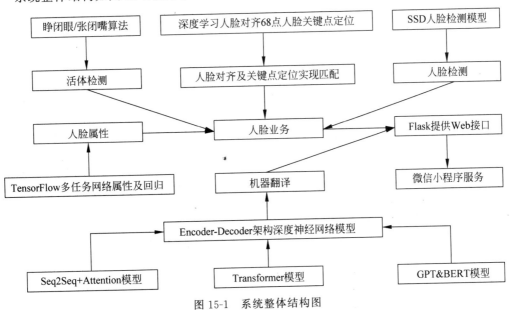

图15-1 系统整体结构图

15.1.2 系统流程图

人脸图像处理流程如图 15-2 所示，机器翻译流程如图 15-3 所示。

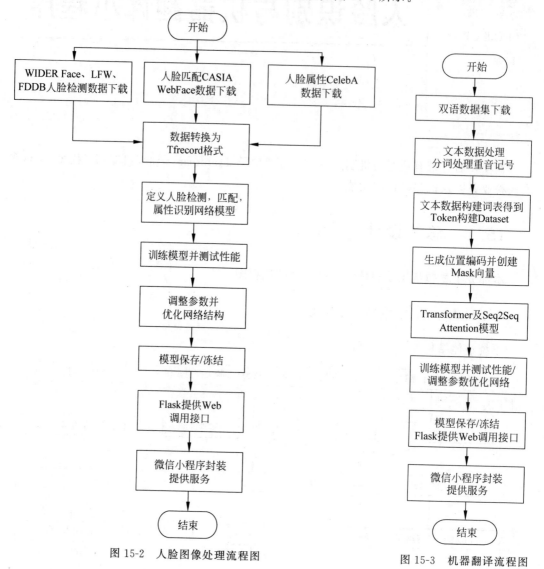

图 15-2 人脸图像处理流程图　　　图 15-3 机器翻译流程图

15.2 运行环境

本部分包括 9 个运行环境。

15.2.1 Python 环境

需要 Python 3.6 及以上配置，在 Windows 环境下推荐下载 Anaconda 完成 Python 所需的配置，下载地址为 https://www.anaconda.com/，也可以下载虚拟机在 Linux 环境下运行代码。需要安装 Numpy、Matplotlib、Pandas、Sklearn 等机器学习常用库。

15.2.2 TensorFlow-GPU/CPU 环境

人脸业务需要使用 1.12.0～1.15.0 版本，由于 TensorFlow 2.0 版本不兼容，而在机器翻译部分使用最新的 TensorFlow 2.0。安装教程参考地址为 https://tensorflow.google.cn/install。

15.2.3 OpenCV2 库

OpenCV 是基于 BSD 许可(开源)发行的跨平台计算机视觉库，可以运行在 Linux、Windows、Android 和 Mac OS 操作系统上。它是轻量级而且高效——由一系列 C 函数和少量 C++ 类构成，同时提供了 Python、Ruby、MATLAB 等语言的接口，实现图像处理和计算机视觉方面的很多通用算法。安装教程参考地址为 https://github.com/opencv/opencv 和 https://opencv.org/。

15.2.4 Dlib 库

安装教程参考地址为 http://dlib.net 和 https://github.com/tensorflow/models/tree/master/research/object_detection。

15.2.5 Flask 环境

Flask 方便提供程序调用所需的 Web 接口，安装教程参考地址为 https://github.com/pallets/flask 和 https://flask.palletsprojects.com/en/1.1.x/。

15.2.6 TensorFlow-SSD 目标(人脸)检测框架

安装教程参考地址为 https://github.com/tensorflow/models/tree/master/research/object_detection，按照 github 教程进行环境安装配置即可。

15.2.7 TensorFlow-FaceNet 人脸匹配框架

安装参考教程地址为 https://github.com/davidsandberg/facenet。

15.2.8 微信小程序开发环境

安装参考教程地址为 https://mp.weixin.qq.com/cgi-bin/wx 和 https://developers.weixin.qq.com/miniprogram/dev/devtools/download.html。

15.2.9 JupyterLab

JupyterLab 基于 Web 的集成开发环境，可以使用它编写 notebook、操作终端、编辑 markdown 文本、打开交互模式、查看 csv 文件及图片等功能。安装参考教程地址为 https://jupyter.org/。

15.3 模块实现

本部分包括 2 个模块——准备预处理和创建模型，下面分别给出各模块的功能介绍和相关代码。

15.3.1 数据预处理

（1）人脸检测算法用到 Widerface 并列转换为 PASCAL VOC 格式的数据集，便于 TensorFlow 进行读取。

VOC 数据集组织包括：

- Annotations 进行 detection 任务时的标签文件，xml 文件形式；
- ImageSets 存放数据集的分割文件，例如 train、val、test；
- JPEGImages 存放 .jpg 格式的图片文件；
- SegmentationClass 按照 class 分割的图片存放；
- SegmentationObject 存放按照 object 分割的图片。

（2）人脸识别中用到 LFW 数据集，其下载地址为 http://vis-www.cs.umass.edu/lfw/。

（3）人脸识别中另一个数据集用 Casia-FaceV5，其下载地址为 https://pan.baidu.com/s/1W-w3raZFtHdls6re3CiuQQ#list/path=%2F。

（4）人脸检测用到 FDDB 数据集，其下载地址为 http://vis-www.cs.umass.edu/fddb/。

（5）人脸匹配身份鉴定中用到 CASIA WebFace 数据集，其下载地址为 https://pgram.com/dataset/casia-webface/。

（6）人脸属性识别用到 CelebA 数据集，其下载地址为 http://mmlab.ie.cuhk.edu.hk/projects/CelebA.html。

（7）机器翻译的 seq2seq＋Attention 中用到双语数据集，其下载地址为 http://www.

manythings.org/anki/，其中 en-spa 的西班牙语、英语数据集需要经过分词等预处理之后构造成 TensorFlow 常用的数据集格式。

```python
def unicode_to_ascii(s):    # 编码转换
    return ''.join(c for c in unicodedata.normalize('NFD', s) if unicodedata.category(c) != 'Mn')
en_sentence = u"May I borrow this book?"
sp_sentence = u"¿Puedo tomar prestado este libro?"
print(unicode_to_ascii(en_sentence))
print(unicode_to_ascii(sp_sentence))
def preprocess_sentence(w):    # 预处理
    w = unicode_to_ascii(w.lower().strip())
    w = re.sub(r"([?.!,¿])", r" \1 ", w)
    w = re.sub(r'[" "]+', " ", w)
    # 用空格替换所有内容,除了(a-z, A-Z, ".", "?", "!", ",")
    w = re.sub(r"[^a-zA-Z?.!,¿]+", " ", w)
    w = w.rstrip().strip()
    # 在句子中添加开始标记和结束标记
    w = '<start> ' + w + ' <end>'
    return w
data_path = './sample_data/spa.txt'
# 去除重音,清理句子,返回单词对：[ENGLISH, SPANISH]
def create_dataset(path, num_examples):
    lines = open(path, encoding='UTF-8').read().strip().split('\n')
    word_pairs = [[preprocess_sentence(w) for w in l.split('\t')]  for l in lines[:num_examples]]
    return zip(*word_pairs)
```

得到分词并增加 token 后的文本结果如图 15-4 所示。

```
<start> if you want to sound like a native speaker , you must be willing to practice saying the same sentence over and over in the same way that banjo players practice the same phrase over and over until they can play it correctly and at the desired tempo . <end>
<start> si quieres sonar como un hablante nativo , debes estar dispuesto a practicar diciendo la misma frase una y otra vez de la misma manera en que un musico de banjo practica el mismo fraseo una y otra vez hasta que lo puedan tocar correctamente y en el tiempo esperado . <end>
```

图 15-4　文本结果

```python
def max_length(tensor):    # 最大长度句子
    return max(len(t) for t in tensor)
def tokenize(lang):    # 单词双向对照表
    lang_tokenizer = tf.keras.preprocessing.text.Tokenizer(filters='')
    lang_tokenizer.fit_on_texts(lang)
    tensor = lang_tokenizer.texts_to_sequences(lang)
    tensor = tf.keras.preprocessing.sequence.pad_sequences(tensor, padding='post')
    return tensor, lang_tokenizer
def load_dataset(path, num_examples=None):    # 创建干净的输入,输出对
    targ_lang, inp_lang = create_dataset(path, num_examples)
    input_tensor, inp_lang_tokenizer = tokenize(inp_lang)
    target_tensor, targ_lang_tokenizer = tokenize(targ_lang)
```

```
            return input_tensor, target_tensor, inp_lang_tokenizer, targ_lang_tokenizer
```

数据集规模如图 15-5 所示。

```
def convert(lang, tensor):    #语言转张量
    for t in tensor:
        if t != 0:
            print ("%d ----> %s" % (t, lang.index_word[t]))
```

```
(24000, 24000, 6000, 6000)
```

图 15-5　数据集规模

处理后得到的 index-word 对应关系结果如图 15-6 所示。

```
Input Language; index to word mapping
1 ----> <start>
5027 ----> arrestadlo
27 ----> !
2 ----> <end>

Target Language; index to word mapping
1 ----> <start>
1308 ----> seize
41 ----> him
37 ----> !
2 ----> <end>
```

图 15-6　对应关系结果

(8) 机器翻译 Transformer 模型用到 TensorFlow 官方提供的 ted_hrlr_translate 数据集。

```
import tensorflow_datasets as tfds    #加载数据
examples, info = tfds.load('ted_hrlr_translate/pt_to_en',
                            with_info = True,
                            as_supervised = True)
train_examples, val_examples = examples['train'], examples['validation']
#需要对数据进行处理
en_tokenizer = tfds.features.text.SubwordTextEncoder.build_from_corpus(
    (en.numpy() for pt, en in train_examples),
    target_vocab_size = 2 ** 13)
pt_tokenizer = tfds.features.text.SubwordTextEncoder.build_from_corpus(
    (pt.numpy() for pt, en in train_examples),
    target_vocab_size = 2 ** 13)    #转换为 TensorFlow 格式
buffer_size = 20000
batch_size = 64
max_length = 40
def encode_to_subword(pt_sentence, en_sentence):    #数据转化为 subwords 格式
    pt_sequence = [pt_tokenizer.vocab_size] \
        + pt_tokenizer.encode(pt_sentence.numpy()) \
        + [pt_tokenizer.vocab_size + 1]
```

```
            en_sequence = [en_tokenizer.vocab_size] \
                + en_tokenizer.encode(en_sentence.numpy()) \
                + [en_tokenizer.vocab_size + 1]
            return pt_sequence, en_sequence
        def filter_by_max_length(pt, en):        #以最大长度进行过滤
            return tf.logical_and(tf.size(pt) <= max_length,
                                  tf.size(en) <= max_length)
        def tf_encode_to_subword(pt_sentence, en_sentence): #TensorFlow运算节点
            return tf.py_function(encode_to_subword,
                                  [pt_sentence, en_sentence],
                                  [tf.int64, tf.int64])
        train_dataset = train_examples.map(tf_encode_to_subword)    #训练集构造
        train_dataset = train_dataset.filter(filter_by_max_length)
        train_dataset = train_dataset.shuffle(
            buffer_size).padded_batch(
            batch_size, padded_shapes = ([-1], [-1]))
        valid_dataset = val_examples.map(tf_encode_to_subword)    #验证集构造
        valid_dataset = valid_dataset.filter(
            filter_by_max_length).padded_batch(
            batch_size, padded_shapes = ([-1], [-1]))
```

15.3.2 创建模型

数据加载进模型之后,需要定义模型结构、损失函数及优化器。

1. 定义模型结构

本部分包括人脸业务、人脸识别、人脸属性识别和机器翻译。

1) 人脸业务

采用 Inception V3 作为特征提取网络的 SSD 模型构建。相关代码如下:

```
class SSDInceptionV3FeatureExtractor(ssd_meta_arch.SSDFeatureExtractor):
    def __init__(self,              #特征提取初始化
                 is_training,
                 depth_multiplier,
                 min_depth,
                 pad_to_multiple,
                 conv_hyperparams_fn,
                 reuse_weights = None,
                 use_explicit_padding = False,
                 use_depthwise = False,
                 num_layers = 6,
                 override_base_feature_extractor_hyperparams = False):
        super(SSDInceptionV3FeatureExtractor, self).__init__(    #实例化参数
            is_training = is_training,
            depth_multiplier = depth_multiplier,
            min_depth = min_depth,
```

```python
            pad_to_multiple = pad_to_multiple,
            conv_hyperparams_fn = conv_hyperparams_fn,
            reuse_weights = reuse_weights,
            use_explicit_padding = use_explicit_padding,
            use_depthwise = use_depthwise,
            num_layers = num_layers,
            override_base_feature_extractor_hyperparams =
            override_base_feature_extractor_hyperparams)
        if not self._override_base_feature_extractor_hyperparams:
            raise ValueError('SSD Inception V3 feature extractor always uses scope returned by 'conv_hyperparams_fn' for both the base feature extractor and the additional layers added since there is no arg_scope defined for the base feature extractor.')    #异常处理
    def preprocess(self, resized_inputs):
        #SSD预处理
        return (2.0 / 255.0) * resized_inputs - 1.0
    def extract_features(self, preprocessed_inputs):
        #从预处理输入中提取特征
        preprocessed_inputs = shape_utils.check_min_image_dim(
            33, preprocessed_inputs)
        feature_map_layout = {    #特征映射
            'from_layer': ['Mixed_5d', 'Mixed_6e', 'Mixed_7c', '', '', ''
                          ][:self._num_layers],
            'layer_depth': [-1, -1, -1, 512, 256, 128][:self._num_layers],
            'use_explicit_padding': self._use_explicit_padding,
            'use_depthwise': self._use_depthwise,
        }
        with slim.arg_scope(self._conv_hyperparams_fn()):    #轻量级训练
            with tf.variable_scope('InceptionV3', reuse = self._reuse_weights) as scope:
                _, image_features = inception_v3.inception_v3_base(
                    ops.pad_to_multiple(preprocessed_inputs, self._pad_to_multiple),
                    final_endpoint = 'Mixed_7c',
                    min_depth = self._min_depth,
                    depth_multiplier = self._depth_multiplier,
                    scope = scope)
            feature_maps = feature_map_generators.multi_resolution_feature_maps(
                feature_map_layout = feature_map_layout,    #特征映射参数输出
                depth_multiplier = self._depth_multiplier,
                min_depth = self._min_depth,
                insert_1x1_conv = True,
                image_features = image_features)
        return feature_maps.values()
```

代码

使用特征提取器进行目标检测模型搭建,bbox生成部分代码,请扫描二维码获取。

参考TensorFlow-models中目标检测部分的代码实现,保留如下部分并进行修改,如图15-7所示,参考地址为https://github.com/tensorflow/models/tree/master/research/object_detection/core。

图 15-7　检测图

2) 人脸识别

使用 Inception V2 和 ResNet 作为特征提取器构建的 FaceNet,存储得到人脸特征如图 15-8 所示,人脸对比如图 15-9 所示。

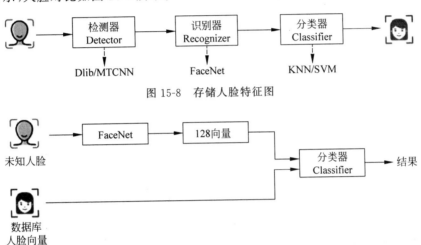

图 15-8　存储人脸特征图

图 15-9　人脸对比图

参考地址为 https://github.com/davidsandberg/facenet/blob/master/src/,代码结构如图 15-10 所示。

特征提取部分模型结构定义代码请扫描二维码获取。

3) 人脸属性识别

采用 ResNet 作为基本结构、残差连接支持深度更大的网络,相关代码如下:

```
def shortcut(input, residual):
# shortcut 连接,也就是身份映射部分
    input_shape = K.int_shape(input)
    residual_shape = K.int_shape(residual)
    stride_height = int(round(input_shape[1] / residual_shape[1]))
    stride_width = int(round(input_shape[2] / residual_shape[2]))
    equal_channels = input_shape[3] == residual_shape[3]
    identity = input
# 如果维度不同,则使用 1 * 1 卷积进行调整
```

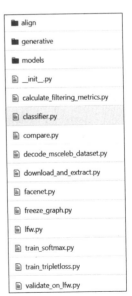

代码

图 15-10　代码结构图

```python
            if stride_width > 1 or stride_height > 1 or not equal_channels:
                identity = Conv2D(filters = residual_shape[3],
                                  kernel_size = (1, 1),
                                  strides = (stride_width, stride_height),
                                  padding = "valid",
                                  kernel_regularizer = regularizers.l2(0.0001))(input)
            return add([identity, residual])
    def basic_block(nb_filter, strides = (1, 1)):
        # 基本的 ResNet 构建模块,适用于 ResNet-18 和 ResNet-34
        def f(input):
            conv1 = conv2d_bn(input, nb_filter, kernel_size = (3, 3), strides = strides)
            residual = conv2d_bn(conv1, nb_filter, kernel_size = (3, 3))
            return shortcut(input, residual)
        return f
    def residual_block(nb_filter, repetitions, is_first_layer = False):
        # 构建每层的残余模块
        def f(input):
            for i in range(repetitions):
                strides = (1, 1)
                if i == 0 and not is_first_layer:
                    strides = (2, 2)
                input = basic_block(nb_filter, strides)(input)
            return input
        return f
    # ResNet 整体模型
    def resnet_18(input_shape = (224,224,3), nclass = 1000):
        # 使用带有 TensorFlow 后端的 Keras 构建 Resnet-18 模型
        input_ = Input(shape = input_shape)
        conv1 = conv2d_bn(input_, 64, kernel_size = (7, 7), strides = (2, 2))
        pool1 = MaxPool2D(pool_size = (3, 3), strides = (2, 2), padding = 'same')(conv1)
        conv2 = residual_block(64, 2, is_first_layer = True)(pool1)
        conv3 = residual_block(128, 2, is_first_layer = True)(conv2) # 卷积
        conv4 = residual_block(256, 2, is_first_layer = True)(conv3)
        conv5 = residual_block(512, 2, is_first_layer = True)(conv4)
        pool2 = GlobalAvgPool2D()(conv5)    # 池化
        output_ = Dense(nclass, activation = 'softmax')(pool2)
        model = Model(inputs = input_, outputs = output_) # 模型输出
        model.summary()
        return model
```

4)机器翻译

本部分包括 Seq2Seq 注意力模型和 Transformer 模型。

(1) Seq2Seq 注意力模型。Seq2Seq 注意力模型相关代码如下。

```python
# 编码部分
class Encoder(tf.keras.Model):
    def __init__(self, vocab_size, embedding_dim, encoding_units, batch_size):    # 初始化
```

```python
        super(Encoder, self).__init__()
        self.batch_size = batch_size
        self.encoding_units = encoding_units
        self.embedding = keras.layers.Embedding(vocab_size, embedding_dim)
        self.gru = keras.layers.GRU(self.encoding_units,
                                    return_sequences = True,
                                    return_state = True,
                                    recurrent_initializer = 'glorot_uniform')
    def call(self, x, hidden):    #定义隐藏状态
        x = self.embedding(x)
        output, state = self.gru(x, initial_state = hidden)
        return output, state
    def initialize_hidden_state(self):    #初始化隐藏状态
        return tf.zeros((self.batch_size, self.encoding_units))
#注意力部分：使用BahdanauAttention的注意力计算方法
class BahdanauAttention(tf.keras.Model):
    def __init__(self, units):    #初始化
        super(BahdanauAttention, self).__init__()
        self.W1 = tf.keras.layers.Dense(units)
        self.W2 = tf.keras.layers.Dense(units)
        self.V = tf.keras.layers.Dense(1)
    def call(self, query, values):    #定义批次和隐藏尺寸
        # hidden shape == (batch_size, hidden size)
        # hidden_with_time_axis shape == (batch_size, 1, hidden size)
        #执行累加分数
        hidden_with_time_axis = tf.expand_dims(query, 1)
        score = self.V(tf.nn.tanh(self.W1(values) + self.W2(hidden_with_time_axis)))
        # attention_weights shape == (batch_size, max_length, 1)
        attention_weights = tf.nn.softmax(score, axis = 1)    #注意力加权
        # context_vector shape after sum == (batch_size, hidden_size)
        context_vector = attention_weights * values    #上下文向量
        context_vector = tf.reduce_sum(context_vector, axis = 1)
        return context_vector, attention_weights
#解码部分
class Decoder(tf.keras.Model):
    def __init__(self, vocab_size, embedding_dim, decoding_units, batch_size):
#初始化参数
        super(Decoder, self).__init__()
        self.batch_size = batch_size
        self.decoding_units = decoding_units
        self.embedding = keras.layers.Embedding(vocab_size, embedding_dim)
        self.gru = keras.layers.GRU(self.decoding_units,
                                    return_sequences = True,
                                    return_state = True,
                                    recurrent_initializer = 'glorot_uniform')
        self.fc = keras.layers.Dense(vocab_size)
        #使用注意力机制
```

```python
        self.attention = BahdanauAttention(self.decoding_units)
    def call(self, x, hidden, encoding_output):
        # enc_output shape == (batch_size, max_length, hidden_size)
        context_vector, attention_weights = self.attention(hidden, encoding_output)
        # 通过嵌入后的 x 为(batch_size, 1, embedding_dim)
        x = self.embedding(x)
        # 通过级联后的 x 为(batch_size, 1, embedding_dim + hidden_size)
        x = tf.concat([tf.expand_dims(context_vector, 1), x], axis = -1)
        # 将连接的向量传递给 GRU
        output, state = self.gru(x)
        # output shape == (batch_size * 1, hidden_size)
        output = tf.reshape(output, (-1, output.shape[2]))
        # output shape == (batch_size, vocab)
        x = self.fc(output)
        return x, state, attention_weights
```

代码

(2) Transformer 模型。

Transformer 模型相关代码请扫描二维码获取。

2. 定义损失函数及优化器

使用 Adam 优化器实现自动调整学习率,调用 Adam 代码实现:

```python
optimizer = keras.optimizers.Adam(learning_rate,
                                   beta_1 = 0.9,
                                   beta_2 = 0.98,
                                   epsilon = 1e-9)
# 在 Transformer 中还使用了自定义的调度器实现学习率自定义调度
class CustomizedSchedule(
    keras.optimizers.schedules.LearningRateSchedule):
    def __init__(self, d_model, warmup_steps = 4000):    # 初始化
        super(CustomizedSchedule, self).__init__()
        self.d_model = tf.cast(d_model, tf.float32)
        self.warmup_steps = warmup_steps
    def __call__(self, step):    # 定义调用参数
        arg1 = tf.math.rsqrt(step)
        arg2 = step * (self.warmup_steps ** (-1.5))
        arg3 = tf.math.rsqrt(self.d_model)
        return arg3 * tf.math.minimum(arg1, arg2)
learning_rate = CustomizedSchedule(d_model)    # 学习率
```

代码

1) 人脸业务中损失计算

相关代码请扫描二维码获取。

2) 机器翻译

本部分包括 Seq2Seq 注意力模型和 Transformer 模型。

(1) Seq2Seq 注意力模型相关代码如下。

```
optimizer = keras.optimizers.Adam()    #优化器
loss_object = keras.losses.SparseCategoricalCrossentropy(from_logits = True, reduction = '
none')
def loss_function(real, pred):  #损失函数
    mask = tf.math.logical_not(tf.math.equal(real, 0))
    loss_ = loss_object(real, pred)
    mask = tf.cast(mask, dtype = loss_.dtype)
    loss_ * = mask
    return tf.reduce_mean(loss_)
```

(2) Transformer 模型相关代码如下。

```
loss_object = keras.losses.SparseCategoricalCrossentropy(    #损失
    from_logits = True, reduction = 'none')
def loss_function(real, pred):  #损失函数
    mask = tf.math.logical_not(tf.math.equal(real, 0))  #掩码
    loss_ = loss_object(real, pred)
        mask = tf.cast(mask, dtype = loss_.dtype)
    loss_ * = mask
        return tf.reduce_mean(loss_)
```

15.4 系统测试

人脸业务小程序初始界面如图 15-11 所示,人脸检测及识别如图 15-12 所示,人脸签到(活体检测+注册+对比)如图 15-13 所示。

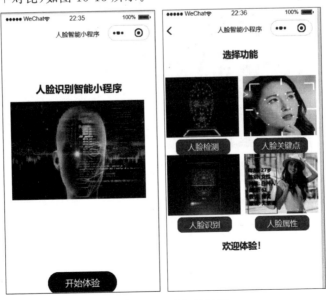

图 15-11　人脸初始界面

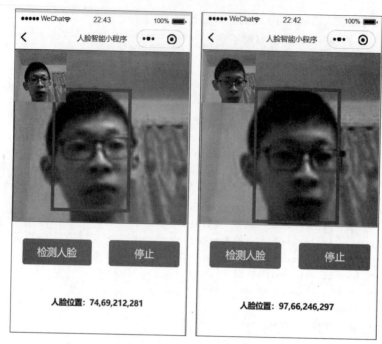

图 15-12 人脸检测及识别

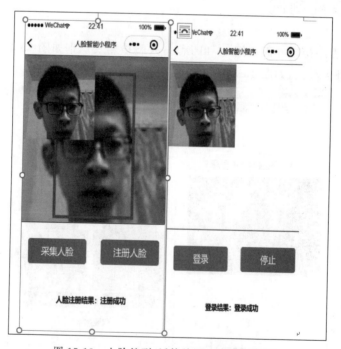

图 15-13 人脸签到(活体检测＋注册＋对比)

项目 16 基于循环神经网络的机器翻译

PROJECT 16

本项目基于 NMT（Neural Machine Translation，神经机器翻译）提供的英语－法语数据集，训练注意力机制的 Seq2Seq 神经网络翻译模型，并将模型移植到 Web 端，实现在线机器翻译。

16.1 总体设计

本部分包括系统整体结构图和系统流程图。

16.1.1 系统整体结构图

系统整体结构如图 16-1 所示。

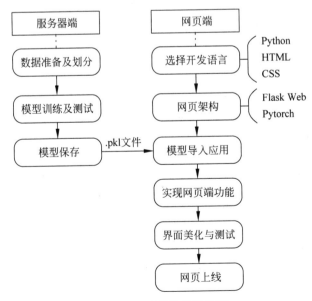

图 16-1 系统整体结构图

16.1.2 系统流程图

系统流程如图 16-2 所示。

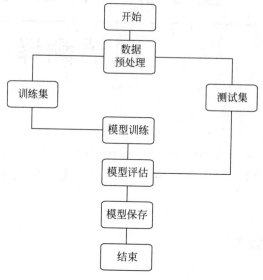

图 16-2 系统流程图

16.2 运行环境

本部分包括 Python 环境、Pytorch 环境和 Flask 环境。

16.2.1 Python 环境

需要 Python 3.7 及以上配置,在 Windows 环境下推荐下载 Anaconda 完成 Python 所需的配置,下载地址为 https://www.anaconda.com/,也可以下载虚拟机在 Linux 环境下运行代码。

16.2.2 PyTorch 环境

打开 Anaconda Prompt,输入清华仓库镜像,输入命令:

```
conda config -- add channels https://mirrors.tuna.tsinghua.edu.cn/anaconda/pkgs/free/
conda config - set show_channel_urls yes
```

创建 Python 3.7 环境,名称为 Pytorch,此时 Python 版本和后面 pytorch 的版本有匹配问题,此步选择 Python 3.7,输入命令:

```
conda create -n pytorch python=3.7
```

需要确认时都输入 y。

在 Anaconda Prompt 中激活 Pytorch 环境,输入命令:

```
activate pytorch
```

安装 CPU 版本的 Pytorch,输入命令:

```
pip install torch
```

安装完毕。

16.2.3 Flask 环境

安装 Flask 库,输入命令 pip install flask。

16.3 模块实现

本项目包括 4 个模块:数据预处理、模型构建、模型训练及保存、模型测试,下面分别给出各模块的功能介绍及相关代码。

16.3.1 数据预处理

数据预处理是为了在原有数据集的基础上进一步统一数据格式,为后面模型训练做准备。

1. 数据集读取

英语及法语数据集网址为 https://tatoeba.org/,共 135842 条英、法语句对。为了读取数据文件,先进行按行分开,将每行分成两对读取文件,添加翻转标志翻转语句对。相关代码如下:

```python
def readLangs(lang1, lang2, reverse=False):
    #读取文件,按行分开
    lines = open('D:/dataset/data/%s-%s.txt' % (lang1, lang2), encoding='utf-8').read().strip().split('\n')
    #将每行划成一对并且做正则化
    pairs = [[normalizeString(s) for s in l.split('\t')] for l in lines]
    #翻转对,改变翻译的顺序
    if reverse:
        pairs = [list(reversed(p)) for p in pairs]
        input_lang = Lang(lang2)
        output_lang = Lang(lang1)
    else:
```

```
        input_lang = Lang(lang1)
        output_lang = Lang(lang2)
    return input_lang, output_lang, pairs
```

2. 数据清洗

文本数据全部采用 Unicode 编码，将 Unicode 字符转换成 ASCII 编码，所有内容小写，并修剪大部分标点符号，相关代码如下：

```python
#Unicode 字符转换成 ASCII 编码
def unicodeToAscii(s):
    return ''.join(
        c for c in unicodedata.normalize('NFD', s)
        if unicodedata.category(c) != 'Mn'
    )
#所有内容小写,并修剪大部分标点符号
def normalizeString(s):
    s = unicodeToAscii(s.lower().strip())
    s = re.sub(r"([.!?])", r" \1", s)
    s = re.sub(r"[^a-zA-Z.!?]+", r" ", s)
    return s
```

由于例句很多，要快速训练模型，需要把数据集修剪为长度相对较短且简单的句子。在这里，最大长度是 10 个单词（包括结尾标点符号），并且对翻译为"I am"或者"He is"形式的句子进行过滤，相关代码如下：

```python
#将数据集修剪为长度相对较短且简单的句子,最长 10 个词
def filterPair(p):
    return len(p[0].split(' ')) < MAX_LENGTH and \
        len(p[1].split(' ')) < MAX_LENGTH and \
        p[1].startswith(eng_prefixes)
def filterPairs(pairs):
    return [pair for pair in pairs if filterPair(pair)]
#过滤后的文本数量从 135842 减少到 10599
```

3. 单词向量化

句子中每个单词对应唯一的索引，将句子向量化。由于多数语言有大量的字，编码向量很大，因此，进行数据修剪以保证每种语言仅仅使用几千字，相关代码如下：

```python
#定义开始,结束的标志
SOS_token = 0
EOS_token = 1
class Lang:
    def __init__(self, name):
        self.name = name
        self.word2index = {} #词到索引
        self.word2count = {}
```

```python
        self.index2word = {0: "SOS", 1: "EOS"} # 索引到词
        self.n_words = 2 # 索引
    # 遍历每一句创建词汇表
    def addSentence(self, sentence):
        for word in sentence.split(' '):
            self.addWord(word)
    # 将不在词汇表里的词加入
    def addWord(self, word):
        if word not in self.word2index:
            self.word2index[word] = self.n_words
            self.word2count[word] = 1
            self.index2word[self.n_words] = word
            self.n_words += 1
        else:
            self.word2count[word] += 1
# 将输入的句子转换成索引表示
def indexesFromSentence(lang, sentence):
    return [lang.word2index[word] for word in sentence.split(' ')]
# 将数值化后的句子转换成 tensor 形式,便于后面进行模型训练
def tensorFromSentence(lang, sentence):
    indexes = indexesFromSentence(lang, sentence)
    indexes.append(EOS_token)
    return torch.tensor(indexes, dtype=torch.long, device=device).view(-1, 1)
# 将每对切分开,一部分作为输入 tensor,一部分作为目标 tensor
def tensorsFromPair(pair):
    input_tensor = tensorFromSentence(input_lang, pair[0])
    target_tensor = tensorFromSentence(output_lang, pair[1])
    return (input_tensor, target_tensor)
```

16.3.2 模型构建

将数据加载进模型之后,需要定义模型结构、优化损失函数。

1. 定义模型结构

Seq2Seq 网络的编码器是 RNN,它为输入序列中的每个单词输出一些值。对每个输入单词,编码器输出一个向量和一个隐状态,并将其用于下一个输入的单词。

构建 GRU 的递归神经网络,并在输入层后面放一个嵌入层。嵌入层的词向量适用之前,将训练好的词向量降维到 hidden_size=256,该层的词向量在训练过程中不断更新。隐藏状态的维度是 256,定义编码器网络,相关代码如下:

```python
class EncoderRNN(nn.Module):
    def __init__(self, input_size, hidden_size):
        super(EncoderRNN, self).__init__()
        # 定义 embedding 层,门单元使用 GRU
        self.hidden_size = hidden_size
```

```python
        self.embedding = nn.Embedding(input_size, hidden_size)
        self.gru = nn.GRU(hidden_size, hidden_size)
    #进行前向传播
    def forward(self, input, hidden):
        embedded = self.embedding(input).view(1, 1, -1)
        output = embedded
        output, hidden = self.gru(output, hidden)
        return output, hidden
    #进行隐藏状态的初始化
    def initHidden(self):
        return torch.zeros(1, 1, self.hidden_size, device=device)
```

解码部分的实现(加入注意力机制):如果仅在编码器和解码器之间传递上下文向量(不使用注意力机制),则该单个向量承担编码整个句子的"负担"。而注意力机制允许解码器网络针对自身输出的每一步聚焦编码器输出不同部分。

计算一组注意力权重,这些将被乘以编码器输出矢量获得加权的组合。结果包含关于输入序列的特定部分信息,从而帮助解码器选择正确的输出单词。

权值的计算是用另一个前馈层进行的,将解码器和隐藏层状态作为输入。由于训练数据中的输入序列(语句)长短不一,为了创建和训练此层,必须选择最大长度的句子,相关代码如下:

```python
class AttnDecoderRNN(nn.Module):
    def __init__(self, hidden_size, output_size, dropout_p=0.1, max_length=MAX_LENGTH):
        super(AttnDecoderRNN, self).__init__()
        self.hidden_size = hidden_size
        self.output_size = output_size
        self.dropout_p = dropout_p
        self.max_length = max_length
        #设置网络的embedding层,gru单元,attention层,dropout机制
        self.embedding = nn.Embedding(self.output_size, self.hidden_size)
        self.attn = nn.Linear(self.hidden_size * 2, self.max_length)
        self.attn_combine = nn.Linear(self.hidden_size * 2, self.hidden_size)
        self.dropout = nn.Dropout(self.dropout_p)
        self.gru = nn.GRU(self.hidden_size, self.hidden_size)
        self.out = nn.Linear(self.hidden_size, self.output_size)
    #定义前向传播函数,设置dropout
    def forward(self, input, hidden, encoder_outputs):
        embedded = self.embedding(input).view(1, 1, -1)
        embedded = self.dropout(embedded)
        #进行注意力权值计算
        attn_weights = F.softmax(
            self.attn(torch.cat((embedded[0], hidden[0]), 1)), dim=1)
        #对编码的隐藏状态加权
        attn_applied = torch.bmm(attn_weights.unsqueeze(0),
                                 encoder_outputs.unsqueeze(0))
```

```python
        output = torch.cat((embedded[0], attn_applied[0]), 1)
        output = self.attn_combine(output).unsqueeze(0)
        # 通过 gru 单元输出 output 和 hidden, output 通过 softmax 输出预测单词
        output = F.relu(output)
        output, hidden = self.gru(output, hidden)
        output = F.log_softmax(self.out(output[0]), dim=1)
        return output, hidden, attn_weights
    # 初始化隐藏状态
    def initHidden(self):
        return torch.zeros(1, 1, self.hidden_size, device=device)
```

2. 优化损失函数

确定模型架构之后进行编译,使用对数概率函数作为损失函数,通过随机梯度下降(Stochastic Gradient Descent,SGD)优化器训练模型,加入丢弃机制,防止过拟合现象,相关代码如下:

```python
# 使用 SGD 优化器训练
encoder_optimizer = optim.SGD(encoder.parameters(), lr=learning_rate)
decoder_optimizer = optim.SGD(decoder.parameters(), lr=learning_rate)
training_pairs = [tensorsFromPair(random.choice(pairs))
                  for i in range(n_iters)]
# 使用对数概率损失函数
criterion = nn.NLLLoss()
```

使用"教师强制",将实际目标输出用做下一个输入概念,而不是将解码器的猜测用做下一个输入。

```python
if use_teacher_forcing:
    # 教师强制:将目标作为下一个输入
    for di in range(target_length):
        decoder_output, decoder_hidden, decoder_attention = decoder(
            decoder_input, decoder_hidden, encoder_outputs)
        loss += criterion(decoder_output, target_tensor[di])
        decoder_input = target_tensor[di]
else:
    # 使用教师强制,将实际输出作为下一个输入,使损失函数收敛更快
    for di in range(target_length):
        decoder_output, decoder_hidden, decoder_attention = decoder(
            decoder_input, decoder_hidden, encoder_outputs)
        topv, topi = decoder_output.topk(1)
        decoder_input = topi.squeeze().detach()  # detach from history as input
        loss += criterion(decoder_output, target_tensor[di])
        if decoder_input.item() == EOS_token:
            break
```

16.3.3　模型训练及保存

在定义模型架构和编译之后,通过训练集训练模型,使模型可以翻译句子,训练结果如图 16-3 所示。

```
10m 40s (- 149m 32s) (5000   6%) 2.8595
21m 29s (- 139m 44s) (10000  13%) 2.2784
32m 29s (- 129m 58s) (15000  20%) 1.9654
43m 21s (- 119m 14s) (20000  26%) 1.7065
54m 29s (- 108m 59s) (25000  33%) 1.5532
65m 36s (-  98m 24s) (30000  40%) 1.3932
76m 41s (-  87m 38s) (35000  46%) 1.2369
87m 41s (-  76m 43s) (40000  53%) 1.1169
98m 35s (-  65m 43s) (45000  60%) 0.9982
109m 25s (- 54m 42s) (50000  66%) 0.8831
120m 18s (- 43m 44s) (55000  73%) 0.8268
131m 5s  (- 32m 46s) (60000  80%) 0.7280
229m 26s (- 35m 17s) (65000  86%) 0.6775
240m 1s  (- 17m 8s)  (70000  93%) 0.6290
250m 45s (- 0m 0s)   (75000 100%) 0.5664
```

图 16-3　训练结果

1. 模型训练

模型训练相关代码如下。

```python
#遍历 n 次数据集,也就是将数据集训练 n 次
for iter in range(1, n_iters + 1):
    training_pair = training_pairs[iter - 1]
    input_tensor = training_pair[0]
    target_tensor = training_pair[1]
    loss = train(input_tensor, target_tensor, encoder,
        decoder, encoder_optimizer, decoder_optimizer, criterion)
    print_loss_total += loss
    plot_loss_total += loss
    #每遍历一定次数就打印损失
    if iter % print_every == 0:
        print_loss_avg = print_loss_total / print_every
        print_loss_total = 0
        print('%s (%d %d%%) %.4f' % (timeSince(start, iter / n_iters),
              iter, iter / n_iters * 100, print_loss_avg))
    #计算平均损失
    if iter % plot_every == 0:
        plot_loss_avg = plot_loss_total / plot_every
        plot_losses.append(plot_loss_avg)
        plot_loss_total = 0
```

通过观察数据集损失函数、准确率大小评估模型的训练程度,进行模型训练的进一步决策。一般来说,模型训练的最佳状态为数据集的损失函数(或准确率)不变且基本相等。

```python
import matplotlib.pyplot as plt
#绘制曲线
plt.switch_backend('agg')
import matplotlib.ticker as ticker
import numpy as np
def showPlot(points):
    plt.figure()
    fig, ax = plt.subplots()
    #该定时器用于定时记录时间
    loc = ticker.MultipleLocator(base=0.2)
    ax.yaxis.set_major_locator(loc)
    plt.plot(points)plt.legend(lns, labels, loc=7)
plt.show()
```

2. 模型保存

为了能够被 Web 端读取，需要将模型文件保存为 .pkl 格式，使用 torch.save() 函数将编码器和解码器模型分别保存到 .pkl 文件中。模型被保存后，可以重用，也可以移植到其他环境中使用。

```
#保存模型
torch.save(encoder1, 'D:/dataset/model/encoder.pkl')
torch.save(attn_decoder1, 'D:/dataset/model/decoder.pkl')
```

3. 模型评估

评估过程与大部分训练过程相同，但没有目标，因此，只是将解码器的每一步预测反馈给它自身。每当预测到一个单词，会添加到输出字符串中，如果预测到停止的 EOS 指令，则预测结束。相关代码如下：

```
decoder_attentions = torch.zeros(max_length, max_length)
        #当预测出 EOS 标志时句子翻译完毕
        for di in range(max_length):
            decoder_output, decoder_hidden, decoder_attention = decoder(
                decoder_input, decoder_hidden, encoder_outputs)
            decoder_attentions[di] = decoder_attention.data
            topv, topi = decoder_output.data.topk(1)
            if topi.item() == EOS_token:
                decoded_words.append('<EOS>')
                break
            else:
                decoded_words.append(output_lang.index2word[topi.item()])
            decoder_input = topi.squeeze().detach()
```

16.3.4 模型测试

本测试实现 Web 端的网页设计、模型导入及调用。

1. 网页设计

网页设计相关代码如下。

```html
# 定义网页主体结构
<!DOCTYPE html>
<html>
<head>
    <title>Home</title>
    <link rel="stylesheet" type="text/css" href="../static/style.css">
    <!--<link rel="stylesheet" type="text/css" href="{{ url_for('static', filename='style.css') }}">-->
</head>
<body>
    <header>
        <div class="container">
        <div id="brandname">
            ML App with Flask
        </div>
        <h2>Neural Machine Translation</h2>
    </div>
    </header>
    <div style="width:100%;text-align:center">
        <form action="{{ url_for('predict')}}" method="POST">
        <p>Enter Your Message Here</p>
        <!-- <input type="text" name="comment"/> -->
        <textarea name="message" rows="4" cols="50"></textarea>
        <br/>
        <input type="submit" class="btn-info" value="translate">
        </form>
    </div>
</body>
</html>
```

```css
# 调用 style.css 文件整理格式问题
body{
    font:15px/1.5 Arial, Helvetica,sans-serif;
    padding: 0px;
    background-color:#f4f3f3;
    # 设置背景图片
    background-image: url(bend-4948376.jpg);
    # 图片适应窗口大小,这里设置的是不进行平铺
    background-repeat:no-repeat;
    # 图片相对于浏览器固定,这里设置背景图片固定,不随内容滚动
    background-attachment: fixed;
    # 从边框区域显示
    background-origin: border-box;
    # 指定图片大小,此时会保持图像的纵横比,并将图像缩放成完全覆盖背景定位区域的大小
```

```css
    background-size:cover;}
.container{
    width:100%;
    margin: auto;
    overflow: hidden;
}
header{
    /* background:#03A9F4;#35434a; */
    border-bottom:3px solid #448AFF;
    height:120px;
    width:100%;
    padding-top:30px;
}
.main-header{
        text-align:center;
        background-color: blue;
        height:100px;
        width:100%;
        margin:0px;
}
#brandname{
    float:left;
    font-size:30px;
    color: #fff;
    margin: 10px;
}
header h2{
    text-align:center;
    color:#fff;
}
.btn-info {background-color: #2196F3;
    height:40px;
    width:100px;
} /* Blue */
.btn-info:hover {
    background: #0b7dda;
}
.resultss{
    border-radius: 15px 50px;
    background: #345fe4;
    padding: 20px;
    width: 200px;
    height: 150px;
}
```

2. 模型导入及调用

用户输入所需翻译的句子,单击"确定"按钮后在 Flask 架构中调用预先训练好的模型,相关代码如下。

```python
app = Flask(__name__)
@app.route('/')
def home():
    return render_template('home.html')
@app.route('/back', methods=['POST'])
def back():
    return render_template('home.html')
@app.route('/predict', methods=['POST'])
def predict():
    #使用存储的模型
    encoder2 = torch.load('./model/encoder.pkl')
    attn_decoder2 = torch.load('./model/decoder.pkl')   #加载模型
    global input_lang, output_lang, pairs
    input_lang, output_lang, pairs = prepareData('eng', 'fra', True)
    print()   #数据准备
    if request.method == 'POST':
        message = request.form['message']
        print(message)
      output_sentence = evaluate_onesetence(encoder2, attn_decoder2, message)
    return render_template('result.html', prediction = output_sentence)
def evaluate_onesetence(encoder, decoder, single_sentence):
    sentence = [normalizeString(s) for s in single_sentence.split('\t')]
    print(sentence[0])   #输出句子
    output_words, attentions = evaluate(encoder, decoder, sentence[0])
    output_sentence = ' '.join(output_words)
    return output_sentence
if __name__ == '__main__':   #主程序
    app.run(debug = True)
private static final String MODEL_FILE = "file:///android_asset/grf.pb";
```

模型重新预测完成后，调用 result.html 文件，展现一个新的页面，其中 result.html 也是使用 HTML 与 CSS 编写的，用户单击 back 按钮重新回到原页面。

```html
<!DOCTYPE html>
<html>
<head>
    <title></title>
    <!--<link rel = "stylesheet" type = "text/css" href = "{{ url_for('static', filename = 'style.css') }}">-->
    <link rel = "stylesheet" type = "text/css" href = "../static/style.css">
</head>
<body>
    <header>
        <div class = "container">
            <div id = "brandname">
```

```html
            ML App
        </div>
        <h2>Neural Machine Translation</h2>
    </div>
</header>
<p style="color: blue;font-size:20;text-align: center;"><b>Results for Comment</b></p>
<br/>
<div style="width:100%;text-align:center">
    <form action="{{ url_for('back') }}" method="POST">
        <div class="results">
            <p style="color: cyan;;text-align: center;font-size: 1000;">{{prediction}}</p>
        </div>
        <br/>
        <input type="submit" class="btn-info" value="back">
    </form>
</div>
</body>
</html>
```

16.4 系统测试

本部分内容包括训练准确率及模型应用。

16.4.1 训练准确率

模型经过75 000次遍历数据集,每遍历5 000次打印进度,优化器的学习率为0.001,设置丢弃率为0.1。随着训练轮次的增加,模型损失减小,说明预测的精准度不断提升,如图16-4所示。

16.4.2 模型应用

本部分包括程序运行、应用使用说明和测试结果。

1. 程序运行

打开Anaconda创建NMT虚拟环境中的命令行,运行程序app.py,界面如图16-5所示。

2. 应用使用说明

打开http://127.0.0.1:5000/,在文本框中输入要翻译的句子,单击translate按钮,如图16-6所示。

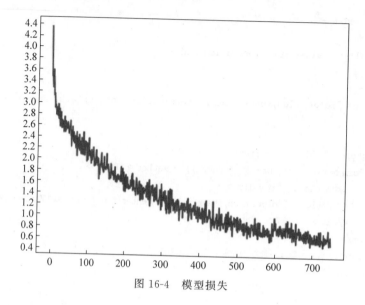

图 16-4 模型损失

```
 * Serving Flask app "app" (lazy loading)
 * Environment: production
   WARNING: This is a development server. Do not use it in a production deployment.
   Use a production WSGI server instead.
 * Debug mode: on
 * Restarting with stat
 * Debugger is active!
 * Debugger PIN: 200-531-087
 * Running on http://127.0.0.1:5000/ (Press CTRL+C to quit)
```

图 16-5 运行界面

图 16-6 应用界面

3. 测试结果

翻译法语句子 je suis pret a tout faire pour toi，如图 16-7 所示，翻译对比如图 16-8 所示。

图 16-7　测试结果示例

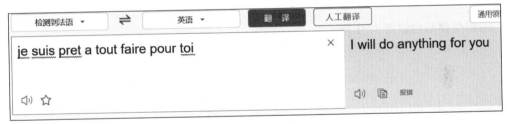

图 16-8　翻译对比

项目 17　基于 LSTM 的股票预测

PROJECT 17

本项目通过调取正弦波、标准普尔 500 股票指数的数据库，进行特征筛选和提取，实现基于 LSTM 时间序列的股票预测。

17.1　总体设计

本部分包括系统整体结构图和系统流程图。

17.1.1　系统整体结构图

系统整体结构如图 17-1 所示。

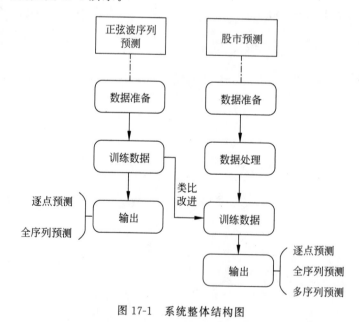

图 17-1　系统整体结构图

17.1.2 系统流程图

系统流程如图 17-2 所示。

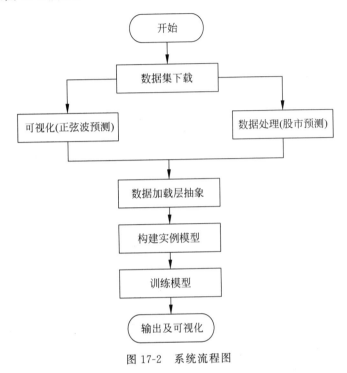

图 17-2 系统流程图

17.2 运行环境

本部分包括 Python 环境、TensorFlow 环境、Numpy 环境、Pandas 环境、Keras 环境和 Matplotlib 环境。

17.2.1 Python 环境

需要 Python 3.6 及以上配置,在 Windows 环境下推荐下载 Anaconda 完成 Python 所需的配置,下载地址为 https://www.anaconda.com/,也可以下载虚拟机在 Linux 环境下运行代码。

17.2.2 TensorFlow 环境

打开 Anaconda Prompt,输入清华仓库镜像,输入命令:

```
conda config -- add channels https://mirrors.tuna.tsinghua.edu.cn/anaconda/pkgs/free/
conda config - set show_channel_urls yes
```

创建 Python 3.5 的环境,名称为 TensorFlow,此时 Python 版本和后面 TensorFlow 的版本有匹配问题,此步选择 python 3.x,输入命令:

```
conda create -n tensorflow python=3.5
```

有需要确认的地方,都输入 y。

在 Anaconda Prompt 中激活 TensorFlow 环境,输入命令:

```
activate tensorflow
```

安装 CPU 版本的 TensorFlow,输入命令:

```
pip install - upgrade -- ignore - installed tensorflow
```

安装完毕。

17.2.3　Numpy 环境

在清华镜像源的环境中直接使用 pip 安装,输入命令:

```
pip install numpy
```

Numpy 选择正确的版本 1.15.0,版本如果有问题会导致配置的代码运行环境出错。

17.2.4　Pandas 环境

在清华镜像源的环境中直接使用 pip 安装,输入命令:

```
pip install pandas
pandas == 0.23.3
```

17.2.5　Keras 环境

在清华镜像源的环境中直接使用 pip 安装,输入命令:

```
pip install keras
keras == 2.2.2
```

17.2.6　Matplotlib 环境

在清华镜像源的环境中直接使用 pip 安装,输入命令:

```
pip install matplotlib
matplotlib == 2.2.2
```

17.3 模块实现

本项目包括 4 个模块：数据预处理、模型构建、模型保存及输出预测和模型测试。下面分别给出各模块的功能介绍及相关代码。

17.3.1 数据预处理

本部分包括正弦波预测所用数据集以及股市预测数据集。下载地址为 https://github.com/Voirtheo/Python-project。

1. 正弦波预测数据集

下载 sinewave.csv 文件，包含 5001 个正弦波时间段，幅度和频率为 1(角频率为 6.28)，时间差值为 0.01，如图 17-3 所示。

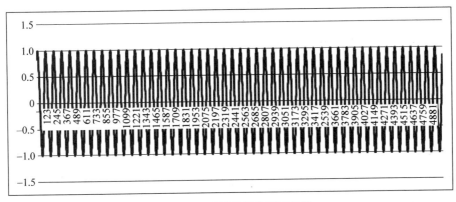

图 17-3　正弦波数据集可视化图

2. 股市预测数据集

数据集 sp500.csv 文件包含 2000 年 1 月～2018 年 9 月的开盘价、最高价、最低价、收盘价以及标准普尔 500 股指的每日交易量。

在股市预测中，与仅在 −1～+1 数值范围的正弦波不同，收盘价是不断变化的股市绝对价格。这意味着，如果不对标准化的情况下进行训练，将永远不会收敛。

为解决这个问题，使用每个大小为 n 的训练/测试数据窗口，并对每个窗口进行标准化处理，以反映该窗口开始时的(百分比)变化。采用如下方程式进行归一化，在预测过程结束时进行反归一化，以从预测中获得真实数字：

$n=$ 价格变化的归一化清单

$p=$ 调整后的每日回报价格的原始清单

归一化：$n_i = \left(\dfrac{p_i}{p_0}\right) - 1$

反归一化：$p_i = p_0(n_i + 1)$

17.3.2 模型构建

首先，将数据转换加载到 Pandas 数据帧；其次，输入 Numpy 数组中；最后，选择 Keras 搭建深度神经网络模型，使用 compile() 方法编译，确定损失函数和优化器。

1. 定义模型结构

Keras LSTM 层的工作方式是采用 3 维(N,W,F)Numpy 数组。其中：N 是训练序列的数量，W 是序列长度，F 是每个序列的特征数量。选择的序列长度（读取窗口大小）为 50，因此，可以在每一个序列看到正弦波的形状，这有利于对正弦波序列进行预测。

为加载此数据，在代码中创建 DataLoader 类，初始化 DataLoader 对象时，将传入文件名、一个拆分变量（该变量确定用于训练与测试的数据百分比）和一个列变量（该变量用于选择一个或多个数据列），用于一维或多维分析。

```python
class DataLoader():
    #用于为 LSTM 模型加载和转换数据的类
    def __init__(self, filename, split, cols):    #初始化
        dataframe = pd.read_csv(filename)    #数据读取
        i_split = int(len(dataframe) * split)
        self.data_train = dataframe.get(cols).values[:i_split]    #训练集
        self.data_test  = dataframe.get(cols).values[i_split:]    #测试集
        self.len_train  = len(self.data_train)
        self.len_test   = len(self.data_test)
        self.len_train_windows = None
```

在拥有加载数据的对象之后，构建深度神经网络模型。本项目的代码框架使用 Model 类以及 config.json 文件来构建模型实例，前提是所需的体系结构存储在 config 文件的超参数中。构建网络的主要功能是 build_model() 函数，该函数接收已解析的配置文件。使用 Keras 中的 Sequential 模型，将一些网络层通过堆叠，构成神经网络模型。相关代码可以进行扩展，以便在更复杂的体系结构上使用。

```python
class Model():
    #一个用于构建和推断 LSTM 模型的类
    def __init__(self):
        self.model = Sequential()
    def load_model(self, filepath):    #加载模型
        print('[Model] Loading model from file %s' % filepath)
        self.model = load_model(filepath)
    def build_model(self, configs):    #构建模型
        timer = Timer()
        timer.start()
        for layer in configs['model']['layers']:    #参数定义
            neurons = layer['neurons'] if 'neurons' in layer else None
```

```python
            dropout_rate = layer['rate'] if 'rate' in layer else None
            activation = layer['activation'] if 'activation' in layer else None
            return_seq = layer['return_seq'] if 'return_seq' in layer else None
            input_timesteps = layer['input_timesteps'] if 'input_timesteps' in layer else None
            input_dim = layer['input_dim'] if 'input_dim' in layer else None
            if layer['type'] == 'dense':  # 类型为 dense
                self.model.add(Dense(neurons, activation = activation))
            if layer['type'] == 'lstm':  # 类型为 LSTM
                self.model.add(LSTM(neurons, input_shape = (input_timesteps, input_dim), return_sequences = return_seq))
            if layer['type'] == 'dropout':  # 丢弃率
                self.model.add(Dropout(dropout_rate))
```

2. 损失函数和优化器

确定模型架构后进行编译,使用 Keras 官网定义的 MSE 均方误差作为损失函数,Adam 作为梯度下降方法来优化模型参数。

```python
# 定义损失函数和优化器
self.model.compile(loss = configs['model']['loss'], optimizer = configs['model']['optimizer'])
# 在 config.json 文件中看到对损失函数和优化模型的配置
"model": {
    "loss": "mse",
    "optimizer": "adam",
    "save_dir": "saved_models",
    "layers": [
        {
            "type": "lstm",
            "neurons": 100,
            "input_timesteps": 49,
            "input_dim": 2,
            "return_seq": true
        },
        {
            "type": "dropout",
            "rate": 0.2
        },
        {
            "type": "lstm",
            "neurons": 100,
            "return_seq": true
        },
        {
            "type": "lstm",
            "neurons": 100,
            "return_seq": false
        },
```

```json
            {
                "type": "dropout",
                "rate": 0.2
            },
            {
                "type": "dense",
                "neurons": 1,
                "activation": "linear"
            }
        ]
    }
```

17.3.3 模型保存及输出预测

在定义模型架构和编译后,使用训练集训练模型,使模型预测曲线走向。这里,将使用训练集和测试集拟合并保存模型。

加载数据并构建模型后,继续使用训练数据对模型进行训练。为此,创建单独的运行模块,该模块利用 Model 和 DataLoader 将它们组合起来进行训练、输出和可视化。

```python
#创建 Model 类中的训练模型
class Model():
    def train(self, x, y, epochs, batch_size, save_dir):  #定义训练参数
        timer = Timer()
        timer.start()
        print('[Model] Training Started')
        print('[Model] %s epochs, %s batch size' % (epochs, batch_size))
        save_fname = os.path.join(save_dir, '%s-e%s.h5' % (dt.datetime.now().strftime('%d%m%Y-%H%M%S'), str(epochs)))  #保存文件
        callbacks = [  #回调函数
            EarlyStopping(monitor='val_loss', patience=2),
            ModelCheckpoint(filepath=save_fname, monitor='val_loss', save_best_only=True)
        ]
#使用 model.fit()方法
        self.model.fit(
            x,
            y,
            epochs=epochs,
            batch_size=batch_size,
            callbacks=callbacks
        )
        self.model.save(save_fname)  #保存模型
        print('[Model] Training Completed. Model saved as %s' % save_fname)
        timer.stop()
#模块实例化
```

```
configs = json.load(open('config.json', 'r'))
data = DataLoader(  #加载数据
    os.path.join('data', configs['data']['filename']),
    configs['data']['train_test_split'],
    configs['data']['columns']
)
model = Model()
#创建模型,传入参数
model.build_model(configs)
x, y = data.get_train_data(
    seq_len = configs['data']['sequence_length'],
    normalise = configs['data']['normalise']
)
#将 json 文件中的参数传入 train 模块中,搭建训练模型
model.train(
    x,
    y,
    epochs = configs['training']['epochs'],
    batch_size = configs['training']['batch_size']
)
x_test, y_test = data.get_test_data(  #获取测试数据
    seq_len = configs['data']['sequence_length'],
    normalise = configs['data']['normalise']
)
```

其中,一个 batch_size 就是在一次前向/后向传播过程用到的训练样例数量,也就是一次用 32 个数据进行训练,共 $2\times32\times124=7936$ 个数据。

1. 模型保存

使用 model.save()方法将模块保存在指定路径中,方便其他程序调用。

```
save_fname = os.path.join(save_dir, '%s-e%s.h5' % (dt.datetime.now().strftime('%d%m%Y-%H%M%S'), str(epochs)))
self.model.save(save_fname)
```

2. 输出预测

对于输出,运行两类预测:一是逐点方式,也就是每次仅预测一个点,将其绘制为预测;二是沿下一个窗口进行预测,具有完整的测试数据,并再次预测下一个点。

本项目进行的第二个预测是完整序列,只用训练数据的第一部分初始化训练窗口一次。第一,该模型将预测下一个点,像逐点方法一样移动窗口。区别在于使用先前预测中的数据进行预测。第二,仅一个数据点(最后一个点)来自先前的预测。第三,最后两个数据点来自先前的预测,以此类推。经过 50 次预测后,模型将进行预测,模块可以使用该模型预测未来的许多时间步长。

```
#点对点预测和全序列预测的代码以及相应的输出
```

```python
def predict_point_by_point(self, data):
    #根据给定真实数据的最后顺序预测每个时间步长,实际上每次仅预测1个步长
    predicted = self.model.predict(data)
    predicted = np.reshape(predicted, (predicted.size,))
    return predicted
def predict_sequence_full(self, data, window_size):
    #每次将窗口移动1个新预测,然后在新窗口上重新运行预测
    curr_frame = data[0]
    predicted = []
    for i in range(len(data)):
        predicted.append(self.model.predict(curr_frame[newaxis,:,:])[0,0])
        curr_frame = curr_frame[1:]
        curr_frame = np.insert(curr_frame, [window_size-2], predicted[-1], axis=0)
    return predicted  #返回预测值
predictions_pointbypoint = model.predict_point_by_point(x_test)
plot_results(predictions_pointbypoint, y_test)
predictions_fullseq = model.predict_sequence_full(x_test, configs['data']['sequence_length'])
#输出预测结果
plot_results(predictions_fullseq, y_test)
```

17.3.4 模型测试

本项目主要基于正弦波的 LSTM 预测方法即神经网络模型运用到股票市场中,并加以改进。

与正弦波不同,股市时间序列不能映射任何特定静态函数,描述股市时间序列运动的最佳属性是随机游走。作为随机过程,真正的随机游走没有可预测的模式。因此,尝试对其建模没有意义。但股票市场是不是纯粹的随机过程仍有争论,可以推断时间序列很可能具有某种隐藏模式。这些隐藏的模式是 LSTM 深度网络可以预测的主要候选对象。

1. 窗口标准化

在 DataLoader 类中添加 normalise_windows() 函数进行转换,在 config 文件中包含一个布尔型归一化标志,用于指示这些窗口的归一化。

```python
def normalise_windows(self, window_data, single_window=False):
    #窗口标准化
    normalised_data = []
    window_data = [window_data] if single_window else window_data
    for window in window_data:
        normalised_window = []
        for col_i in range(window.shape[1]):
            normalised_col = [((float(p) / float(window[0, col_i])) - 1) for p in window[:, col_i]]
            normalised_window.append(normalised_col)
            #重塑数组并将其转置为原始多维格式
```

```
        normalised_window = np.array(normalised_window).T
        normalised_data.append(normalised_window)
return np.array(normalised_data)
```

2. 模型改进

窗口标准化后，针对正弦波数据运行。但是，在运行时进行了更改：未使用框架的 model.train() 方法，而是使用已创建的 model.train_generator() 方法。这是因为尝试训练大型数据集时很容易耗尽内存，model.train() 函数将整个数据集加载到内存中，规范化应用于内存中的每个窗口，容易引起内存溢出。因此，改为使用 Keras 的 fit_generator() 函数，以便使用 Python 生成器绘制数据来动态训练数据集，降低内存利用率。

```
configs = json.load(open('config.json', 'r'))
data = DataLoader(  #加载数据
    os.path.join('data', configs['data']['filename']),
    configs['data']['train_test_split'],
    configs['data']['columns']
)
model = Model()
model.build_model(configs)  #构建模型
x, y = data.get_train_data(
    seq_len = configs['data']['sequence_length'],
    normalise = configs['data']['normalise']
)
#生成训练
steps_per_epoch = math.ceil((data.len_train - configs['data']['sequence_length']) / configs
['training']['batch_size'])
model.train_generator(
    data_gen = data.generate_train_batch(  #训练数据参数
        seq_len = configs['data']['sequence_length'],
        batch_size = configs['training']['batch_size'],
        normalise = configs['data']['normalise']
    ),
    epochs = configs['training']['epochs'],
    batch_size = configs['training']['batch_size'],
    steps_per_epoch = steps_per_epoch
)
x_test, y_test = data.get_test_data(  #测试数据
    seq_len = configs['data']['sequence_length'],
    normalise = configs['data']['normalise']
)
predictions_multiseq = model.predict_sequences_multiple(x_test, configs['data']['sequence_
length'], configs['data']['sequence_length'])
predictions_fullseq = model.predict_sequence_full(x_test, configs['data']['sequence_length'])
predictions_pointbypoint = model.predict_point_by_point(x_test)
plot_results_multiple(predictions_multiseq, y_test, configs['data']['sequence_length'])
```

```
#预测输出
plot_results(predictions_fullseq, y_test)
plot_results(predictions_pointbypoint, y_test)
```

在引入正弦波逐点预测和完整序列预测方法后，尝试在股票市场预测中结合两种方法改进预测值的过拟合与欠拟合问题，即引入多序列预测。

从某种意义上说，这是全序列预测的混合，因为它使用测试数据来初始化测试窗口，预测该窗口的下一个点，并创建一个新窗口。但是，如果到达输入窗口完全由过去的预测组成的点，将停止向前移动一个完整的窗口长度，使用真实的测试数据重置窗口，再次开始该过程。从本质上讲，这为测试数据提供了多个类似趋势线的预测，从而分析模型如何适应未来的动量趋势。

```
def predict_sequences_multiple(self, data, window_size, prediction_len):
    #在预测向前移动 50 步之前,预测 50 步的顺序
    print('[Model] Predicting Sequences Multiple...')
    prediction_seqs = []
    for i in range(int(len(data)/prediction_len)):
        curr_frame = data[i*prediction_len] #当前帧
        predicted = []
        for j in range(prediction_len):
            predicted.append(self.model.predict(curr_frame[newaxis,:,:])[0,0])#预测
            curr_frame = curr_frame[1:]
            curr_frame = np.insert(curr_frame, [window_size-2], predicted[-1], axis=0)
        prediction_seqs.append(predicted) #预测序列输出
    return prediction_seqs
```

3. 相关代码

本部分包括数据加载及窗口搭建类、模型预测类和主活动类代码。

1) 数据加载及窗口搭建类

数据加载及窗口搭建类相关代码如下。

```
import math
import numpy as np
import pandas as pd
class DataLoader():
    #用于为 LSTM 模型加载和转换数据的类
    def __init__(self, filename, split, cols): #初始化参数
        dataframe = pd.read_csv(filename)
        i_split = int(len(dataframe) * split)
        self.data_train = dataframe.get(cols).values[:i_split]
        self.data_test  = dataframe.get(cols).values[i_split:]
        self.len_train  = len(self.data_train)
        self.len_test   = len(self.data_test)
        self.len_train_windows = None
    def get_test_data(self, seq_len, normalise): #获取测试数据
```

```python
        #创建x,y测试数据窗口
        data_windows = []
        for i in range(self.len_test - seq_len):
            data_windows.append(self.data_test[i:i+seq_len])
        data_windows = np.array(data_windows).astype(float)
        data_windows = self.normalise_windows(data_windows, single_window=False) if normalise else data_windows
        x = data_windows[:, :-1]
        y = data_windows[:, -1, [0]]
        return x,y
    def get_train_data(self, seq_len, normalise): #获取训练数据
        #创建x,y训练数据窗口''
        data_x = []
        data_y = []
        for i in range(self.len_train - seq_len):
            x, y = self._next_window(i, seq_len, normalise)
            data_x.append(x)
            data_y.append(y)
        return np.array(data_x), np.array(data_y)
    def generate_train_batch(self, seq_len, batch_size, normalise):
        #从给定的列表文件名中生成训练数据的生成器,以进行训练/测试
        i = 0
        while i < (self.len_train - seq_len):
            x_batch = []
            y_batch = []
            for b in range(batch_size):
                if i >= (self.len_train - seq_len):
                    #如果数据未平均分配,则终止条件可用于较小的最终批次
                    yield np.array(x_batch), np.array(y_batch)
                    i = 0
                x, y = self._next_window(i, seq_len, normalise)
                x_batch.append(x)
                y_batch.append(y)
                i += 1
            yield np.array(x_batch), np.array(y_batch)
    def _next_window(self, i, seq_len, normalise):
        #从给定的索引位置i生成下一个数据窗口
        window = self.data_train[i:i+seq_len]
        window = self.normalise_windows(window, single_window=True)[0] if normalise else window
        x = window[:-1]
        y = window[-1, [0]]
        return x, y
    def normalise_windows(self, window_data, single_window=False):
        #归一化窗口,基值为零
        normalised_data = []
        window_data = [window_data] if single_window else window_data
```

```python
        for window in window_data:
            normalised_window = []
            for col_i in range(window.shape[1]):
                normalised_col = [((float(p) / float(window[0, col_i])) - 1) for p in window[:, col_i]]
                normalised_window.append(normalised_col)
            normalised_window = np.array(normalised_window).T
            #重塑数组并将其转置为原始多维格式
            normalised_data.append(normalised_window)
        return np.array(normalised_data)
```

2)模型预测类

模型预测类相关代码如下。

```python
import os
import math
import numpy as np
import datetime as dt
from numpy import newaxis
from core.utils import Timer
from keras.layers import Dense, Activation, Dropout, LSTM
from keras.models import Sequential, load_model
from keras.callbacks import EarlyStopping, ModelCheckpoint
class Model():
    #用于构建和推断 LSTM 模型的类
    def __init__(self):  #初始化
        self.model = Sequential()
    def load_model(self, filepath):  #加载模型
        print('[Model] Loading model from file %s' % filepath)
        self.model = load_model(filepath)
    def build_model(self, configs):  #构建模型
        timer = Timer()
        timer.start()
        for layer in configs['model']['layers']:
            neurons = layer['neurons'] if 'neurons' in layer else None
            dropout_rate = layer['rate'] if 'rate' in layer else None
            activation = layer['activation'] if 'activation' in layer else None
            return_seq = layer['return_seq'] if 'return_seq' in layer else None
            input_timesteps = layer['input_timesteps'] if 'input_timesteps' in layer else None
            input_dim = layer['input_dim'] if 'input_dim' in layer else None
            if layer['type'] == 'dense':  #类型为 dense
                self.model.add(Dense(neurons, activation=activation))
            if layer['type'] == 'lstm':  #类型为 LSTM
                self.model.add(LSTM(neurons, input_shape=(input_timesteps, input_dim), return_sequences=return_seq))
            if layer['type'] == 'dropout':  #类型为丢弃率
```

```python
        self.model.add(Dropout(dropout_rate))
        self.model.compile(loss = configs['model']['loss'], optimizer = configs['model']['optimizer'])  # 模型编译
        print('[Model] Model Compiled')
        timer.stop()
    def train(self, x, y, epochs, batch_size, save_dir):  # 训练模型
        timer = Timer()
        timer.start()
        print('[Model] Training Started')
        print('[Model] %s epochs, %s batch size' % (epochs, batch_size))
        save_fname = os.path.join(save_dir, '%s-e%s.h5' % (dt.datetime.now().strftime('%d%m%Y-%H%M%S'), str(epochs)))
        callbacks = [  # 回调函数
            EarlyStopping(monitor = 'val_loss', patience = 2),
            ModelCheckpoint(filepath = save_fname, monitor = 'val_loss', save_best_only = True)
        ]
        self.model.fit(  # 模型拟合
            x,
            y,
            epochs = epochs,
            batch_size = batch_size,
            callbacks = callbacks
        )
        self.model.save(save_fname)  # 模型保存
        print('[Model] Training Completed. Model saved as %s' % save_fname)
        timer.stop()
    def train_generator(self, data_gen, epochs, batch_size, steps_per_epoch, save_dir):  # 训练引擎
        timer = Timer()
        timer.start()
        print('[Model] Training Started')
        print('[Model] %s epochs, %s batch size, %s batches per epoch' % (epochs, batch_size, steps_per_epoch))
        save_fname = os.path.join(save_dir, '%s-e%s.h5' % (dt.datetime.now().strftime('%d%m%Y-%H%M%S'), str(epochs)))  # 保存模型
        callbacks = [
            ModelCheckpoint(filepath = save_fname, monitor = 'loss', save_best_only = True)
        ]
        self.model.fit_generator(  # 模型拟合引擎
            data_gen,
            steps_per_epoch = steps_per_epoch,
            epochs = epochs,
            callbacks = callbacks,
            workers = 1
        )
        print('[Model] Training Completed. Model saved as %s' % save_fname)
```

```python
            timer.stop()
        def predict_point_by_point(self, data):
            # 根据给定真实数据的最后顺序预测每个时间步长,实际上每次仅预测1个步长
            print('[Model] Predicting Point-by-Point...')
            predicted = self.model.predict(data)
            predicted = np.reshape(predicted, (predicted.size,))
            return predicted
        def predict_sequences_multiple(self, data, window_size, prediction_len):
            # 在将预测向前移动50步之前,预测50步的顺序
            print('[Model] Predicting Sequences Multiple...')
            prediction_seqs = []
            for i in range(int(len(data)/prediction_len)):
                curr_frame = data[i*prediction_len]
                predicted = []
                for j in range(prediction_len):        # 预测序列
                    predicted.append(self.model.predict(curr_frame[newaxis,:,:])[0,0])
                    curr_frame = curr_frame[1:]
                    curr_frame = np.insert(curr_frame, [window_size-2], predicted[-1], axis=0)
                prediction_seqs.append(predicted)
            return prediction_seqs
        def predict_sequence_full(self, data, window_size):
            # 每次将窗口移动1个新预测,在新窗口上会重新运行预测
            print('[Model] Predicting Sequences Full...')
            curr_frame = data[0]
            predicted = []
            for i in range(len(data)):  # 预测序列        predicted.append(self.model.predict(curr_frame[newaxis,:,:])[0,0])
                curr_frame = curr_frame[1:]
                curr_frame = np.insert(curr_frame, [window_size-2], predicted[-1], axis=0)
            return predicted
```

3)主活动类

主活动类相关代码如下。

```python
import os   # 导入模块
import json
import time
import math
import matplotlib.pyplot as plt
from core.data_processor import DataLoader
from core.model import Model
def plot_results(predicted_data, true_data):  # 结果绘制参数
    fig = plt.figure(facecolor='white')
    ax = fig.add_subplot(111)
    ax.plot(true_data, label='True Data')
    plt.plot(predicted_data, label='Prediction')
    plt.legend()
```

```python
        plt.show()
def plot_results_multiple(predicted_data, true_data, prediction_len):
    fig = plt.figure(facecolor = 'white')    #多结果绘制
    ax = fig.add_subplot(111)
    ax.plot(true_data, label = 'True Data')
    #填充预测列表将其在图表中移动到正确的起点
    for i, data in enumerate(predicted_data):
        padding = [None for p in range(i * prediction_len)]
        plt.plot(padding + data, label = 'Prediction')  #数据及填充
        plt.legend()
    plt.show()
def main():    #主函数
    configs = json.load(open('config.json', 'r'))
    if not os.path.exists(configs['model']['save_dir']): os.makedirs(configs['model']['save_dir'])
    data = DataLoader(    #加载数据
        os.path.join('data', configs['data']['filename']),
        configs['data']['train_test_split'],
        configs['data']['columns']
    )
    model = Model()
    model.build_model(configs)    #构建模型
    x, y = data.get_train_data(
        seq_len = configs['data']['sequence_length'],
        normalise = configs['data']['normalise']
    )
    '''
    #模型训练
    model.train(
        x,
        y,
        epochs = configs['training']['epochs'],
        batch_size = configs['training']['batch_size'],
        save_dir = configs['model']['save_dir']
    )
    '''
    #生成训练
    steps_per_epoch = math.ceil((data.len_train - configs['data']['sequence_length']) / configs['training']['batch_size'])
    model.train_generator(    #训练引擎
        data_gen = data.generate_train_batch(
            seq_len = configs['data']['sequence_length'],
            batch_size = configs['training']['batch_size'],
            normalise = configs['data']['normalise']
        ),
        epochs = configs['training']['epochs'],
        batch_size = configs['training']['batch_size'],
```

```
            steps_per_epoch = steps_per_epoch,
            save_dir = configs['model']['save_dir']
    )
    x_test, y_test = data.get_test_data(    # 测试数据
            seq_len = configs['data']['sequence_length'],
            normalise = configs['data']['normalise']
    )
    # predictions = model.predict_sequences_multiple(x_test, configs['data']['sequence_length'], configs['data']['sequence_length'])
    predictions = model.predict_sequence_full(x_test, configs['data']['sequence_length'])
    #模型预测序列
    # predictions = model.predict_point_by_point(x_test)
    # plot_results_multiple(predictions, y_test, configs['data']['sequence_length'])
    plot_results(predictions, y_test)
if __name__ == '__main__':    #主函数
    main()
```

17.4 系统测试

本部分包括训练准确率及模型效果。

17.4.1 训练准确率

损失函数在一轮训练的过程中已经降低到接近 0 的值，说明训练模型状态较佳，如图 17-4 所示。

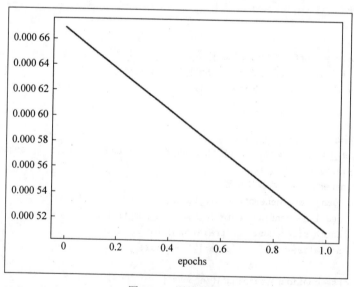

图 17-4 训练结果

17.4.2 模型效果

将预测数据与测试数据可视化,完整序列预测效果如图 17-5 所示,可以得到验证:模型欠拟合,基本无法得到正确的预测模型。

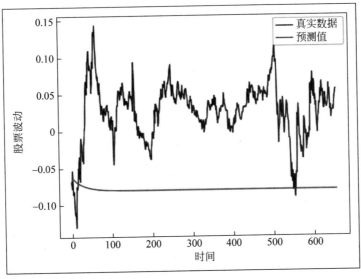

图 17-5 完整序列预测效果

逐点序列预测模型:由图 17-6 可得,在单个逐点预测上运行数据接近匹配返回的内容。经过仔细检查,发现预测线由"奇异"的预测点组成,这些预测点在后面具有整个先前的

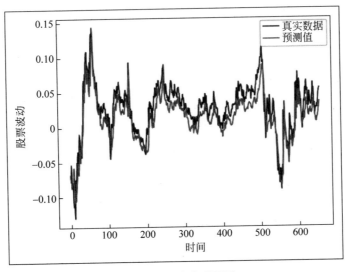

图 17-6 逐点序列预测

真实历史窗口。因此，网络不需要了解时间序列本身，除了下一个点很可能不会离最后一点太远。即使得到错误点的预测，下一个预测也将考虑真实的历史并忽略不正确的预测，然后再次允许产生错误。

此信息可用于波动率预测等应用（能够预测市场中高或低波动的时段，这对于特定交易策略非常有利），或远离交易（这也可用作良好指标的异常检测）。通过预测下一个点，将其与真实数据进行比较来实现异常检测，如果真实数据值与预测点不同，则可以针对该数据点标出异常标记。多序列预测如图 17-7 所示。

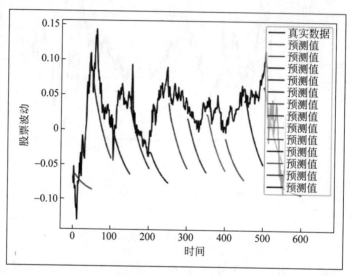

图 17-7　多序列预测

项目 18 基于 LSTM 的豆瓣影评分类情感分析

PROJECT 18

本项目基于 Word2Vec 模型，采用 LSTM 架构搭建情感分类，结合 Python 原生的 GUI 库 Tkinter，将分析结果和生成的词云通过 Tkinter 界面进行显示。

18.1 总体设计

本部分包括系统整体结构图和系统流程图。

18.1.1 系统整体结构图

系统整体结构如图 18-1 所示。

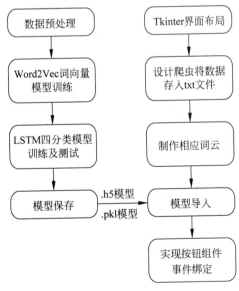

图 18-1 系统整体结构图

18.1.2 系统流程图

系统流程如图 18-2 所示。

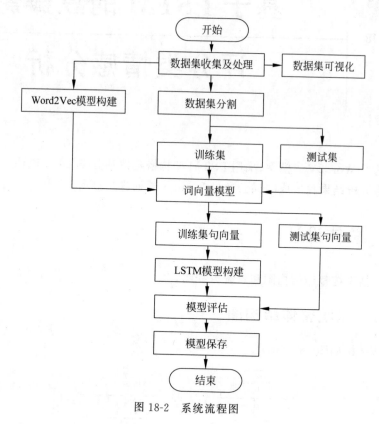

图 18-2 系统流程图

18.2 运行环境

本部分包括 Python 环境、TensorFlow 环境和 Keras 环境。

18.2.1 Python 环境

在 Windows 64 位操作系统下,下载 Anaconda 完成 Python 3.7 的配置。

18.2.2 TensorFlow 环境

打开 Anaconda Prompt,以管理员权限运行,安装 Python 3.7,输入命令:

```
conda install python = 3.7
```

建立名为 TensorFlow 的 conda 计算环境，输入命令：

```
conda create -n tensorflow python=3.7
```

激活 TensorFlow 环境，输入命令：

```
activate tensorflow
```

查看是否切换到 Python 3.7 工作环境，输入命令：

```
python -version
```

在 ANACONDA NAVIGATOR 中的 Environments 里可以看到新的环境，如图 18-3 所示。

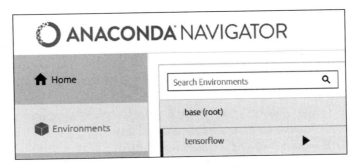

图 18-3　新环境 TensorFlow

安装 TensorFlow，输入命令：

```
pip install tensorflow
```

调用 TensorFlow，若运行成功则安装完毕。输入命令：

```
python
import tensorflow as tf
```

18.2.3　Keras 环境

打开 Anaconda Prompt，激活 TensorFlow 环境，输入命令：

```
activate tensorflow
```

安装 Keras，输入命令：

```
pip install keras
```

安装完毕。

18.3 模块实现

本项目包括 6 个模块：数据收集、数据处理、Word2Vec 模型、LSTM 模型、完整流程和模型测试。下面分别给出各模块的功能介绍及相关代码。

18.3.1 数据收集

数据集下载地址为 https://download.csdn.net/download/turkan/9181661。包含了"开心"(happy)、"愤怒"(angry)、"不喜欢"(dislike)、"沮丧"(upset)四种情感的语料集。

以下是豆瓣评论的爬取代码：

```python
import urllib.request    #导入模块
from bs4 import BeautifulSoup
import re
def get_Html(url):
    #获取 url 页面
    headers = {'User-Agent':'Mozilla/5.0 (Windows NT 10.0; WOW64) AppleWebKit/537.36 (KHTML, like Gecko) Chrome/62.0.3202.94 Safari/537.36'}
    req = urllib.request.Request(url,headers = headers)
    req = urllib.request.urlopen(req)
    content = req.read().decode('utf-8')
    return content
def get_Comment(url):
    #解析 HTML 页面
    html = get_Html(url)
    soupComment = BeautifulSoup(html, 'html.parser')
    comments = soupComment.findAll('div', {'class':'comment-item'})
    onePageComments = []
    for comment in comments:
        shortjudge = comment.find('span','short')
        star = comment.find('span',{'class':re.compile(r'^allstar(.*?)')})
        #舍弃无评分的无效数据
        if star is None:
            continue
        #三星-五星的评分标记为 1(积极),否则标记为 0(消极)
        if int(star['class'][0][7])> 2:target = 1
        else:target = 0
        content = shortjudge.get_text().strip()
        #正则表达式匹配中文
        pattern = re.compile(r'[\u4e00-\u9fa5]+')
        filterdata = re.findall(pattern, content)
        cleaned_comments = ''.join(filterdata)
        print(str(target) + cleaned_comments + '\n')
```

```
        onePageComments.append(str(target) + cleaned_comments + '\n')
    return onePageComments
if __name__ == '__main__':
    #将爬取的数据写入.txt文件
    f = open('data.txt', 'w', encoding = 'utf-8')
    for page in range(10):
        url = 'https://movie.douban.com/subject/26752088/comments?start = ' + str(20 * page) + '&limit = 20&sort = new_score&status = P'
        for i in get_Comment(url):
            f.write(i)
            print(i)
    print('\n')
```

18.3.2 数据处理

数据处理包括数据清洗、构建词向量、构建词索引。原始数据中有很多无用的特殊符号,需要去掉这些符号及停用词。分词把句子分成独立的词语以便构建词向量。构建词索引可以理解成一种对高维向量的降维,在下一步处理中,Word2Vec 模型将每个词构建成了 150 维的向量,再引入分词后,整个数据集将非常庞大,这为项目带来了很大的运算量。构建词索引即给每个词赋予一个数值,用于简化运算。

1. 数据清洗

使用正则表达式将中文匹配出来的数据保存到新的.txt 文件中:

```
def clean_data(rpath, wpath):
    #正则表达式匹配中文字符
    pchinese = re.compile('([\u4e00-\u9fa5]+)+?')
    f = open(rpath, encoding = 'UTF-8')
    fw = open(wpath, "w", encoding = 'UTF-8')
    #取 f 文件中的前 50000000 个字符写入文件,其中 readlines()函数会读取整行
    for line in f.readlines(50000000):
        m = pchinese.findall(str(line))
        if m:
            str1 = ''.join(m)
            str2 = str(str1)
            fw.write(str2)
            fw.write("\n")
    f.close()
    fw.close()
```

2. 构建词向量

本项目选择了最好的中文分词工具 jieba,参数设置为精准匹配,并使用隐马尔可夫模型提升分词准确率,标记好四类情感并将四个语料库合并成一个完整的数据集,相关代码如下。

```python
def loadfile():    # 加载文件
    happy = []
    angry = []
    dislike = []
    upset = []
    # jieba分词使用精准匹配并启用隐马尔可夫模型
    with open('happy.txt', encoding='UTF-8') as f:        # 高兴
        for line in f.readlines():
            happy.append(list(jieba.cut(line, cut_all=False, HMM=True))[:-1])
    with open('angry.txt', encoding='UTF-8') as f:        # 愤怒
        for line in f.readlines():
            angry.append(list(jieba.cut(line, cut_all=False, HMM=True))[:-1])
        f.close()
    with open('dislike.txt', encoding='UTF-8') as f:      # 不喜欢
        for line in f.readlines():
            dislike.append(list(jieba.cut(line, cut_all=False, HMM=True))[:-1])
        f.close()
    with open('upset.txt', encoding='UTF-8') as f:        # 难过
        for line in f.readlines():
            upset.append(list(jieba.cut(line, cut_all=False, HMM=True))[:-1])
        f.close()
    # 合并数据集
    X_Vec = np.concatenate((happy, angry, dislike, upset))
    # 标记四类情感
    y = np.concatenate((np.zeros(len(happy), dtype=int),
                        np.ones(len(angry), dtype=int),
                        2 * np.ones(len(dislike), dtype=int),
                        3 * np.ones(len(upset), dtype=int)))
    return X_Vec, y
```

3. 构建词索引

相关代码如下。

```python
# 对数据集分词得到的每个词构建词索引
def data2inx(w2indx, X_Vec):
    data = []
    for sentence in X_Vec:
        new_txt = []
        for word in sentence:
            try:
                new_txt.append(w2indx[word])
            except:    # 异常处理
                new_txt.append(0)
        data.append(new_txt)
    return data
```

18.3.3 Word2Vec 模型

Word2Vec 是由 Google 公司的工程师和机器学习专家所提出的一种算法，主要完成词语和高维向量的映射。Gensim 是开源的第三方 Python 工具包，用于从原始非结构化的文本中，无监督地学习到文本隐层主题向量表达。本项目使用的 Word2Vec 是 Gensim 中的一个模型。同时考虑模型的准确性和计算成本，将词向量的长度设为 150、滑动窗口的大小设为 7，并将出现频数不超过 4 的词语都编为零向量。

```python
def word2vec_train(X_Vec):
    # 使用 gensim 库里的 Word2Vec 模型构建词向量
    model_word = Word2Vec(size = voc_dim,          # 词向量长度
                          min_count = min_out,      # 被编码词语的最小频数
                          window = window_size,     # 滑动窗口大小
                          workers = cpu_count,      # 控制训练的并行数为 GPU 核数
                          iter = 100)
    model_word.build_vocab(X_Vec)
    model_word.train(X_Vec, total_examples = model_word.corpus_count, epochs = model_word.iter)
    model_word.save('new_Word2Vec.pkl')
    print(len(model_word.wv.vocab.keys()))
    # 频数小于阈值的词语编码为 0
    input_dim = len(model_word.wv.vocab.keys()) + 1
    # 初始化权重矩阵
    embedding_weights = np.zeros((input_dim, voc_dim))
    # 从 Word2Vec 中提取权重矩阵，即词向量
    w2dic = {}
    for i in range(len(model_word.wv.vocab.keys())):
        embedding_weights[i + 1, :] = model_word[list(model_word.wv.vocab.keys())[i]]
        w2dic[list(model_word.wv.vocab.keys())[i]] = i + 1
    return input_dim, embedding_weights, w2dic
```

18.3.4 LSTM 模型

Keras 是一个高层神经网络 API，由 Python 编写并基于 TensorFlow、Theano 以及 CNTK 后端，它支持 RNN 网络，有两种不同的构建模型方法。本项目使用 Sequential 模型构建 LSTM 网络。Sequential 模型使用多个网络层的线性叠加构建整个神经网络，将各层参数列表传递给 Sequential 的构造函数，来创建分类使用的 LSTM 模型，同时 add() 方法也可以将新层添加到模型中。构建后使用合适的损失函数训练模型，最后进行评估并保存模型。

1. 数据可视化和预分析

先进行简单的可视化观察数据特征，以便调整模型参数：

```python
import pandas as pd
```

```python
# 读取正则化处理过后的各个数据集文件，合并为统一的数据集
f_angry = open(r"angry.txt", 'r', encoding = 'utf-8')
reviews_angry = f_angry.readlines()
f_dislike = open(r"dislike.txt", 'r', encoding = 'utf-8')
reviews_dislike = f_dislike.readlines()
f_upset = open(r"upset.txt", 'r', encoding = 'utf-8')
reviews_upset = f_upset.readlines()
f_happy = open(r"happy.txt", 'r', encoding = 'utf-8')
reviews_happy = f_happy.readlines()
data = []
for r in reviews_dislike:     # 不喜欢
    data.append(r)
for r in reviews_upset:       # 难过
    data.append(r)
for r in reviews_angry:       # 愤怒
    data.append(r)
for r in reviews_happy:       # 高兴
    data.append(r)
d1 = pd.DataFrame(data)
pd.set_option('max_colwidth', 1000)
d1.columns = ['comment']
# 使用 jieba.lcut() 函数将分词结果存入, list 统计每条语句包含的词语数量
import jieba
numWords = []
for r in data:
    counter = len(jieba.lcut(r))
    numWords.append(counter)
# 作图
import matplotlib.pyplot as plt
from matplotlib.font_manager import FontProperties
font = FontProperties(fname = r"c:\windows\fonts\simsun.ttc", size = 14)
plt.hist(numWords, 20)
plt.xlabel("语句长度", fontproperties = "SimHei")
plt.ylabel("频率", fontproperties = "SimHei")
plt.axis([0, 200, 0, 80000])
plt.show()
```

大部分语句的词汇量在 100 以内，因此，将 LSTM 的输入序列长度定为 100，如图 18-4 所示。

2. 定义模型结构

使用 Sequential 模型，首先，加入嵌入层；其次，连接 LSTM 层。用一个全连接层和激活函数以构建整个 LSTM 神经网络。

```python
def lstm(input_dim, embedding_weights):
    model = Sequential()
    # 嵌入层参数设置
```

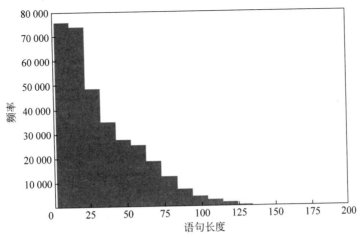

图 18-4 语句长度-频率图

```
model.add(Embedding(output_dim = voc_dim,
                    input_dim = input_dim,
                    mask_zero = True,
                    weights = [embedding_weights],
                    input_length = lstm_input))
# LSTM 参数设置,丢弃正则化防止过拟合
model.add(LSTM(128, activation = 'softsign', dropout = 0.2, recurrent_dropout = 0.2))
# 全连接层
model.add(Dense(4))
# 多分类任务,激活函数为 softmax
model.add(Activation('softmax'))
return model
```

3. 模型训练、评估及保存

确定模型架构后进行编译,这是多类别的分类问题。因此,使用交叉熵作为损失函数。由于所有标签都带有相似的权重,本项目使用精确度作为性能指标,同时还计算了通用的平均绝对误差 mae。Adam 是常用的梯度下降方法,使用它来优化模型参数。设定整体架构,使用训练集训练模型,测试集测试模型,输出测试结果并保存模型为 .h5 文件以便后续调用。

```
def train_lstm(model, x_train, y_train, x_test, y_test):
    print('Compiling the Model...')
    # 使用交叉熵作为损失函数,adam 优化,输出准确率和平均绝对误差
    model.compile(loss = 'binary_crossentropy',
                  optimizer = 'adam', metrics = ['mae', 'acc'])
    # 训练模型
    print("Train...")
    model.fit(x_train, y_train, batch_size = batch_size, epochs = epoch_time, verbose = 1)
```

```python
#测试模型
print("Evaluate...")
print(model.predict(x_test))
score = model.evaluate(x_test, y_test,
                    batch_size=batch_size)
#保存 yaml 文件
yaml_string = model.to_yaml()
with open('new_lstm.yml', 'w') as outfile:
    outfile.write(yaml.dump(yaml_string, default_flow_style=True))
#保存权重文件
model.save('new_total.h5')
print('Test score:', score)
```

18.3.5 完整流程

在代码中频繁使用 print()显示程序进程:

```python
print("开始清洗数据................")
clean_data('0_simplifyweibo.txt','happy.txt')
clean_data('1_simplifyweibo.txt','angry.txt')
clean_data('2_simplifyweibo.txt','dislike.txt')
clean_data('3_simplifyweibo.txt','upset.txt')
print("清洗数据完成................")
print("开始下载数据................")
X_Vec, y = loadfile()
print("下载数据完成................")
print("开始构建词向量................")
input_dim,embedding_weights,w2dic = word2vec_train(X_Vec)
print("构建词向量完成................")
#词索引构建
index = data2inx(w2dic,X_Vec)
#词索引在数据集上的映射
index2 = sequence.pad_sequences(index, maxlen=lstm_input)
#分割数据集,其中验证集占 20%
x_train,x_test,y_train,y_test = train_test_split(index2, y, test_size=0.2)
#把类标签转为独热编码
y_train = keras.utils.to_categorical(y_train, num_classes=4)
y_test = keras.utils.to_categorical(y_test, num_classes=4)
#模型加载和训练
model = lstm(input_dim, embedding_weights)
train_lstm(model, x_train, y_train, x_test, y_test)
```

通过观察训练集、测试集的损失函数和准确率来调整参数,进行模型训练的进一步决策。一般来说,训练集和测试集的损失函数不变且基本相等时,模型训练达到理想状态。运行训练进度如图 18-5 所示,可以看出,训练过程有较好的可视化效果。

```
开始清洗数据...............
Building prefix dict from the default dictionary ...
Loading model from cache C:\Users\MONOLO~1\AppData\Local\Temp\jieba.cache
清洗数据完成...............
开始下载数据...............
Loading model cost 0.725 seconds.
Prefix dict has been built succesfully.
下载数据完成...............
开始构建词向量...............
C:/Users/Monologue/newanalysis.py:87: DeprecationWarning: Call to
deprecated `iter` (Attribute will be removed in 4.0.0, use self.epochs
instead).
  model_word.train(X_Vec, total_examples=model_word.corpus_count,
epochs=model_word.iter)
96186
C:/Users/Monologue/newanalysis.py:95: DeprecationWarning: Call to
deprecated `__getitem__` (Method will be removed in 4.0.0, use
self.wv.__getitem__() instead).
  embedding_weights[i+1, :] = model_word[list(model_word.wv.vocab.keys())
[i]]
构建词向量完成...............
```

图 18-5　训练进度

18.3.6　模型测试

有两种方式用于模型测试：一是使用 Python 原生 GUI 库 Tkinter，设计一个简单合理的界面；二是对按钮进行事件绑定，通过单击按钮触发爬虫，将相应评论存入.txt 文件。单击第二个按钮，加载模型对文件中的评论进行情感分析，并返回分析结果和生成的词云。

1. 界面设置

首先，设置两个文本框分别用来输入电影 ID 信息和输出分析结果；其次，设置两个按钮获取文本框输入内容和情感分析；最后，设置标签用于显示词云，组件的放置使用 pack 布局方式。

```
#设置窗口
window = tk.Tk()
window.title('情感分析')
window.geometry('500x500')
#设置输入窗口
tk.Label(window, text = "请输入想了解的电影ID:").pack()
e = tk.Entry()
e.pack()
#设置两个插入按钮
b1 = tk.Button(text = '获取信息', width = 20, height = 2, command = show)
b1.pack()
b2 = tk.Button(text = '分析结果', width = 20, height = 2, command = analysis)
b2.pack()
#设置文本显示框
t = tk.Text(width = 20, height = 2)
t.pack()
window.mainloop()
```

2. 评论的获取

（1）定义 getReview() 函数，使用 get() 方法获取文本框输入的电影 ID，通过分析豆瓣电影链接变化，构造短评网页的 URL。

(2) 在 getReview() 中调用 getComment() 函数解析获取的 HTML 页面，对获取的评论数据进行简单的清洗——去掉标点及特殊符号，只保留汉字，并将清洗后的数据写入 review.txt 文件中。下面列出获取并保存评论用到的 3 个函数。

```python
def getReview():                                                    # 获取评论
    global ni
    ni = e.get()                                                    # 使用 get() 方法获取输入文本框内容
    print("输入内容为:%s" % ni)
    f = open('review.txt', 'w', encoding = 'utf-8')
    for page in range(10):                                          # 爬取相应电影的 10 页评论
        url = 'https://movie.douban.com/subject/' + str(ni) + '/comments?start = ' + str(20 *
page) + '&limit = 20&sort = new_score&status = P'                   # 构造短评网页的 URL
        for i in getComment(url):                                   # 调用 getComment 函数
            f.write(i)
            print(i)
            print('\n')
def getHtml(url):
    # 获取 url 页面
    headers = {'User - Agent':'Mozilla/5.0 (Windows NT 10.0; WOW64) AppleWebKit/537.36
(KHTML, like Gecko) Chrome/62.0.3202.94 Safari/537.36'}
    req = urllib.request.Request(url, headers = headers)
    req = urllib.request.urlopen(req)
    content = req.read().decode('utf-8')
    return content
def getComment(url):
    # 解析 HTML 页面
    html = getHtml(url)
    soupComment = BeautifulSoup(html, 'html.parser')
    comments = soupComment.findAll('div', {'class':'comment - item'})
    # 获取评论区标签
    onePageComments = []
    for comment in comments:
        shortjudge = comment.find('span','short')                   # 获取短评
        star = comment.find('span',{'class':re.compile(r'^allstar(.*?)')})  # 获取短评对应的星级
        if star is None:
            continue
        if int(star['class'][0][7]) > 2:target = 1                  # 将 3～5 星标记为 1
        else:target = 0                                             # 将 1～2 星标记为 0
        # 对数据进行简单的清洗
        content = shortjudge.get_text().strip()
        # 清除所有 html 标签元素,移除字符串头尾的空格或换行符
        pattern = re.compile(r'[\u4e00 - \u9fa5] + ')
        filterdata = re.findall(pattern, content)                   # 只保留汉字
        cleaned_comments = ''.join(filterdata)
        print(str(target) + cleaned_comments + '\n')
        onePageComments.append(str(target) + cleaned_comments + '\n')
```

3. 分析结果的返回

(1) 定义 analysis 函数，加载 Word2Vec_java.pkl 和 lstm_java_total.h5 模型，定义标签。
(2) 读取评论文件，将内容按行划分，并连接成一个字符串，转换成高维向量进行连接。
(3) 调用 wordcloud 函数，根据评论内容形成个性化词云。

```python
def analysis():
    #加载模型
    model_word = Word2Vec.load('Word2Vec_java.pkl')
    input_dim = len(model_word.wv.vocab.keys()) + 1
    embedding_weights = np.zeros((input_dim, voc_dim))
    w2dic = {}
    for i in range(len(model_word.wv.vocab.keys())):
            embedding_weights[i+1, :] = model_word[list(model_word.wv.vocab.keys())[i]]
            w2dic[list(model_word.wv.vocab.keys())[i]] = i+1
    model = load_model('lstm_java_total.h5')
    pchinese = re.compile('([\u4e00-\u9fa5]+)+?')
    label = {0:"happy",1:"angry",2:"dislike",3:"upset"}
    #读取文件，将内容按行划分并连接成一个长字符串
    in_str = open('review.txt',encoding = 'utf-8')
    lines = []
    for i in in_str:
        lines.append(i)
        txt = ''.join(lines)
    in_stc = ''.join(pchinese.findall(txt))
    wordcloud()       #形成个性化词云
    in_stc = list(jieba.cut(in_stc,cut_all = True, HMM = False))
    #使用全模式进行分词，不采用 HMM 模型
    new_txt = []
    data = []
    #将词语转换为高维向量并连接
    for word in in_stc:
        try:
            new_txt.append(w2dic[word])
        except:    #异常处理
            new_txt.append(0)
    data.append(new_txt)
    data = sequence.pad_sequences(data, maxlen = voc_dim )    #数据序列
    pre = model.predict(data)[0].tolist()
    result = label[pre.index(max(pre))]
    t.insert('insert', result)
def wordcloud():     #词云
    text = open("review.txt","rb").read()
    #jieba 分词
    wordlist = jieba.cut(text,cut_all = True)
    wl = " ".join(wordlist)
    wc = WordCloud(background_color = "white",     #设置背景颜色
```

```
                    mask = imageio.imread('original.jpg'),    #设置背景图片
                    max_words = 2000,       #设置最大显示的字数
                    stopwords = ["的", "这种", "这样", "还是", "就是", "这个"],
                    #设置停用词
                    font_path = "C:\Windows\Fonts\simkai.ttf",    #设置为楷体
           #设置中文字体,使得词云可以显示(词云默认字体是"DroidSansMono.ttf 字体库")
                    max_font_size = 60, #设置字体最大值
                    random_state = 30, #设置有多少种随机生成状态,即有多少种配色方案
)
myword = wc.generate(wl)#生成词云
wc.to_file('result.jpg')
global img_png
#设置标签的文字
Img = Image.open('C:\\Users\\15671\\result.jpg')
img_png = ImageTk.PhotoImage(Img)
label_Img = tk.Label(window, image = img_png)
label_Img.pack()
#展示词云图
plt.imshow(myword)
plt.axis("off")
plt.show()
```

18.4 系统测试

本部分包括训练准确率及应用效果。

18.4.1 训练准确率

训练准确率在进行二分类情感分析时达到了90%以上,如图18-6所示。但在进行四分类时仅为85%左右,模型过拟合情况良好,如图18-7所示,说明模型有很大提升空间。进行四分类时准确率低也与下载的数据集规范度等情况有关。

18.4.2 应用效果

运行程序初始界面如图18-8所示,其中"电影 ID"在豆瓣页面的 URL 中。

界面从上至下,分别是一个输入文本框、两个按钮、一个输出文本框(显示结果)、一个标签(用于显示图片)。输入电影 ID 后,单击第一个按钮"获取信息",可触发爬虫,待爬虫结束后,单击第二个按钮"分析结果",可以看到输出文本框显示对短评文件进行情感分析后的结果(happy、upset、dislike、angry),同时在文本框下方显示生成的词云,预测结果显示界面如图18-9所示,移动端测试结果如图18-10所示。

```
Epoch 1/10
13388/13388 [==============================] - 76s 6ms/step - loss: 0.5433 - mae: 0.3615 - acc: 0.7365
Epoch 2/10
13388/13388 [==============================] - 85s 6ms/step - loss: 0.4280 - mae: 0.2667 - acc: 0.8225
Epoch 3/10
13388/13388 [==============================] - 93s 7ms/step - loss: 0.3222 - mae: 0.1928 - acc: 0.8768
Epoch 4/10
13388/13388 [==============================] - 90s 7ms/step - loss: 0.2689 - mae: 0.1543 - acc: 0.9015
Epoch 5/10
13388/13388 [==============================] - 96s 7ms/step - loss: 0.2529 - mae: 0.1446 - acc: 0.9055
Epoch 6/10
13388/13388 [==============================] - 102s 8ms/step - loss: 0.2118 - mae: 0.1221 - acc: 0.9241
Epoch 7/10
13388/13388 [==============================] - 101s 8ms/step - loss: 0.1703 - mae: 0.0945 - acc: 0.9423
Epoch 8/10
13388/13388 [==============================] - 108s 8ms/step - loss: 0.1502 - mae: 0.0822 - acc: 0.9494
Epoch 9/10
13388/13388 [==============================] - 103s 8ms/step - loss: 0.1170 - mae: 0.0637 - acc: 0.9612
Epoch 10/10
13388/13388 [==============================] - 95s 7ms/step - loss: 0.0993 - mae: 0.0530 - acc: 0.9683
Evaluate...
3347/3347 [==============================] - 7s 2ms/step
Test score: [0.2696365864108775, 0.1031867042183876, 0.915297269821167]
```

图 18-6　二分类训练结果

```
Epoch 1/5
265603/265603 [==============================] - 2047s 8ms/step - loss: 0.4082 - mae: 0.2539 - acc: 0.8290
Epoch 2/5
265603/265603 [==============================] - 1940s 7ms/step - loss: 0.3844 - mae: 0.2372 - acc: 0.8412
Epoch 3/5
265603/265603 [==============================] - 1870s 7ms/step - loss: 0.3687 - mae: 0.2265 - acc: 0.8480
Epoch 4/5
265603/265603 [==============================] - 1865s 7ms/step - loss: 0.3554 - mae: 0.2176 - acc: 0.8539
Epoch 5/5
265603/265603 [==============================] - 1866s 7ms/step - loss: 0.3428 - mae: 0.2094 - acc: 0.8595
Evaluate...
[[0.65534014 0.11793894 0.0867953  0.13992563]
 [0.8730366  0.02846943 0.04689741 0.05159654]
 [0.95720625 0.01716856 0.0174463  0.00817891]
 ...
 [0.0056415  0.08688713 0.87967455 0.02779683]
 [0.2594925  0.11196107 0.3630051  0.2655413 ]
 [0.6930258  0.14605646 0.1259604  0.03495736]]
66401/66401 [==============================] - 69s 1ms/step
Test score: [0.3880039076225522, 0.22213022410869598, 0.8423178791999817]
```

图 18-7　四分类训练结果

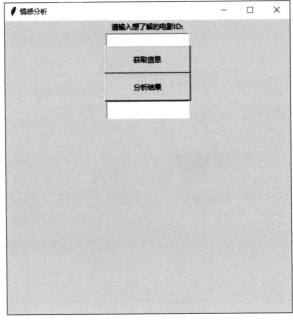

图 18-8　应用初始界面

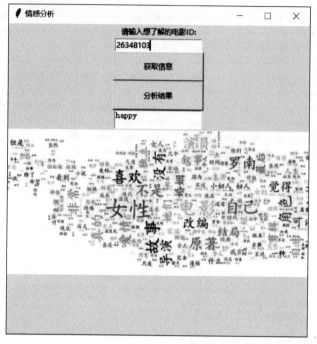

图 18-9　预测结果显示界面

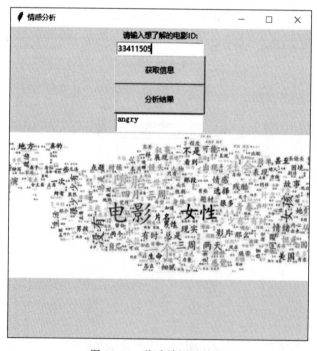

图 18-10　移动端测试结果

项目 19　AI 写诗机器人

PROJECT 19

本文通过 GitHub 中文数据集，基于 TensorFlow 的两层门控循环神经网络 LSTM 模型，实现根据指定输入写诗的功能，在 Qt 上搭建界面进行可视化操作。

19.1　总体设计

本部分包括系统整体结构图和系统流程图。

19.1.1　系统整体结构图

系统整体结构如图 19-1 所示。

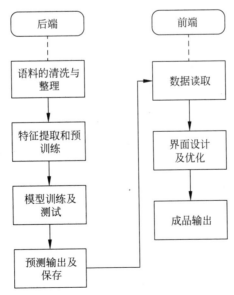

图 19-1　系统整体结构图

19.1.2 系统流程图

系统流程如图 19-2 所示。

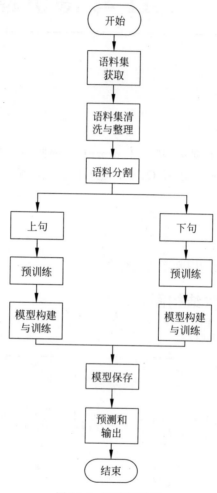

图 19-2 系统流程图

19.2 运行环境

本部分包括 Python 环境、TensorFlow 环境和 Qt Creator 环境。

19.2.1 Python 环境

下载 Anaconda 完成 Python 所需的配置，下载地址为 https://www.anaconda.com/。

19.2.2 TensorFlow 环境

打开 Anaconda Prompt,输入清华仓库镜像,输入命令:

```
conda config -- add channels https://mirrors.tuna.tsinghua.edu.cn/anaconda/pkgs/free/
conda config - set show_channel_urls yes
```

创建 Python 3.5 环境,名称为 TensorFlow,此时 Python 版本和后面 TensorFlow 的版本有匹配问题,此步选择 Python 3.x,输入命令:

```
conda create - n tensorflow python = 3.5
```

有需要确认的地方,都输入 y。
在 Anaconda Prompt 中激活 TensorFlow 环境,输入命令:

```
activate tensorflow
```

安装 CPU 版本的 TensorFlow,输入命令:

```
pip install - upgrade -- ignore - installed tensorflow
```

安装完毕。

19.2.3 Qt Creator 下载与安装

下载 Qt Creator:进入 http://download.qt.io/网站,选择 official_releases 或 archive,选择后者。进入 archive,此时有四个选项,选择 qt/,选择符合自己需要版本的 Qt 安装包(此代码使用 5.9.0 版本)。

安装 Qt Creator:单击下一步或 skip 按钮直到选择好安装路径,其他选项默认即可。

选择需要的组件:在 Qt 5.9 中,如果使用 MinGW 进行编译,选中 MinGW 模块。如果调用 VS 的编译器,则需要勾选 VS 模块。

Tools 工具项的选择:第一项是 CDB 的调试器,如果仅使用 MinGW 进行编译,则此项可以不选。第二项虽然名字带 MinGW,但只是用于交叉编译(即在某一平台上用其他平台的程序),如果不需要,也可以不选。第三项用于 Perl。如果没有安装 Perl,无法操作,继续选择"下一步"直到安装完成。

19.3 模块实现

本项目包括 7 个模块:语料获取和整理、特征提取与预训练、构建模型、模型训练、结果预测、设置诗句评分标准和界面设计。下面分别给出各模块的功能介绍及相关代码。

19.3.1 语料获取和整理

语料集下载地址为 https://github.com/jinfagang/tensorflow_poems/tree/master/data。首先，去掉英文字母特殊符号，并且筛选得到符合律诗结构的语料；其次，将语料集分解成上下两句，统计每一个汉字出现的频率后按顺序、排列保存在字典里；最后，提取上下句的句子向量，并与之前得到的字典整理成诗歌矩阵备用。相关代码请扫描二维码获取。

代码

19.3.2 特征提取与预训练

对于全唐诗结构特点的调研结果，在数据加载到模型训练之前，提取句子向量中的一些额外特征，如平仄、押韵，进行 Embedding 和 additional_Embedding 作为预训练模块。相关代码如下：

```python
skiptoken = "_((«[E{"
for mname in ["qijue-all", "qilv-all", "wujue-all", "wulv-all"]:
    with open("./data/%s.txt" % mname, encoding='utf8') as f:
        lines = f.read().split('\n')
        lines = [i for i in lines if ':' in i and not any([c in i for c in skiptoken])]
        corpus = [[i for i in poem.split(':')[1]] for poem in lines]
        model = Word2Vec(corpus, size=58, min_count=0, hs=1)
        model.save("./data/embedding/%s.dat" % mname)
#加载预训练模型代码如下:
def getEmbedding(vocabularies: list, add_dim: dict, data_name) -> np.ndarray:
    #从 gensim::Word2Vec 中嵌入
    model = Word2Vec.load('./data/embedding/%s.dat' % data_name)
    #使用笛卡尔乘积生成(-1, 0, 1)组成的五位三进制数
    vdim = add_dim["vowel"]
    ls = [i for i in product(*([range(-1, 2)] * vdim))]
    vdic = {k: ls[v] for k, v in getVowel().items()}
    tdim = add_dim["tune"]
    ls = [[i] * tdim for i in range(-1, 2)]
    tdic = {k: ls[v + 1] for k, v in getTune().items()}
    del ls
    embeddings = []
    for dim, dic in zip([64 - vdim - tdim, vdim, tdim], [model, vdic, tdic]):
        z = [0] * dim
        add_embedding = [ #添加嵌入
            np.array([dic[c] if c in dic else z for c in vocab])
            for vocab in vocabularies
        ]
        embeddings.append(add_embedding)
    return [np.concatenate(tuple(e[i] for e in embeddings), axis=1) for i in range(len(vocabularies))]
```

19.3.3 构建模型

数据加载进模型之后，需要定义模型结构、优化损失函数。

1. 定义模型结构

封装 2 层的 RNN 模型用于兼容两种不同的 LSTM，相关代码如下：

```python
class MultiFusedRNNCell(FusedRNNCell):
    # 多融合 RNN 单元类
    def __init__(self, cells: list):    # 初始化
        for i in cells: assert isinstance(i, LSTMBlockFusedCell)
        self._cells = cells
    def __call__(self, inputs, initial_state = None, dtype = None, sequence_length = None, scope = None):    # 调用函数参数设置
        new_states = []
        cur_inp = inputs
        if scope is None: scope = "MultiFusedRNNCell"
        with tf.name_scope(scope):    # 设定范围
            for i, cell in enumerate(self._cells):
                cur_inp, cur_stat = cell(
                    inputs = cur_inp, initial_state = None if initial_state is None else initial_state[i],
                    dtype = dtype, sequence_length = sequence_length
                )    # 当前输入和状态
                new_states.append(cur_stat)
        return cur_inp, tuple(new_states)
    @property
    def state_size(self):    # 状态值大小
        return [i.state_size for i in self._cells]
    @property
    def output_size(self):    # 输出值大小
        return [i.output_size for i in self._cells]
    def zero_state(self, batch_size, dtype):    # 零状态定义
        return tuple([
            tuple([
                tf.zeros((batch_size, cell.num_units), dtype = dtype) for i in range(2)
            ]) for cell in self._cells
        ])
class MyLSTMAdapter(base_layer.Layer):    # LSTM 适配器类
    def __init__(self, GPU, num_layers, num_units, dropout = 0., dtype = tf.dtypes.float32, name = None):    # 初始化参数
        base_layer.Layer.__init__(self, dtype = dtype, name = name)
        self.GPU = GPU
        self.dropout = dropout
        if GPU:    # GPU 情况
            self.model = CudnnLSTM(num_layers, num_units, dtype = self.dtype, name = name)
```

```python
        else: #其他情况
            self.model = MultiFusedRNNCell(
                [LSTMBlockFusedCell(num_units, dtype = self.dtype, name = '%s_%d' % (name, i)) for i in range(num_layers)]
            )
    def __call__(self, inputs, initial_state = None, training = True):
        if self.GPU:     #定义调用模型 GPU
            return self.model(
                inputs, training = training, initial_state = initial_state
            )
        else:    #其他情况
            inputs = tf.transpose(inputs, [1, 0, 2])
            time_len, batch_size, hidden_size = inputs.shape
            output, states = self.model(inputs, initial_state = initial_state, dtype = self.dtype)
            output = tf.transpose(output, [1, 0, 2])
            output = tf.layers.dropout(output, rate = self.dropout, training = training)
#丢弃处理
            return output, states
    def zero_state(self, batch_size):    #零状态处理
        if self.GPU:
            return self.model._zero_state(batch_size)
        else:
            return self.model.zero_state(batch_size, self.dtype)
#封装模型类和公共的中间代码
class RNNModel:
    #构建 RNN Seq2Seq 模型
    num_layers: int
    rnn_size: int
    batch_size: int
    vocab_size: int
    time_len: int
    add_dim: dict
    def __init__(self, name: str, num_layers, rnn_size, batch_size, vocabularies, add_dim: dict, substr_len: int):     #初始化参数
        assert rnn_size % 2 == 0
        self.model_name = name
        self.num_layers = num_layers
        self.rnn_size = rnn_size
        self.batch_size = batch_size
        self.up_vocab = len(vocabularies[0])
        self.down_vocab = len(vocabularies[1])
        self.time_len = substr_len
        self.add_dim = sum(add_dim["sentense"].values())
        self.up_model = MyLSTMAdapter(
            GPU = GPU, num_layers = self.num_layers, num_units = self.rnn_size // 2 + self.add_dim, name = 'up_lstm'
```

```python
        )
        self.down_model = MyLSTMAdapter(
            GPU = GPU, num_layers = self.num_layers, num_units = self.rnn_size, name = "down_lstm"
        )
        # 取得 embedding 层参数
        embedding = getEmbedding(vocabularies, add_dim["word"], self.model_name)
        self.up_embedding, self.down_embedding = [
            tf.constant(
                embedding[i], dtype = tf.float32,
                name = '%s_embedding' % name
            ) for i, name in enumerate(["up", "down"])
        ]
        # 增加 embedding
        self.add_embedding = tf.get_variable(
            'add_embedding',
            initializer = tf.ones([self.time_len, self.add_dim]), dtype = tf.float32
        )
    def __middleware(self, input_data: tf.Tensor, add_data: tf.Tensor, up: bool, is_training):
        # middleware 是一步中间操作,不对外暴露
        if up:   # 中间层向上操作
            prefex = "up/"
            vocab_embedding = self.up_embedding
            add_embedding = self.add_embedding
            model = self.up_model
            rnn_size = self.rnn_size // 2 + self.add_dim
            vocab_size = self.up_vocab
            add_data = add_data[:, 0, :]
            input_data = input_data[:, 0, :]
        else:  # 中间层向下操作
            prefex = "down/"
            vocab_embedding = self.up_embedding
            add_embedding = self.down_embedding
            model = self.down_model
            rnn_size = self.rnn_size
            vocab_size = self.down_vocab
            t = add_data
            add_data = input_data[:, 1, :]
            input_data = t[:, 0, :]
            del t
        inputs = tf.nn.embedding_lookup(vocab_embedding, input_data)
        addmat = tf.nn.embedding_lookup(add_embedding, add_data)
        inputs = tf.concat([inputs, addmat], axis = 2)
        # 建立一个 RNN 网络
        # inputs:神经网络的输入
        initial_state = None if is_training else model.zero_state(1)
        outputs, last_state = model(inputs, initial_state = initial_state, training = is_
```

```
training)
        output = tf.reshape(outputs, [-1, rnn_size])
            #网络参数 weights:用截断正态分布初始化
        weights = tf.Variable(tf.truncated_normal([rnn_size, vocab_size + 1]), name =
prefex + "Weights")
            #网络参数 bias:初始化为 0
        bias = tf.Variable(tf.zeros(shape = [vocab_size + 1]), name = prefex + "Bias")
        logits = tf.nn.bias_add(tf.matmul(output, weights), bias = bias)
        return initial_state, logits, last_state
```

2. 优化损失函数

使用交叉熵作为损失函数,Adam 优化模型参数,相关代码如下:

```
with tf.name_scope(prefex + "prepare_label"):
        #输出数据必须是一位有效编码
    with tf.device('/cpu:0'):
        labels = tf.one_hot(
            tf.reshape(label_data[:, 1 - int(up), :], [-1]),
            depth = (self.up_vocab if up else self.down_vocab) + 1
        )
with tf.name_scope(prefex + "cal_loss"):
    #使用 softmax 交叉熵计算损失
    loss = tf.nn.softmax_cross_entropy_with_logits(labels = labels, logits = logits)
    #所有损失的均值
    total_loss = tf.reduce_mean(loss)
    tf.summary.scalar("loss", total_loss)
with tf.name_scope(prefex + "optimize"):
    #Adam 优化
    train_op = tf.train.AdamOptimizer(learning_rate).minimize(total_loss)
```

19.3.4 模型训练

模型训练相关代码如下:

```
def train(self, input_data, add_data, label_data, learning_rate = 0.01) -> list:
    ops = [] #训练函数
    for up in [True, False]:
        initial_state, logits, last_state = self.__middleware(input_data, add_data if up
else label_data, up, True)
        prefex = "up/" if up else "down/"
        #计算损失
        with tf.name_scope(prefex + "prepare_label"):
            #输出数据必须是一位有效编码
            with tf.device('/cpu:0'):
                labels = tf.one_hot(
                    tf.reshape(label_data[:, 1 - int(up), :], [-1]),
```

```python
                depth = (self.up_vocab if up else self.down_vocab) + 1
            )
        with tf.name_scope(prefex + "cal_loss"):
            #使用softmax交叉熵计算损失
            loss = tf.nn.softmax_cross_entropy_with_logits(labels = labels, logits = logits)
            #所有损失的均值
            total_loss = tf.reduce_mean(loss)
            tf.summary.scalar("loss", total_loss)
        with tf.name_scope(prefex + "optimize"):
            #Adam 优化
            train_op = tf.train.AdamOptimizer(learning_rate).minimize(total_loss)
#在定义模型架构和编译模型之后，使用语料集训练并保存模型
    def run_training():
        if not os.path.exists(FLAGS.model_dir):
            os.makedirs(FLAGS.model_dir)
        #语料矩阵，每层为一行诗，分上下句，其中每个字用对应的序号表示
        #字到对应序号的映射
        #单词表出现频率由高到低
    poems_vector, word_to_int, vocabularies = process_poems(FLAGS.file_path)
        _, _, substr_len = poems_vector.shape
    #语料矩阵按 batch_size 分为若干块
    #batches_inputs:四维ndarray，每块中每层为一个数据(2 * substr_len)
    #batches_outputs:四维ndarray，batches_inputs向左平移一位得到
    batches_inputs, batches_outputs = generate_batch(FLAGS.batch_size, poems_vector, word_to_int)
    graph = tf.Graph()
    with graph.as_default():
        #声明占位符(batch_size, 2, substr_len)
        input_data = tf.placeholder(tf.int32, [FLAGS.batch_size, 2, substr_len], name = "left_word")
        output_targets = tf.placeholder(tf.int32, [FLAGS.batch_size, 2, substr_len], name = "right_word")
        add_mat = tf.placeholder(tf.int32, [FLAGS.batch_size, 2, substr_len], name = "additional_feature")
        #取得模型
        rnn = RNNModel(
            model_name, num_layers = 2, rnn_size = 64, batch_size = 64, vocabularies = vocabularies,
            add_dim = add_feature_dim, substr_len = substr_len
        )
        #获得2个端点
        endpoints = rnn.train(
            input_data = input_data, add_data = add_mat, label_data = output_targets,
            learning_rate = FLAGS.learning_rate
        )
        #只保存一个文件
```

```python
        saver = tf.train.Saver(tf.global_variables(), max_to_keep = 1)
        init_op = tf.group(tf.global_variables_initializer(), tf.local_variables_initializer())
    # session 配置
    config = tf.ConfigProto()
    config.gpu_options.allow_growth = True
    with tf.Session(config = config, graph = graph) as sess:
        # 初始化
        sess.run(init_op)
        summary_writer = tf.summary.FileWriter(FLAGS.log_path, graph = graph)
        # start_epoch,训练完的轮数
        start_epoch = 0
        # 建立检查点
        checkpoint = tf.train.latest_checkpoint(FLAGS.model_dir)
        os.system('cls')
        if checkpoint:
            # 从检查点中恢复
            saver.restore(sess, checkpoint)
            print("## restore from checkpoint {0}".format(checkpoint))
            start_epoch += int(checkpoint.split('-')[-1])
        print('## start training...')
        print("## run 'tensorboard --logdir %s', and view localhost:6006." % (os.path.abspath("./log/train/%s" % model_name)))
        # n_chunk, chunk 块的大小
        n_chunk = len(poems_vector) // FLAGS.batch_size
        tf.get_default_graph().finalize()
        for epoch in range(start_epoch, FLAGS.epochs):
            bar = Bar("epoch %d" % epoch, max = n_chunk)
            for batch in range(n_chunk):
                # 训练两个模型
                summary = easyTrain(
                    sess, endpoints,
                    inputs = (input_data, batches_inputs[batch]), label = (output_targets, batches_outputs[batch]),
                    pos_data = (add_mat, generate_add_mat(batches_inputs[batch], 'binary'))
                )
                if batch % 16 == 0:
                    summary_writer.add_summary(summary, epoch * n_chunk + batch)
                bar.next(16)
            # 每轮结束保存
            saver.save(sess, os.path.join(FLAGS.model_dir, FLAGS.model_prefix), global_step = epoch)
            bar.finish()
        # 保存退出
        saver.save(sess, os.path.join(FLAGS.model_dir, FLAGS.model_prefix), global_step = epoch)
        print('## Last epoch were saved, next time will start from epoch {}.'.format(epoch))
```

19.3.5 结果预测

预测图构造相关代码如下:

```python
def predict(self, input_data, add_data) -> list: #定义预测
    ops = []
    for up in [True, False]:
        initial_state, logits, last_state = self.__middleware(input_data, add_data, up, False)
        prefex = "up/" if up else "down/"
        #将logits传入softmax得到最后输出
        with tf.name_scope(prefex + "predict"):
            prediction = tf.nn.softmax(logits)
#预测过程封装代码
#从begin_word开始推导length个字符
        assert length <= substr_len
        if not begin_char: begin_char = start_token
        #奇数行(下句)
        odd = idx & 1
        #当前的端点
        end = self.endpoints[odd]
        #当前的word_map
        word_map = self.word_int_map[odd]
        #当前的输入
        inputs = np.full((1, 2, 1), word_map[begin_char], dtype=np.int32)
        if odd: at = lambda s, i: s[i] if i < len(s) else ','
        #设置位置函数
        if pos_mode == "linear":
            pos_func = lambda i: i % substr_len
        elif pos_mode == "binary":
            pos_func = lambda i: int(i == substr_len - 1)
        else: raise ValueError("illegal pos_mode: %s" % pos_mode)
        #初始化
        pos_mat = np.zeros((1, 2, 1), np.int32)
        #当前状态
        cur_state = self.state[odd]
        feed_dict = {}
        #生成的字符序列
        s = ''
        for i in range(substr_len - length, substr_len):
            #如果下句: pos_mat 由上句中的相应字符填充
            #如果上句: pos_mat 由 pos 函数中填充
            pos_mat[:] = self.word_int_map[1 - odd][at(self.poem[idx - 1], i)] if odd else pos_func(i)
            feed_dict[self.pos_mat], feed_dict[self.input_data] = (inputs, pos_mat) if odd else (pos_mat, inputs)
```

```python
        #预测
        predict, cur_state = self.sess.run(
            [end.prediction, end.last_state],
            feed_dict = feed_dict
        )
        feed_dict[end.initial_state] = cur_state
        #转变为字符
        w = self.to_word(predict, self.vocabularies[odd])
        #作为下一个输入
        inputs[:] = word_map[w]
        #与"s"拼接
        s += w
    self.state[odd] = cur_state
    return s
```

19.3.6 设置诗句评分标准

设置诗句评分标准的相关代码如下:

```python
class Rater: #评分类
    sml_model = None
    tune_patterns: np.ndarray
    def __init__(self, model_name, substr_len):           #初始化模型参数
        self.model = model_name
        assert substr_len == 5 or substr_len == 7
        self.__gen_tune_pattern(substr_len)
    def __gen_tune_pattern(self, substr_len):             #产生音调模式
        p1 = [-1, -1, 1, 1, -1, -1, 1]
        p3 = [1, 1, -1, -1, -1, 1, 1]
        p1 = np.array(p1, np.int8)[-substr_len:]
        p3 = np.array(p3, np.int8)[-substr_len:]
        p1 = np.hstack((p1, -p1))
        p3 = np.hstack((p3, -p3))
        self.tune_patterns = np.vstack((p1, -p1, p3, -p3))
    def rate(self, poems, subjects = None):               #评价
        if subjects is None:
            subjects = [self.similarity, self.perplexity, self.vowel_score, self.tune_score]
        for subject in subjects:
            print("%s: %.3f" % (subject.__name__, subject(poems)))
    def similarity(self, poems) -> float:
        #计算上/下句的相似度
        if self.sml_model is None:
            self.sml_model = Word2Vec.load("./data/embedding/qijue-all.dat")
        l = [self.sml_model.wv.similarity(u, d) for i in range(0, len(poems), 2) for u, d in zip(*poems[i: i + 2])]
```

```python
        return np.mean(l)
    def perplexity(self, poems, exp = False) -> float:
        # 这不是标准的困惑算法,仅用于比较
        with open("./data/%s.txt" % self.model, encoding = 'utf8') as f:
            lines = f.read().split('\n')
        # 获取所有内容
        contents = [i.split(':')[1] for i in lines if ':' in i and not any([c in i for c in skiptoken])]
        f = '\n'.join(contents)
        chain = [
            # 二元条件概率
            (len(re.findall(p[i: i + 2], f, re.S)) + 1) / len(re.findall(p[i], f, re.S))
            
            for p in poems for i in range(len(p) - 1)
        ]
        # log P(sentense) ^ -1/N = - log sum ( p(w_i | w_i-1 ) ) / N
        p = np.sum(np.log(np.array(chain)))
        p = -p / sum([len(i) for i in poems])
        if exp: p = np.exp(p)
        return p
    def vowel_score(self, poems = False) -> float:
        # 诗歌中的统计元音类,未知元音跳过
        d = getVowel()
        c = [d.get(p[-1], 0) for p in poems]
        n = np.count_nonzero(c)
        c = set(c)
        if 0 in c: c.remove(0)
        c = max(1, len(c))
        return (n - c) / (n - 1)
    def tune_score(self, poems = False) -> float:
        # 诗歌中的统计音类,未知的音调跳过
        d = getTune()
        score = 0
        n = 0
        for i in range(0, len(poems), 2):
            seq = [d.get(i, 0) for i in ''.join(poems[i: i + 2])]
            seq = np.array(seq, np.int8)
            n += np.count_nonzero(seq)
            # 计算每种模式的内积
            score += max([np.sum(i * seq) for i in self.tune_patterns])
        return score / n
```

19.3.7 界面设计

界面与前后端连接的相关代码请扫描二维码获取。

代码

19.4 系统测试

上下句训练模型的损失函数值在逐渐降低并且趋于平稳,如图 19-3 所示;模型初步训练效果如图 19-4 所示;最终成品展示如图 19-5～图 19-8 所示。

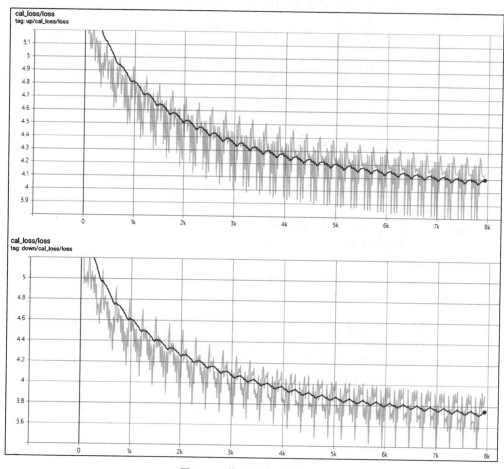

图 19-3 模型损失函数曲线

图 19-4 模型初步训练效果

图 19-5　成品展示图 1

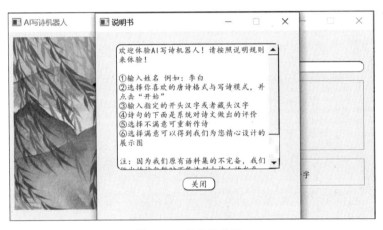

图 19-6　成品展示图 2

图 19-7　成品展示图 3

图 19-8　成品展示图 4

项目 20 基于 COCO 数据集的自动图像描述

PROJECT 20

本项目通过 NLTK(Natural Language Toolkit,自然语言处理工具包)自带的 jieba 分词器构建词向量,使用 CNN 提取图形特征,RNN 完成对词向量的构建,实现对图像的自动描述,并在 Python 自带的 Tkinter 上完成图像界面的展示。

20.1 总体设计

本部分包括系统整体结构图和系统流程图。

20.1.1 系统整体结构图

系统整体结构如图 20-1 所示。

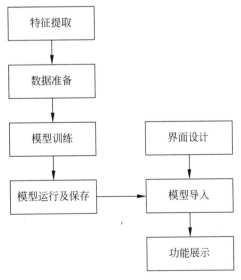

图 20-1 系统整体结构图

20.1.2 系统流程图

系统流程如图 20-2 所示。

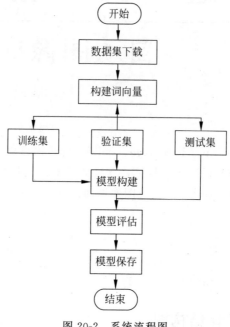

图 20-2 系统流程图

20.2 运行环境

需要 Python 3.6 及以上配置,在 Windows 环境下推荐下载 Anaconda 完成 Python 所需的配置,下载地址为 https://www.anaconda.com/,也可以下载虚拟机在 Linux 环境下运行代码。版本为 1.4.0,需要在 pip install torch 环境中。

20.3 模块实现

本项目包括 4 个模块:数据准备、模型创建及保存、模型训练及保存、界面设置及演示。下面分别给出各模块的功能介绍及相关代码。

20.3.1 数据准备

数据集下载地址为 http://cocodataset.org/#home。包括 91 类、328 000 个影像和 2 500 000 个标签。到目前为止有语义分割的最大数据集,提供 80 类和 33 万张图片,其中

20万张有标注,整个数据集中的数目超过 150 万个。

1. 词语向量表的创建

使用 NLTK 自带的分词工具 jieba,实现对 COCO 数据集标注的分词并保存。

```
import nltk
import pickle
import argparse
from collections import Counter
from pycocotools.coco import COCO
# 导入需要的 package 包
class Vocabulary(object):                    # 定义 Vocabulary 类
    def __init__(self):                      # 初始化
        self.word2idx = {}
        self.idx2word = {}
        self.idx = 0
    def add_word(self, word):                # 定义单词写入函数
        if not word in self.word2idx:
            self.word2idx[word] = self.idx
            self.idx2word[self.idx] = word
            self.idx += 1
    def __call__(self, word):                # 定义调用
        if not word in self.word2idx:
            return self.word2idx['<unk>']
        return self.word2idx[word]
    def __len__(self):                       # 定义长度
        return len(self.word2idx)
def build_vocab(json, threshold):            # 创建词向量和函数
    """Build a simple vocabulary wrapper."""
    coco = COCO(json)
    counter = Counter()
    ids = coco.anns.keys()
    for i, id in enumerate(ids):
        caption = str(coco.anns[id]['caption'])
        tokens = nltk.tokenize.word_tokenize(caption.lower())
        counter.update(tokens)
        if i % 1000 == 0:
            print("[%d/%d] Tokenized the captions." %(i, len(ids)))
    # 如果单词的频率小于阈值,则将单词舍去
    words = [word for word, cnt in counter.items() if cnt >= threshold]
    # 创建词向量抓取器,并设置一些特殊的词向量
    vocab = Vocabulary()
    vocab.add_word('<pad>')
    vocab.add_word('<start>')
    vocab.add_word('<end>')
    vocab.add_word('<unk>')
    # 将单词放入 vocab 中
```

```python
        for i, word in enumerate(words):
            vocab.add_word(word)
        return vocab
    def main(args):
        vocab = build_vocab(json = args.caption_path,
                            threshold = args.threshold)
        vocab_path = args.vocab_path
        with open(vocab_path, 'wb') as f:
            pickle.dump(vocab, f, pickle.HIGHEST_PROTOCOL)
        print("Total vocabulary size: %d" % len(vocab))
        print("Saved the vocabulary wrapper to '%s'" % vocab_path)
    if __name__ == '__main__':                  #主函数
        parser = argparse.ArgumentParser()      #解析参数
        parser.add_argument('--caption_path', type = str, default = 'J:\谢明熹\pytorch_image_
caption-master\data\captions_train_val2014/annotations\captions_train2014.json',
                            help = 'path for train annotation file')
        parser.add_argument('--vocab_path', type = str, default = './data/vocab.pkl',
                            help = 'path for saving vocabulary wrapper')
        parser.add_argument('--threshold', type = int, default = 4,
                            help = 'minimum word count threshold')
        args = parser.parse_args()
        main(args)
```

2. 图片格式设置

图片格式相关代码如下：

```python
    def resize_image(image, size):                          #将图片统一成一个格式
        return image.resize(size, Image.ANTIALIAS)
    def resize_images(image_dir, output_dir, size):         #提示存储的文件夹以及位置
        if not os.path.exists(output_dir):
            os.makedirs(output_dir)
        images = os.listdir(image_dir)
        num_images = len(images)
        for i, image in enumerate(images):                  #枚举图片
            with open(os.path.join(image_dir, image), 'r+b') as f:
                with Image.open(f) as img:                  #打开图片
                    img = resize_image(img, size)
                    img.save(os.path.join(output_dir, image), img.format)   #保存
            if i % 100 == 0:
                print ("[%d/%d] Resized the images and saved into '%s'."
                       % (i, num_images, output_dir))
    def main(args):                                         #定义图片的格式
        splits = ['train', 'val']
        for split in splits:
            image_dir = args.image_dir
            output_dir = args.output_dir
```

```python
            image_size = [args.image_size, args.image_size]
            resize_images(image_dir, output_dir, image_size)
if __name__ == '__main__':                              #主函数
    parser = argparse.ArgumentParser()                  #参数解析
    parser.add_argument('--image_dir', type = str, default = './data/train2014/',
                        help = 'directory for train images')
    parser.add_argument('--output_dir', type = str, default = './data/resized2014/',
                        help = 'directory for saving resized images')
    parser.add_argument('--image_size', type = int, default = 256,
                        help = 'size for image after processing')
    args = parser.parse_args()
    main(args)
```

20.3.2 模型创建及保存

创建用于编码的 CNN 模型以及解码的 RNN 模型，相关代码如下：

```python
class EncoderCNN(nn.Module):                            #创建 CNN 模型
    def __init__(self, embed_size):
        #加载预训练的 ResNet-152 并替换顶层
        super(EncoderCNN, self).__init__()
        self.resnet = models.resnet152(pretrained = True)
        for param in self.resnet.parameters():
            param.requires_grad = False
        self.resnet.fc = nn.Linear(self.resnet.fc.in_features, embed_size)
        self.bn = nn.BatchNorm1d(embed_size, momentum = 0.01)
        self.init_weights()
    def init_weights(self):                             #定义权重
        self.resnet.fc.weight.data.normal_(0.0, 0.02)
        self.resnet.fc.bias.data.fill_(0)
    def forward(self, images):                          #提取图像中的特征向量
        features = self.resnet(images)
        features = self.bn(features)
        return features
class DecoderRNN(nn.Module):                            #定义 RNN 模型
    def __init__(self, embed_size, hidden_size, vocab_size, num_layers):
        #设置超参数和构建层
        super(DecoderRNN, self).__init__()
        self.embed = nn.Embedding(vocab_size, embed_size)
        self.lstm = nn.LSTM(embed_size, hidden_size, num_layers, batch_first = True)
        self.linear = nn.Linear(hidden_size, vocab_size)
        self.init_weights()
        def init_weights(self):                         #设置权重
            self.embed.weight.data.uniform_(-0.1, 0.1)
            self.linear.weight.data.uniform_(-0.1, 0.1)
            self.linear.bias.data.fill_(0)
```

```python
def forward(self, features, captions, lengths):
    #解码图形特征向量并创建解释语句
    embeddings = self.embed(captions)
    embeddings = torch.cat((features.unsqueeze(1), embeddings), 1)
    packed = pack_padded_sequence(embeddings, lengths, batch_first = True)
    hiddens, _ = self.lstm(packed)
    outputs = self.linear(hiddens[0])
    return outputs
def sample(self, features, states):     #展示给出图像的解释句子
    sampled_ids = []
    inputs = features.unsqueeze(1)
    for i in range(20):
        hiddens, states = self.lstm(inputs, states)
        outputs = self.linear(hiddens.squeeze(1))
        predicted = outputs.max(1)[1]
        sampled_ids.append(predicted)
        inputs = self.embed(predicted.unsqueeze(1))
    sampled_ids = torch.cat(sampled_ids, 0)
    return sampled_ids.squeeze()
```

20.3.3 模型训练及保存

训练结果如图 20-3 所示。其中 Epoch 表示训练的轮数；step 表示每轮训练的样本数；loss 表示损失值；perplexity 表示训练的准确值，其值越低准确度越高。

```
Epoch [0/5], Step [24/3236], Loss: 5.1719, Perplexity: 176.2451
Epoch [0/5], Step [25/3236], Loss: 5.0953, Perplexity: 163.2583
Epoch [0/5], Step [26/3236], Loss: 5.1062, Perplexity: 165.0352
Epoch [0/5], Step [27/3236], Loss: 5.0297, Perplexity: 152.8909
Epoch [0/5], Step [28/3236], Loss: 5.0122, Perplexity: 150.2372
Epoch [0/5], Step [29/3236], Loss: 4.9624, Perplexity: 142.9325
Epoch [0/5], Step [30/3236], Loss: 4.9649, Perplexity: 143.3006
Epoch [0/5], Step [31/3236], Loss: 5.0632, Perplexity: 158.0939
Epoch [0/5], Step [32/3236], Loss: 5.0136, Perplexity: 150.4489
Epoch [0/5], Step [33/3236], Loss: 4.7827, Perplexity: 119.4222
Epoch [0/5], Step [34/3236], Loss: 4.9245, Perplexity: 137.6145
Epoch [0/5], Step [35/3236], Loss: 4.9026, Perplexity: 134.6366
Epoch [0/5], Step [36/3236], Loss: 4.8871, Perplexity: 132.5665
Epoch [0/5], Step [37/3236], Loss: 4.8226, Perplexity: 124.2845
Epoch [0/5], Step [38/3236], Loss: 4.7812, Perplexity: 119.2498
Epoch [0/5], Step [39/3236], Loss: 4.7593, Perplexity: 116.6664
Epoch [0/5], Step [40/3236], Loss: 4.7911, Perplexity: 120.4357
Epoch [0/5], Step [41/3236], Loss: 4.7548, Perplexity: 116.1424
Epoch [0/5], Step [42/3236], Loss: 4.7774, Perplexity: 118.7946
Epoch [0/5], Step [43/3236], Loss: 4.7690, Perplexity: 117.8004
Epoch [0/5], Step [44/3236], Loss: 4.6573, Perplexity: 105.3512
Epoch [0/5], Step [45/3236], Loss: 4.7147, Perplexity: 111.5708
Epoch [0/5], Step [46/3236], Loss: 4.6788, Perplexity: 107.6408
Epoch [0/5], Step [47/3236], Loss: 4.5700, Perplexity: 96.5462
Epoch [0/5], Step [48/3236], Loss: 4.7630, Perplexity: 117.0930
Epoch [0/5], Step [49/3236], Loss: 4.6824, Perplexity: 108.0304
Epoch [0/5], Step [50/3236], Loss: 4.5935, Perplexity: 98.8393
Epoch [0/5], Step [51/3236], Loss: 4.5954, Perplexity: 99.0260
Epoch [0/5], Step [52/3236], Loss: 4.6079, Perplexity: 100.2731
```

图 20-3　训练结果

相关代码如下:

```python
def main(args):
    # 创建模型目录
    if not os.path.exists(args.model_path):
        os.makedirs(args.model_path)
    # 图片预处理
    transform = transforms.Compose([
        transforms.RandomCrop(args.crop_size),
        transforms.RandomHorizontalFlip(),
        transforms.ToTensor(),
        transforms.Normalize((0.5, 0.5, 0.5), (0.5, 0.5, 0.5))])
    # 加载单词抓取器
    with open(args.vocab_path, 'rb') as f:
        vocab = pickle.load(f)
    # 创建数据加载函数
    data_loader = get_loader(args.image_dir, args.caption_path, vocab,
                             transform, args.batch_size,
                             shuffle = True, num_workers = args.num_workers)
    # 建立模型
    encoder = EncoderCNN(args.embed_size)
    decoder = DecoderRNN(args.embed_size, args.hidden_size,
      len(vocab), args.num_layers)
    if torch.cuda.is_available():
      encoder.cuda()
      decoder.cuda()
    # 损失函数以及优化器
    criterion = nn.CrossEntropyLoss()
    params = list(decoder.parameters()) + list(encoder.resnet.fc.parameters())
    optimizer = torch.optim.Adam(params, lr = args.learning_rate)
    # 模型训练
    total_step = len(data_loader)
    for epoch in range(args.num_epochs):
        for i, (images, captions, lengths) in enumerate(data_loader):
            # 设置小批量数据集
            images = Variable(images)
            captions = Variable(captions)
            if torch.cuda.is_available():
                images = images.cuda()
                captions = captions.cuda()
            targets = pack_padded_sequence(captions, lengths, batch_first = True)[0]
            # 期望值、损失值以及优化
            decoder.zero_grad()
            encoder.zero_grad()
            features = encoder(images)
            outputs = decoder(features, captions, lengths)
            loss = criterion(outputs, targets)
```

```python
                loss.backward()
                optimizer.step()
                # 打印日志信息
                if i % args.log_step == 0:
                    print('Epoch [%d/%d], Step [%d/%d], Loss: %.4f, Perplexity: %5.4f'
                          % (epoch, args.num_epochs, i, total_step,
                             loss.item(), np.exp(loss.item())))
                # 保存模型
                if (i+1) % args.save_step == 0:
                    torch.save(decoder.state_dict(),
                               os.path.join(args.model_path,
                                            'decoder-%d-%d.pkl' % (epoch+1, i+1)))
                    torch.save(encoder.state_dict(),
                               os.path.join(args.model_path,
                                            'encoder-%d-%d.pkl' % (epoch+1, i+1)))
if __name__ == '__main__':  # 主函数
    parser = argparse.ArgumentParser()  # 参数解析
    parser.add_argument('--model_path', type=str, default='./models/',
                        help='path for saving trained models')
    parser.add_argument('--crop_size', type=int, default=224,
                        help='size for randomly cropping images')
    parser.add_argument('--vocab_path', type=str, default='./data/vocab.pkl',
                        help='path for vocabulary wrapper')
    parser.add_argument('--image_dir', type=str, default='./data/resized2014',
                        help='directory for resized images')
    parser.add_argument('--caption_path', type=str,
                        default='./data/captions_train_val2014/annotations/captions_train2014.json',
                        help='path for train annotation json file')
    parser.add_argument('--log_step', type=int, default=1,
                        help='step size for prining log info')
    parser.add_argument('--save_step', type=int, default=100,
                        help='step size for saving trained models')
    # 设置模型参数
    parser.add_argument('--embed_size', type=int, default=256,
                        help='dimension of word embedding vectors')
    parser.add_argument('--hidden_size', type=int, default=512,
                        help='dimension of lstm hidden states')
    parser.add_argument('--num_layers', type=int, default=1,
                        help='number of layers in lstm')
    parser.add_argument('--num_epochs', type=int, default=5)
    parser.add_argument('--batch_size', type=int, default=128)
    parser.add_argument('--num_workers', type=int, default=2)
    parser.add_argument('--learning_rate', type=float, default=0.001)
    args = parser.parse_args()
    print(args)
    main(args)
```

20.3.4 界面设置及演示

选择图片地址并输出,当按下按钮时显示当前图片的描述。

```python
def test(args):
    # 图像预处理
    transform = transforms.Compose([
        transforms.Scale(args.crop_size),
        transforms.CenterCrop(args.crop_size),
        transforms.ToTensor(),
        transforms.Normalize((0.5, 0.5, 0.5), (0.5, 0.5, 0.5))])
    # 加载词向量抓取器
    with open(args.vocab_path, 'rb') as f:
        vocab = pickle.load(f)
    # 创建模型
    encoder = EncoderCNN(args.embed_size)
    encoder.eval() # 评估模式
    decoder = DecoderRNN(args.embed_size, args.hidden_size,
                         len(vocab), args.num_layers)
    encoder.load_state_dict(torch.load(args.encoder_path))
    decoder.load_state_dict(torch.load(args.decoder_path))
    # 准备图片
    image = Image.open(args.image)
    image_tensor = Variable(transform(image).unsqueeze(0))
    # 设置初始值
    state = (Variable(torch.zeros(args.num_layers, 1, args.hidden_size)),
             Variable(torch.zeros(args.num_layers, 1, args.hidden_size)))
    if torch.cuda.is_available():
        encoder.cuda()
        decoder.cuda()
        state = [s.cuda() for s in state]
        image_tensor = image_tensor.cuda()
    # 根据图像创建相关对图像理解的句子
    feature = encoder(image_tensor)
    sampled_ids = decoder.sample(feature, state)
    sampled_ids = sampled_ids.cpu().data.numpy()
    # 将单词编码解码成相应的单词
    sampled_caption = []
    for word_id in sampled_ids:
        word = vocab.idx2word[word_id]
        sampled_caption.append(word)
        if word == '<end>':
            break
    sentence = ' '.join(sampled_caption)
    # 输出图片和相应图片理解的句子
    print(sentence)
    plt.imshow(np.asarray(image))
    return sentence
class Application(Frame):
```

```python
# 创建图形界面
    def __init__(self, master = None):  # 设置初始值
        Frame.__init__(self, master, bg = 'black')
        self.pack(expand = YES, fill = BOTH)
        self.window_init()
        self.createWidgets(image_address, sentence)
    def window_init(self):  # 设置窗口的初始值
        self.master.title('welcome to IMAGE - captioning system')
        self.master.bg = 'black'
        width, height = self.master.maxsize()
        self.master.geometry("{}x{}".format(width, height))
    def createWidgets(self, address, sentence):  # 定义各种部件的参数
        # 第一部分
        self.fm1 = Frame(self, bg = 'black')
        self.titleLabel = Label(self.fm1, text = "video - captioning system", font = ('微软雅黑', 64), fg = "white", bg = 'black')
        self.titleLabel.pack()
        self.fm1.pack(side = TOP, expand = YES, fill = 'x', pady = 20)
        # 第二部分
        self.fm2 = Frame(self, bg = 'black')
        self.fm2_left = Frame(self.fm2, bg = 'black')
        self.fm2_right = Frame(self.fm2, bg = 'black')
        self.fm2_left_top = Frame(self.fm2_left, bg = 'black')
        self.fm2_left_bottom = Frame(self.fm2_left, bg = 'black')
        self.predictEntry = Entry(self.fm2_left_top, font = ('微软雅黑', 24), width = '72', fg = '#FF4081')
        self.predictButton = Button(self.fm2_left_top, text = 'predict sentence', bg = '#FF4081', fg = 'white',
                                    font = ('微软雅黑', 36), width = '16',
                                    command = self.output_predict_sentence)
        self.predictButton.pack(side = LEFT)
        self.predictEntry.pack(side = LEFT, fill = 'y', padx = 20)
        self.fm2_left_top.pack(side = TOP, fill = 'x')
        self.fm2_left_bottom.pack(side = TOP, pady = 10, fill = 'x')
        self.fm2_left.pack(side = LEFT, padx = 60, pady = 20, expand = YES, fill = 'x')
        self.fm2_right.pack(side = RIGHT, padx = 60)
        self.fm2.pack(side = TOP, expand = YES, fill = "x")
        # 第三部分
        self.fm3 = Frame(self, bg = 'black')
        load = Image.open(address)
        initIamge = ImageTk.PhotoImage(load)
        self.panel = Label(self.fm3, image = initIamge)
        self.panel.image = initIamge
        self.panel.pack()
        self.fm3.pack(side = TOP, expand = YES, fill = BOTH, pady = 10)
    def output_predict_sentence(self):
        predicted_sentence_str = sentence
        self.predictEntry.delete(0, END)
        self.predictEntry.insert(0, predicted_sentence_str)
if __name__ == '__main__':  # 主函数
```

```python
parser = argparse.ArgumentParser()  # 参数解析
parser.add_argument('-- image', type = str, default = './data/2.jpg',
                    help = 'input image for generating caption')
parser.add_argument('-- encoder_path', type = str, default = './models/encoder-1-900.pkl',
                    help = 'path for trained encoder')
parser.add_argument('-- decoder_path', type = str, default = './models/decoder-1-900.pkl',
                    help = 'path for trained decoder')
parser.add_argument('-- vocab_path', type = str, default = './data/vocab.pkl',
                    help = 'path for vocabulary wrapper')
parser.add_argument('-- crop_size', type = int, default = 224,
                    help = 'size for center cropping images')
# 模型参数设置
parser.add_argument('-- embed_size', type = int, default = 256,
                    help = 'dimension of word embedding vectors')
parser.add_argument('-- hidden_size', type = int, default = 512,
                    help = 'dimension of lstm hidden states')
parser.add_argument('-- num_layers', type = int, default = 1,
                    help = 'number of layers in lstm')
args = parser.parse_args()
image_address = './data/2.jpg'
sentence = test(args)
app = Application()
app.mainloop()
```

20.4 系统测试

页面上方显示项目名称 image-captioning system，下方显示选择的图片，单击 predict sentence 按钮将显示输出文本结果，如图 20-4 所示。

图 20-4　界面展示

图书资源支持

感谢您一直以来对清华大学出版社图书的支持和爱护。为了配合本书的使用，本书提供配套的资源，有需求的读者请扫描下方的"书圈"微信公众号二维码，在图书专区下载，也可以拨打电话或发送电子邮件咨询。

如果您在使用本书的过程中遇到了什么问题，或者有相关图书出版计划，也请您发邮件告诉我们，以便我们更好地为您服务。

我们的联系方式：

地　　址：北京市海淀区双清路学研大厦 A 座 701

邮　　编：100084

电　　话：010-83470236　010-83470237

资源下载：http://www.tup.com.cn

客服邮箱：tupjsj@vip.163.com

QQ：2301891038（请写明您的单位和姓名）

用微信扫一扫右边的二维码，即可关注清华大学出版社公众号。

教学资源·教学样书·新书信息

人工智能科学与技术
人工智能|电子通信|自动控制

资料下载·样书申请

书圈